PLANT SCIENCE RESEARCH AND PRACTICES

RECENT ADVANCES IN PLANT RESEARCH

PLANT SCIENCE RESEARCH AND PRACTICES

Additional books and e-books in this series can be found on Nova's website under the Series tab.

PLANT SCIENCE RESEARCH AND PRACTICES

RECENT ADVANCES IN PLANT RESEARCH

BEATRICE WELCH

AND

MICHEAL WILKERSON

EDITORS

nova
science publishers
New York

Library of Congress Cataloging-in-Publication Data

ISBN: 978-1-53614-170-2

Published by Nova Science Publishers, Inc. † New York

CONTENTS

PREFACE

Recent Advances in Plant Research begins by providing a summary of work performed on the effects of environmental factors and growth treatments on secondary metabolite production in both conventionally and *in vitro* cultivated plant material, focusing on different classes of volatile and non-volatile secondary metabolites.

Following this, the authors set out to determine the diversity of *Fabaceae* in vegetational fragments of Cerrado in the state of Maranhão. Monthly expeditions were carried out in the period between September 2016 and June 2017 for the observation, collection, and identification of botanical material.

The great potential of the most important by-products of apples, grapes and berry fruits processing as a source of antioxidants is presented in one chapter. Some novel and practical aspects of extraction of natural antioxidants are also discussed.

Also discussed is gray mold rot caused by *Botrytis cinerea* Persoon: Fries [teleomorph *Botryotinia fuckeliana* (de Bary) Whetzel]. It is a serious disease causing severe damage to fruits and vegetables at both pre- and post-harvest periods. The authors aim to highlight try to highlight the disease distribution and hosts range, pathogenesis of the pathogen, symptoms of the disease, disease cycle and epidemiology of the pathogen as well as the current status of the disease in Malaysia.

Plants produce an array of toxins and defensive proteins through various metabolic pathways for their defence against insects. To cope with these defences, herbivores have developed counter-defences which the authors examine in this compilation. Furthermore, insects have employed a diverse array of strategies that enable them to bypass defensive barriers, or to metabolise these chemicals after ingestion.

Next, angiosperm flowers that have been found in mid-Cretaceous Myanmar (Burmese) amber are analyzed. Flowers in amber often show details of their reproductive parts that are not evident in blooms preserved in sedimentary deposits. Two new genera are described, *Chenocybus allodapus* gen et sp. nov. and *Diaphoranthus burmensis* gen et sp. nov.

Japan, which extends north and south, has four clearly defined seasons that provide a variety of unique habitats for the growth of various aromatic and medicinal plants. Recent focus on these traditional plants has begun to revitalize local communities, providing diversification to alternative medicines and the development of other products, as well as suggesting increased and use of plants.

Aiming to better understand the phytodiversity of a vegetation area in the Piauí state, Brazil, in addition to aspects such as the geographical distribution of species and their uses, the concluding study was carried out in municipality of Bom Princípio, northern Piauí. The diverse flora was represented by trees, shrubs and herbaceous species, distributed in 54 families, 114 generas, and 146 species.

Chapter 1 - Secondary metabolites production, accumulation and translocation are dependent on the presence of highly specialized anatomical structures within the plant organism. Therefore secondary metabolites content is strongly affected by the developmental patterns and morphogenesis of the plant individual.

The present chapter summarizes work performed on the effects of environmental factors and growth treatments on secondary metabolite production in both conventionally and *in vitro* cultivated plant material. Different classes of volatile and non-volatile secondary metabolites have been concerned.

The author's own experience in the impact of tissue culture treatments on the developmental patterns and morphological features *in vitro* and the subsequent modification of secondary metabolite production of species of the genera *Hypericum*, *Sideritis* and *Artemisia* have been summarized and discussed.

Chapter 2 - *Fabaceae* is one of the largest families among the angiosperms, with representatives widely distributed throughout the globe, where 770 genera and 19.500 species are recognized and divided into six subfamilies (*Caesalpinioideae*, *Cercidoideae*, *Detarioideae*, Dialioideae, Duparquetioideae, and *Papilionoideae*). Floristic and taxonomic works indicate *Fabaceae* as the largest species family in Brazil and most representative in forest formations. The Cerrado has a significant representation of the family, where the state of Maranhão covers a broad area. In this way, the research had as goal to know the diversity of *Fabaceae* in vegetational fragments of Cerrado in the state of Maranhão. Monthly expeditions were carried out in the period between September 2016 and June 2017 for observation, collection, and identification of botanical material. The Cerrado area sampled belongs to the municipality of São João do Sóter, in the Maranhense´s East. After the collections, the specimens were herborized, analyzed with the help of taxonomic keys, specialized literature for identification, and determined by area expert. A total of 68 specimens, 31 genera, and 45 species were cataloged. Of all the specimens collected, the subfamilies *Papilionoideae* and *Caesalpinioideae* were the most representative of 21 species each. As for the Life form, was observed that the prevailing growth habit was the shrub type (20). Concerning to the physiognomies of the Cerrado, was observed the predominance of the species in the gallery forest environment (28).

Were obtained three new records to the Maranhão, the species: *Desmodium subsecundum* Vogel. *Aeschynomene viscidula* Michx and *Vigna lasiocarpa* (Mart. ex Benth.) Verdc. being *V. lasiocarpa* a new record for the Northeast region of Brazil. Taxonomic keys, descriptions and photo plates were elaborated with all the studied species, composing a taxonomic treatment. In this way, it can be established that the Cerrado of Maranhão possesses a diversity of species for the *Fabaceae* family and that the research carried out has provided a basis for later studies, since these are few for Maranhão.

Chapter 3 - The awareness of consumers regarding the functional food led to an increasing demand for functional foods especially for foods rich in antioxidants, which have a wide range of activities such as antitumoral, antiviral, cardioprotective and antimutagenic activities. Phenolic compounds are an essential part of the human diet and are of considerable interest due to their antioxidant activities. Wastes from fruits represent a major disposal problem for the industry concerned. The by-products, which are results of the fruit processing, are rich in valuable compounds and represent a novel, natural and economic sources of flavorings, colorants, polyphenols and natural antioxidants, which can be used in the food industry as a source of natural food additives. This chapter is underlying the great potential of the most important by-products of apples, grapes and berry fruits processing as a source of antioxidants. Some novel and practical aspects of extraction of natural antioxidants will be discussed in this chapter.

Chapter 4 - The gray mold rot is caused by *Botrytis cinerea* Persoon: Fries [teleomorph *Botryotinia fuckeliana* (de Bary) Whetzel]. It is a serious disease causing severe damages on fruits and vegetables at both pre and post-harvest periods. *Botrytis cinerea* is an ascomycetous fungus as well as a necrotrophic plant pathogen that was ranked second in the top 10 most devastating fungal plant pathogens worldwide. Due to its broad host range, the fungus infects more than 200 plant species of agricultural importance such as cucumbers, strawberries, grapes, tomatoes, cut flowers and ornamental plants. The fungus occurs mostly where its host plants are cultivated, under climatic conditions ranging from cool temperate zones to subtropical regions. The fungus is highly concentrated in the temperate areas between latitudes 25 and 30. In Malaysia, the occurrence of this disease has been widespread in the greenhouses of the commercial farms in the Cameron Highland area of Pahang, Malaysia. This hindered the growing industries, hence detection, and characterization of the pathogen using molecular approaches provide an essential tool for establishing disease management programs. Since the pathogen is highly destructive, in this book chapter the authors will try to highlight the disease distribution and hosts range, pathogenesis of the pathogen (*B. cinerea*), symptoms of the disease, disease cycle and epidemiology of the pathogen (*B. cinerea*) as well as the current status of the disease (gray mold rot) in Malaysia.

Chapter 5 - Insect herbivores use different feeding strategies to obtain nutrients and energy from plants for their growth and reproduction. Plants produce an array of toxins and defensive proteins through various metabolic pathways for their defence against insects. To cope with these defences, herbivores have developed counter defences. These responses of insect herbivores are important in insect plant interactions. Among these interactions, host plant selection and feeding behaviour are important factors and play a significant role in habitat selection and discrimination mechanisms between host species. Furthermore, insects have employed a diverse array of strategies that enable them to bypass defensive barriers, or to metabolise these chemicals after ingestion.

Chapter 6 - Amberization is a unique type of fossilization that preserves both animal and plant parts for millions of years through a unique fixation-dehydration process. Flowers in amber often show details of their reproductive parts that are not evident in blooms preserved in sedimentary deposits. The present chapter discusses angiosperm flowers that have been found in mid-Cretaceous Myanmar (Burmese) amber. Two new genera are described, *Chenocybus allodapus* gen et sp. nov. and *Diaphoranthus burmensis* gen et sp. nov. These two new genera, along with other flowers described from this amber source, namely, *Antiquifloris latifibris*, *Cascolaurus burmitis*, *Endobeuthos paleosum*, *Eoepigynia burmensis*, *Jamesrosea burmensis*, *Lachnociona terriae*, *Micropetasos burmensis*, *Palaeoanthella huangii*, *Programinis burmitis*, *Programinis laminatus*, *Tropidogyne pikei* and *Tropidogyne pentaptera*, are considered extinct at the generic level. However, the level of extinction of the various lineages varies and they can be arranged in roughly three groups. Group one consists of flowers with no obvious affinities to any extant lineages. Group 2 are flowers with possible affinities to extant lineages above the family level. Group 3 are flowers that can be placed in extant families. Comments on possible pollinators in the Myanmar amber forest and extinction events regarding Myanmar amber flowers are included.

Chapter 7 - Japan, which extends north and south, has four clearly defined seasons that provide a variety of unique habitats for the growth of various aromatic and medicinal plants. Recent focus on these traditional plants has begun to revitalize local communities, providing diversification to alternative medicines and the development of other products. Aromatic plants are widely used in Japanese foods such as Washoku, the cuisine of Japan, which is recognized on the UNESCO Heritage list. The traditional Japanese "Ko-do" (an elegant incense ceremony) uses several aromatic plants that provide graceful aromas. Indigenous aromatic and medicinal plants continue the connection of plants with life and with spirit for the Japanese. Due to global warming, a change may limit plant growth and alter constituents, endangering both domestic and wild species. Thus, conservation of genetic resources is necessary to insure that the unique aromatic plant species of Japan, such as *Eutrema japonicum* (Wasabi) and *Cryptomeria japonica* (Sugi), are maintained. While universities, companies, institutes, and homes own a number of gardens and conservatories, the most important concerns are the use, conservation, and return of natural resources in a daily life. Various locations in Japan practice a circulation-type forestry. This method creates new jobs that eventually lead to regional revitalization. Recently, the introduction of profiles and advantages of domestic Japanese medicinal and aromatic plants have developed markets with remarkable growth in Japan, suggesting increased interest and use of plants for better health, education, and environment in Japan.

Chapter 8 - Aiming to better understand the phytodiversity of a vegetation area in the Piauí state, Brazil, in addition to aspects such as the geographical distribution of species and their uses, this study was carried out in municipality of Bom Princípio, northern Piauí. The diverse flora was represented by trees, shrubs and herbaceous species, distributed in 54 families, 114 generas, and 146 species. The families of richest generas were *Fabaceae* (22 generas), followed by Malvaceae and Rubiaceae (seven generas) and Poaceae (five generas). Approximately 64,81% (35) of families were represented by a single genre at this location. It was found, according to the literature, that 29 species belonging to 16 families have economic potential. Most species have medicinal potential, with 22 species; followed by 15 timbers species, with 14 honey producer species; 12 ornamental species, food with eight species, forage nine and use for fuel, eight species. Of the species recorded in area 97 (66.4%) are not endemic in Brazil and 47 (32,2%) are endemic in this country. As consultations held in the collections of several Virtual Herbarium and Herbarium "HDelta," checked that 17 species did not have occurrence records in Piauí state. Analyzing the geographical distribution of species recorded in the study area and comparing with other surveys, it was found that 13 species found in survey's "cerrado" vegetation and 18 species in "caatinga" vegetation areas. The floristic survey in this semideciduous vegetation area in northern Piauí showed that in the area there are a number of species belonging to the plant formations of "cerrado" and "caatinga" vegetation, according to data of floristic lists made these vegetational formations, thus suggesting, treat itself to a transition area between these two types of vegetation.

In: Recent Advances in Plant Research
Editors: B. Welch and M. Wilkerson
ISBN: 978-1-53614-170-2

Chapter 1

Roles of Developmental Patterns and Morphogenesis in the Secondary Metabolite Production of Conventionally and Biotechnologically Cultivated Medicinal and Aromatic Plants

Kalina Danova
Institute of Organic Chemistry with Centre of Phytochemistry,
Bulgarian Academy of Sciences, Bulgaria

Abstract

Secondary metabolites production, accumulation and translocation are dependent on the presence of highly specialized anatomical structures within the plant organism. Therefore secondary metabolites content is strongly affected by the developmental patterns and morphogenesis of the plant individual.

The present chapter summarizes work performed on the effects of environmental factors and growth treatments on secondary metabolite production in both conventionally and *in vitro* cultivated plant material. Different classes of volatile and non-volatile secondary metabolites have been concerned.

The author's own experience in the impact of tissue culture treatments on the developmental patterns and morphological features *in vitro* and the subsequent modification of secondary metabolite production of species of the genera *Hypericum*, *Sideritis* and *Artemisia* have been summarized and discussed.

1. Introduction

Growth and development are determined by the complex interrelations of the plant with its surrounding environment. Thus, the combined effects of climatic and geographic factors, as well as the plant phenology have been shown to strongly affect secondary metabolites productivity of medicinal and aromatic plants in nature. In addition, agro-technique has been shown to be non-less decisive, than the genotype for the productivity of these plants when being transferred from the natural habitats to the field.

The controlled environment of plant cell tissue and organ culture breeding makes plant biotechnology a flexible approach for targeting valuable phytopharmaceuticals through modification of growth conditions. The extent of differentiation/de-differentiation has been shown to play vital role in secondary metabolite levels in the different types of *in vitro* cultured cells and tissues. This phenomenon being determined by plant growth regulators treatments makes them one of the most widely utilized tools in modification of developmental patterns, morphogenesis and hence secondary metabolite production *in vitro*. The comprehension of the role of developmental and morphogenetic features for the biogenesis of plant secondary metabolites is essential for optimization of secondary metabolite productivity *in vitro*. In addition, identification of key factors responsible for the productivity of certain secondary metabolites in plant cell tissue and organ culture could be useful to also develop suitable agro-technique approaches for the enhanced productivity of these metabolites also on the farm.

2. Metabolite Production in the Plant Organism

2.1. Primary Plant Metabolites

Primary metabolites (such as carbohydrates, amino acids, fatty acids and organic acids, nucleotides, sugars and lipids) are widely distributed in the plant kingdom. They are produced as a result of primary cell metabolism and are vital for the plant survival (Figure 1). They are involved in processes such as growth and development, respiration and photosynthesis, as well as protein and hormone synthesis. Humankind has been using plant primary metabolites as food and raw materials since ancient times (Balandrin & Kloecke 1988; Hounsome et al. 2008).

2.2. Secondary Plant Metabolites

During its growth and development, the plant organism exists in a dynamic equilibrium with the numerous stimuli of its surrounding environment. Each ecosystem is characterized by numerous factors such as bacteria, viruses, fungi, nematodes, insects and herbivores. In addition, each habitat is characterized by its specific climatic factors and seasonal fluctuations, soil and water parameters, etc. Given their attached way of life, plants cannot directly avoid unfavorable impacts by simply changing their location at the moment of the unfavorable interaction. Although plants are capable of migrating, this process is complicated and time consuming, and occurs on the population level taking several seasons to be realized. In addition this process depends strongly on the physiological adaptability capacity of the individual plants constituting the population. Therefore the intermediate survival of the plant individual is bound to other mechanisms encompassing a number of complex biochemical adaptations which have evolved during evolution. Plants produce a wide array of chemical compounds which seemingly are not directly related to their growth and development (Figure 1). These compounds, designated secondary metabolites (natural compounds, biologically active compounds), do not have a direct widely accepted role in the processes of photosynthesis, respiration, mineral assimilation and transportation, differentiation, protein and lipids biosynthesis (Taiz and Zeiger 2006). A distinctive feature for secondary metabolites is their limited distribution in the plant kingdom. Often certain secondary metabolites are found only in certain species, genus or family, being a chemotaxonomic feature of this respective taxon. A vital parameter for secondary metabolite production and especially translocation and accumulation is the presence of the respective structural prerequisites, related to the morphological development of the plant.

2.2.1. Classification of Secondary Metabolites According to Their Functional Role for the Plant Organism

Secondary metabolites could functionally be classified as (Taiz & Zeiger 2006):

- *Attractants* – their function is to attract pollinating insects or animals whose biological role is to disperse plant diaspore.
- *Factors of chemical adaptation* towards the unfavorable environmental conditions – compounds with protective, repelling and warning functions.
- *Growth regulators and signaling molecules* – having in mind their vital role for synchronizing all physiological processes within the integral plant organism their functions are close to the ones of primary metabolites.

It is worth mentioning that unlike primary metabolites, plant secondary metabolites are "expensive to produce" for plant cellular energetics. Therefore in order to avoid the unnecessary waste of resources, plants have evolved complicated early signal systems, enabling them to distinguish between a purely mechanical damage and the attack of an insect/herbivore (Fürstenberg-Hägg et al. 2013). Thus, for example plants are able to evaluate the extent of damage, caused by herbivore feeding guilds, insect oral secretions, oviposition fluids, etc. These events lead to a succession of activities, leading to the production of certain metabolites: metabolic changes, followed by gene activation, jasmonic acid (JA) changes, kinase cascades, hydrogen peroxide production, cytosolic calcium ion fluxes, as well as membrane potential changes (Fürstenberg-Hägg et al. 2013 and references cited within).

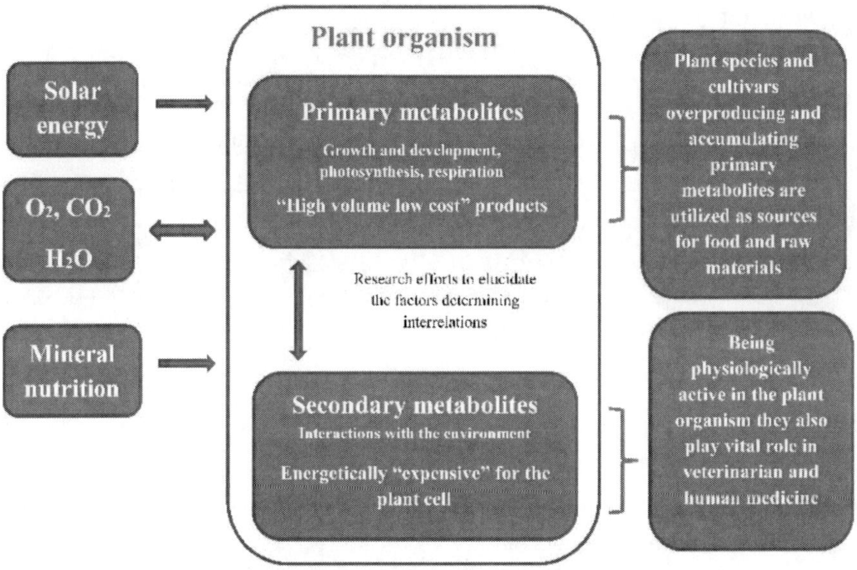

Figure 1. Schematic overview of primary and secondary plant metabolites.

Many of the models of biotic/abiotic stress response of the plants in their indigenous environment underline the experimental set-up of secondary metabolite elicitation of plant cell tissue and organ culture. The good knowledge and understanding of the physiological role of given secondary metabolite for the plant organism of interest is the key to its successful production by any breeding technique (i.e., by different types of cultivation techniques in soil, hydroponics, etc. as well as by the tools of plant biotechnology).

2.2.2. Main Biosynthetic Pathways and Regulation of Secondary Metabolism in the Plant Organism

The building blocks of secondary metabolites are derived from the primary metabolism (Figure 2). With the advancement of plant sciences, more light has been shed on the transcriptional regulation of secondary metabolites biosynthesis. It is now clear that the expression of activators and repressors comes as a complex response to phytohormones and different environmental signals, resulting in a dynamic regulatory network which determines the timing, amplitude and tissue specific expression of pathway genes and the subsequent accumulation of secondary metabolites (Patra et al. 2013). Transcription factor families such as MYC, MYB, WRKY and AP2/ERF have been identified to be responsible for secondary metabolite regulation and Jasmonic acid (JA) has been found to be an inducer in these processes (Afrin et al. 2015).

2.2.3. Main Classes of Secondary Metabolites and Their Production from the Cellular (Different Plant Organelles) to the "Supra-Cellular" (Secretory Glands, Trichromes, etc.) Level. Structural Prerequisites, Responsible for Their Production, Translocation and Accumulation

Secondary metabolites can be categorized into three major groups: *(i)* terpenes (Figure 3) *(ii)* phenolic compounds (Figure 4) and *(iii)* nitrogen-containing. The physiological roles of the respective compounds for the survival of the plant organism in its indigenous environment define also the pharmacological role which these substances play in the mammalian organism and hence their applicability in the veterinarian and humanitarian medicinal practices. Data from literature sources on the two first classes of compounds are summarized in Figures 3 and 4 (Gersbach 2002; Taiz and Zeiger 2006; Hassan and Mathesius 2012; Li et al. 2012; Gupta and Chakrabarty 2013; Agati et al. 2013; Singh and Sharma 2015).

The biogenesis, translocation, secretion and accumulation of the different secondary metabolites are bound to specific cellular organelles and require strictly the presence of the respective anatomic structures.

Plants respond by secondary metabolite production in dependence of the following factors:

- *Environmental impact*: water, nutritional elements, altitude, light, temperature, pathogens and herbivore pressure, etc. affect secondary metabolite production through affecting plant growth and development, morphology and primary metabolism (photosynthetic efficiency, carbon fixation and sugar accumulation) (Table 1).

Thus, herbivore attack signal is being perceived by undamaged tissues and low molecular messenger compounds are being released, followed by a cascade of defense reactions at the wounding site such as alteration of calcium ion fluxes, phosphorylation and systematic and jasmonate signaling. The bioactive compounds released as a result function to repel and intoxicate insects. The volatiles released repel herbivores, attract predators or serve for signaling between the different plant leaves and induce defense responses. Other morphological features as waxes, trichomes and lattices also serve to deter insects and herbivores (Fürstenberg-Hägg et al. 2013).

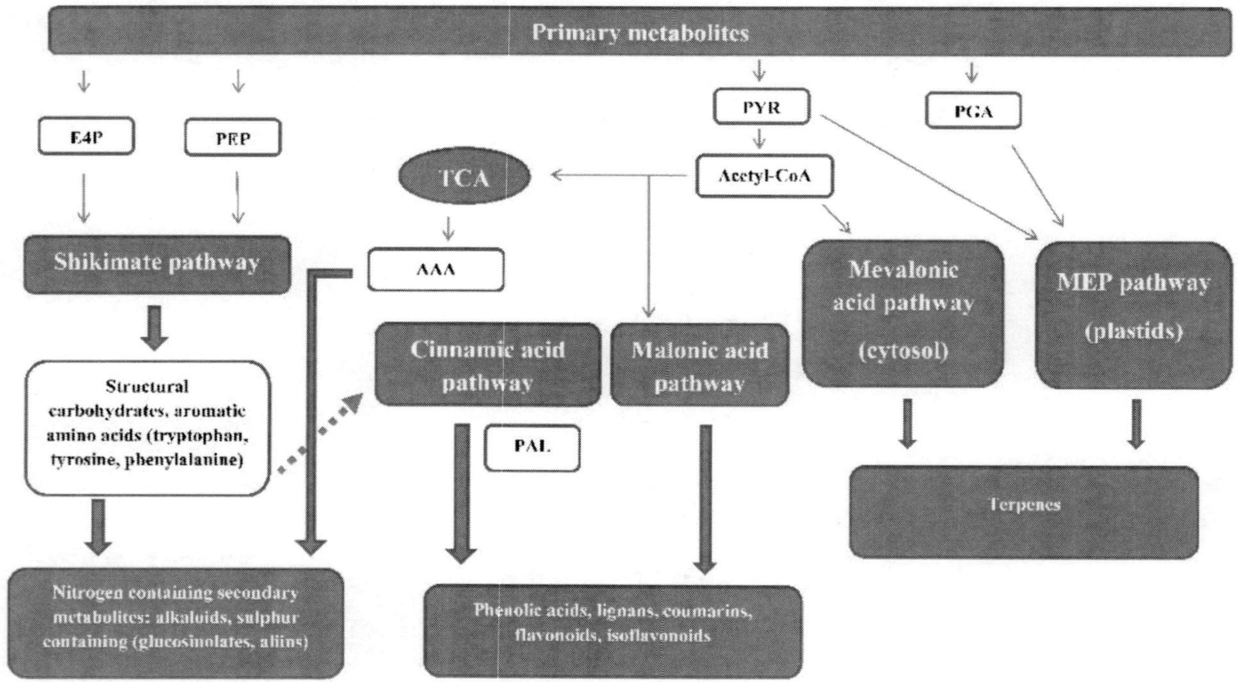

Figure 2. Schematic representation of the secondary metabolites biosynthetic pathways (Bajaj et al. 1998; Bouvier et al. 2000; Kasahara et al. 2004; Taiz and Zeiger 2006; Kougan et al. 2013). *PAL – Phenylalanine ammonia lyase (EC 4.3.1.24); E4P – erythrose-4-phosphate; PEP –phosphoenolpyruvate; PYR – pyruvate; PGA – 3-phospho-glycerate; Acetyl-CoA – acetyl coenzyme a; TCA – tricarboxylic acid cycle; AAA – aliphatic amino acids; MEP – 2-C-methyl-D-erythritol 4-phosphate.*

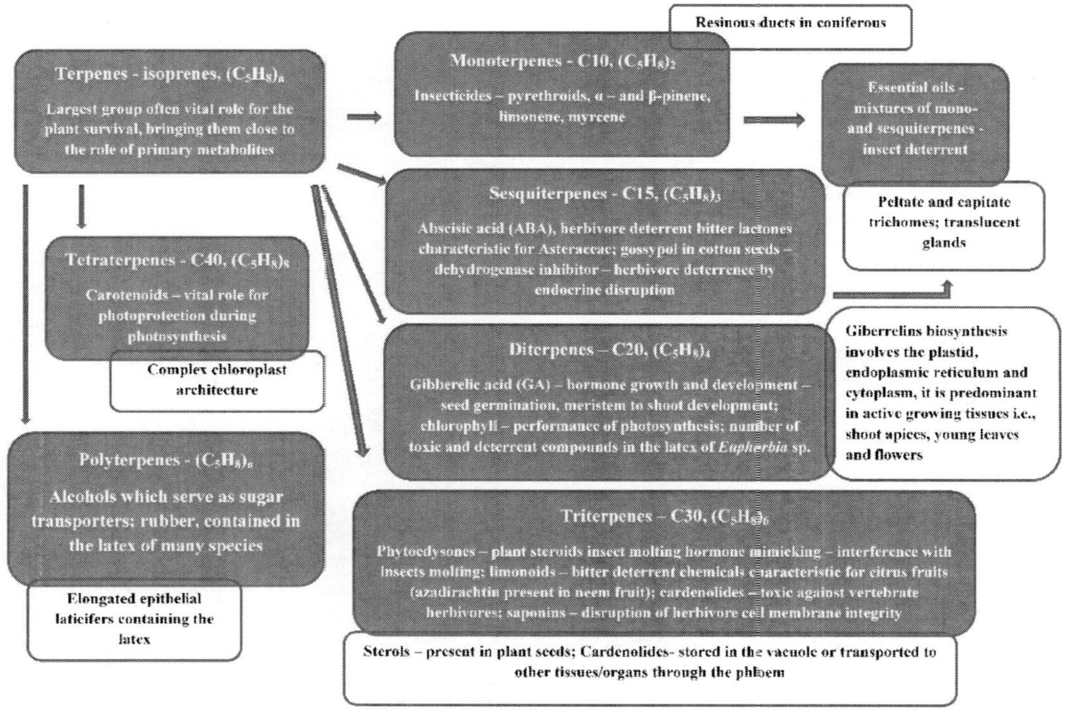

Figure 3. Schematic representation of some of the main roles of terpenes for the adaptive response and survival of the plant organism in its environment. Examples of anatomical structures for their biogenesis, accumulation secretion are denoted.

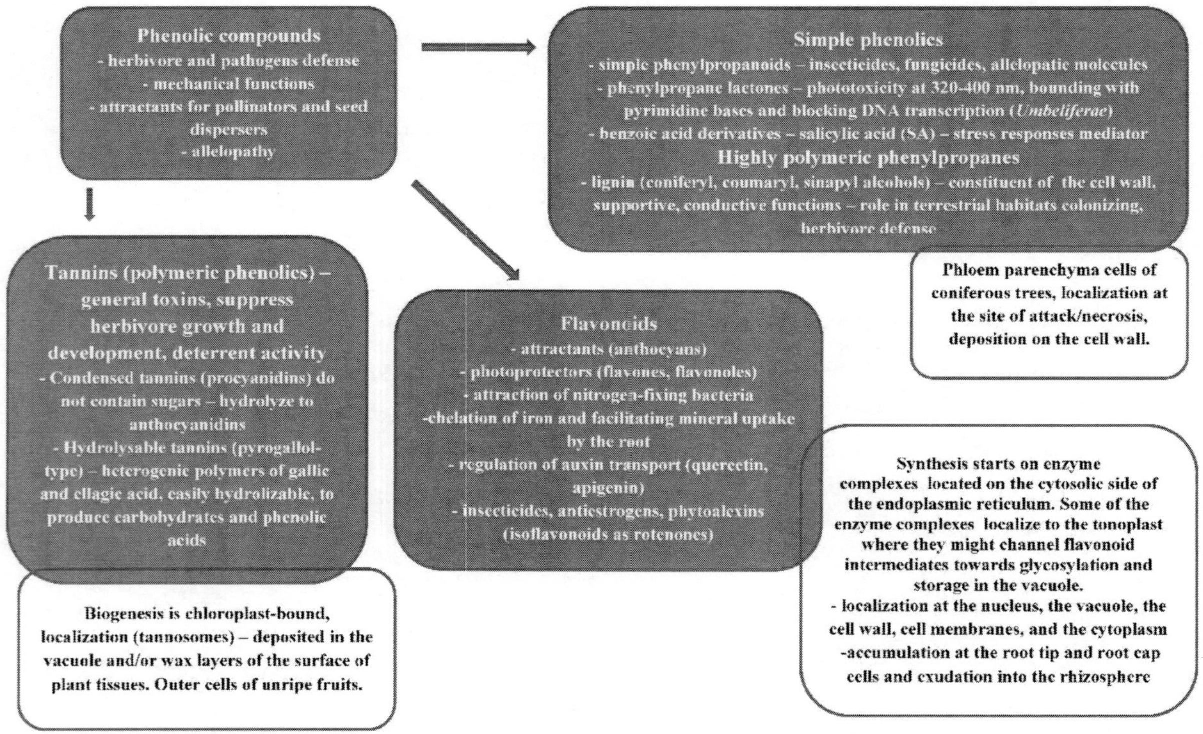

Figure 4. Schematic representation of some of the main roles of phenolics for the adaptive response and survival of the plant organism in its environment. Examples of anatomical structures for their biogenesis, accumulation and secretion are denoted.

Table 1. Effect of diverse environmental factors on the secondary metabolite levels in different species

Plant species	Secondary metabolites produced	Effect of environmental factors	Reference
Nitrogen containing compounds			
Papaver somniferum	Alkaloid levels	Morphine reaches maximum value of 1.1% in terminal and lateral capsules six weeks after fool bloom, then levels drop with 10% by the stage of dry harvest; in leaves the maximal 0.1% is at the same stage but drops by 50% at harvest.	Laughlin 1980
Pinus ponderosa *Pinus contorta*	2,6-disubstituted Piperidine alkaloids	Geographical dependence of alkaloid levels	Gerson and Kelsey 1998
Coffea arabica L. *C. canephora* P.	Caffeine and trigonelline of wild accessions (*C. arabica* - 38 genotypes originating from Ethiopia and Kenya and *C. canephora* - 38 genotypes originating from Côte d'Ivoire, Guinea, Congo, Cameroon and Central African Republic)	Higher trigonelline content for *C. arabica*, higher caffeine content and genetic diversity for *C. canephora* Both alkaloids content varied in *C. arabica* between the accessions	Ky et al. 2001
Conium maculatum	Piperidine alkaloids	Comparison of three locations with different infestation levels. Higher alkaloid production and less herbivore damage by the caterpillar *Agonopterix alstroemeriana* in locations with higher overall infestation levels.	Castells et al. 2005
Survey on the distribution of alkaloid-rich plant species in shortgrass steppe vegetation	Identification of alkaloid containing species	All alkaloid-positive taxa occurred in "water enriched" habitats, which composed less than 20% of the study area	Hazlett and Sawyer 2008
Catharanthus roseus	Vindoline, catharanthine and vinblastine	Highest vindoline and catharanthine in leaves; no vindoline in roots; highest vinblastine in blossoms; vindoline and catharanthine reach maximum between August and September; vinblastine reaches maximum after September; highest biomass in the beginning of September	Yu et al. 2010
Duboisia myoporoides	Alkaloid levels	69% higher in leaves harvested in February as compared with July or October	Ram et al. 2010
Camelia sinensis	L-theanine and caffeine	Effect of leaf age, bud and first two leaves, and shade levels: the content decreased as leaf age increased moving from bud to first and then second leaf	Song et al. 2012
Camellia sinensis	Theanine and theobromine	Highest levels in March, as compared with May and September	Fang et al. 2017
Thalictrum foliolosum	Berberine in root samples	Levels dropped with increasing the altitude	Pandey et al. 2017
Phenolic compounds			
Review on different plant species	Flavonoid compounds	Solar radiation and specifically UV-B wavelengths (280 – 320 nm) increase flavonoid levels	Caldwell et al. 1983 and references cited within

Table 1. (Continued)

Plant species	Secondary metabolites produced	Effect of environmental factors	Reference
Coffea arabica L. C. canephora P.	Chlorogenic acids of wild accessions (*C. arabica* - 38 genotypes originating from Ethiopia and Kenya and *C. canephora* - 38 genotypes originating from Côte d'Ivoire, Guinea, Congo, Cameroon and Central African Republic)	Higher chlorogenic acids levels and genetic diversity for *C. canephora.* Levels varied significantly between accessions, being dependent on the country of origin	Ky et al. 2001
Lycopersicon esculentum Mill.	Total phenolics and phenolic acids in the fruit	Comparison of effect of blocking UV solar radiation (290-400 nm). The studied metabolites were 20% higher in irradiated plants, as compared with plants which were protected from wavelengths < 380 nm.	Luthria et al. 2006
Pteridium arachnoideum	Phenolics and condensed tannins	High altitude related to higher secondary metabolite levels, attributed to sunlight exposure, elevation, and water restriction	Alonso-Amelot et al. 2007
Camelia sinensis	(-)-epigallocatechin gallate, (-)-epigallocatechin, (-)-epicatechin, and (-)-epicatechin gallate	Effect of leaf age, bud and first two leaves, and shade levels: the content increased from the bud to first and second leaves	Song et al. 2012
Camellia sinensis (L.) O. Kuntze	Proanthocyanins and O-glycosylated flavonols, phenolic acids, lignin	Shaded leaves: proanthocyanins and O-glycosylated flavonols decreased 53.37% and 43.26%, respectively; marked increase in phenolic acids and increase of lignin accumulation	Wang et al. 2012
Vitis vinifera L	Flavonols, flavanols, phenolic acids,	Effect of UV-B radiation on specific signalling pathway in the skin of fruits leading to elevated flavonols, stilbenes, protocatechuic and p-coumaric acids; phenolic acids from the methanol-insoluble fraction, flavanols, and anthocyanins did not show any significant variation	Carbonell-Bejerano et al. 2014
Camellia sinensis	Phenolic acids and flavonoid content	Sampling in March, May and September. Higher afzelechin gallate, theogallin, kaempferol coumaroyl hexoside, abundance of kaempferol derivatives and lower epicatechin and catechin in March	Fang et al. 2017
Thalictrum foliolosum	Total phenolic and flavonoid contents of root samples	Levels increased at higher altitudes	Pandey et al. 2017
Camellia sinensis	Flavonoids and tannins	Sampling at different growth stages (from April to August). Flavonols and theaflavins increased gradually; gallotannins and flavan-3-ols decreased continuously depending on the growth stage	Ryu et al. 2017

Plant species	Secondary metabolites produced	Effect of environmental factors	Reference
Terpenes			
Artemisia alba Turra	*Artemisia alba* Turra	*Artemisia alba* Turra	*Artemisia alba* Turra
Artemisia alba Turra (*A. lobelia* All.)	*Artemisia alba* Turra (*A. lobelia* All.)	*Artemisia alba* Turra (*A. lobelia* All.)	*Artemisia alba* Turra (*A. lobelia* All.)
Artemisia alba Turra	*Artemisia alba* Turra	*Artemisia alba* Turra	*Artemisia alba* Turra
Artemisia alba Turra	Essential oil (EO) terpenoid profile	Sesquiterpenoid domination in the oils of samples from Montenegro, Ljevište	Jović et al. 2001
Artemisia alba Turra	Essential oil (EO) terpenoid profile	Monoterpenoid domination - generalization of data on majority of studied populations in different parts of Europe and grown in different natural environments	Radulović and Blagojević 2010 and ref. cited within
Artemisia alba Turra	Essential oil (EO) terpenoid profile	Monoterpenoid domination – samples from calcareous soils; sesquiterpenoid domination - Serpentinite (water and Ca deficient, supraoptimal Al, Fe, Mg, Ni, Co, Cr, low plant nutrition elements, pH basic to ultrabasic) – 4 times lower oil yields	Radulović and Blagojević 2010
Artemisia alba Turra	Essential oil (EO) terpenoid profile	Four Italian populations compared – in one of them – over 60% of the oil – sesquiterpenoid domination; in the other three – comparable levels of the mono- and sesquiterpenoids	Maggio et al. 2012
Lavandula angustifolia cv. Etherio	Essential oil (EO) profile	The EO content positively regulated by temperature and flowering stage development and negatively - by rainfall during flowering period. Rainfalls remarkably decreased linalool production, which was restored within 10 days period	Hassiotis et al. 2014
Origanum compactum Benth.	Essential oil profile	Two major and one intermediate type of populations according to their chemical composition: *(i)* high levels of thymol, α-terpineol, linalool, and carvacryl methyl oxide of samples growing in regions with humid climate, clayey, sandy, and alkaline soils; *(ii)* high levels of carvacrol, α-thujene, α-terpinene, and myrcene content in plants from semi-arid climate, and growing at high altitudes and silty soils; *(iii)* intermediate type populations exposed to sub-humid climate, appeared less homogeneous and belong mainly either to the first or second group	Aboukhalid et al. 2017

Table 2. Effect of diverse agro technique treatments on the secondary metabolite levels

Plant species	Secondary metabolites produced	Effect of agro technique treatments	Reference
Nitrogen containing compounds			
Papaver somniferum L.	Alkaloid production	Experiment of light and temperature – optimal growth at 1.6 x 10^4 lux light intensity and "low temperature" (daily rhythm of 12.5/7.5°C increasing to 18.5/11.5°C during vegetation); maximal dry weight at "low temperature" and higher lux intensity of 3.2 x 10^4 lux; "low temperature" hindered development by 10-15 days and modified the ratio of organs: proportion of leaves increased, and that of the stems diminished; highest alkaloid content at "high temperatures" (daily rhythm of 12.5/7.5°C reaching to 26.0/16.0°C) and at 3.2 x 10^4 lux illumination	Bernáth and Tétányi 1981
Papaver somniferum L.	Morphine	Morphine content in the capsules increased with the increasing supply of N from 0.85 to 1.01%	Lošák and Richter 2004
Lolium perenne L.	Pyrrolopyrazine alkaloid peramine, ergot alkaloid ergovaline and mycotoxin lolitrem B.	Endophyte-free plants did not produce peramine, ergovaline or lolitrem B. Both alkaloids declined with increasing N at ambient CO_2 but remained roughly constant across N levels at elevated CO_2	Hunt et al. 2005
Camellia sinensis	Caffeine	Nitrogen, potassium and micronutrients (Zinc and Copper) stimulated caffeine production	Sedaghathoor et al. 2009
Erythroxylum novogranatense var. *novogranatense*	Truxillines	Increased alkaloid levels with increased UV exposure	Lydon et al. 2009
Duboisia myoporoides	Alkaloid levels	1 m x 1 m spacing optimal for maximizing biomass harvest and alkaloid content	Ram et al. 2010
Papaver somniferum L.	Alkaloid content	Pre-anthesis water deficit caused a reduction in the biomass production without apparent reduction in capsule yield, which implies an increased harvest index for plots under water deficit	Mahdavi-Damghani et al. 2010
Spilanthes oleracea cv. Jambuarana	spilanthol	No significant differences of spilanthol when comparing organic and mineral (urea) fertilization	Borges et al. 2012
Withania somnifera Dunal.	Alkaloid levels	Highest alkaloid yield at 3000 ppm CCC (chlormequat chloride – GA biosynthesis inhibitor and growth retardant), at 30 x 20 cm planting distance (as compared with IAA and GA treatments and 30 x 30 cm plot).	Shukla and Shukla 2012
Colonization of *Claviceps purpurea* on winter rye (*Secale cereale* L.)	Ergot alkaloids	Application of gametocidal agents - maleic hydrazide and 2-chloroethylphosphonic acid to enhance *C. purpurea* infestation on unfertilized ovaries	Hanosová et al. 2015

Plant species	Secondary metabolites produced	Effect of agro technique treatments	Reference
Lupinus angustifolius	Sum of angustifoline, isolupanine, lupanine, and 13-hydroxylupanine	Alkaloid content was significantly influenced by the growing system, year and genotype. Higher N-content in the soil was related to higher concentrations of alkaloids; thus organically grown crop displayed lower productivity than the conventionally grown one	Jansen et al. 2015
Piper longum Linn.	Assay of total alkaloid content	"Protected cultivation system" experiments. Highest alkaloid production was recorded by planting in hanging pots and fertilization through mist	Jayanth et al. 2015
Ephedra sinica	Total alkaloids, ephedrine and pseudoephedrine	Achieved stabilization of high alkaloid levels by selective breeding and stolon propagation; selection criteria after 3 years of cultivation – total alkaloid content, high ephedrine/pseudoephedrine ratio, minimal variation between lowest and highest total alkaloid values, stolon formation capacity of the individuals	Hiyama et al. 2017
Phenolic compounds			
Geranium thunbergii Sieb. Et Zucc.	*Geranium thunbergii* Sieb. Et Zucc.	*Geranium thunbergii* Sieb. Et Zucc.	*Geranium thunbergii* Sieb. Et Zucc.
Camellia sinensis	*Camellia sinensis*	*Camellia sinensis*	*Camellia sinensis*
Vanilla planifolia A.	*Vanilla planifolia* A.	*Vanilla planifolia* A.	*Vanilla planifolia* A.
Brassica campestris L. ssp. *chinensis* var. *communis*	Flavonoids and hydroxycinnamic acids	Pre-harvest UV-B irradiation resulted in elevated hydroxycinnamic acid derivatives at low temperature (9°C) and of flavonoids at ambient temperature (22°C)	Harbaum-Piayda et al. 2010
Camellia sinensis	Optimal biomass formation and meeting market specifications of "Light leaf" and "Dark leaf" tea.	Production of hybride, biclonal and polyclonal seeds; spacing, pruning, mineral nutrition, shading, drainage, irrigation, pest management, soil rehabilitation (growing Guatemala grass, *Mimosa invisa*, Pusa Hybrid Napier for 18-24 months).	Barbora 2013
Lactuca sativa *(red leaf lettice)*	Cyanidin -3-O-(6"-O-malonyl)-glucoside and caffeoylmalic acid	Cultivation in greenhouse at low temperature stimulates anthocyanin biosynthesis	Becker et al. 2014
Calendula officinalis L.	Flavonoids	Blossom-mass and flavonoid levels related to the developmental stage - highest yield of flavonoids was obtained three days after anthesis	Honório et al. 2016
Lactuca sativa L. var. *crispa* L. cv. *Eventai* RZ and L. *sativa* L. var. *crispa* L., cv. *Satine* (Red Oak Leaf and Red Lollo lettuce)	Phenolic acids, flavonol and flavone glycosides, anthocyanidin glycosides	CO_2 enrichment resulted in high yields of red leaf lettuce rich in phenolic compounds (most flavonoid glycosides and only some caffeic acid derivatives)	Becker and Kläring, 2016

Table 2. (Continued)

Plant species	Secondary metabolites produced	Effect of agro technique treatments	Reference
Morus sp.	Tannins, phenolics	Elevated CO_2 increased biomass formation, primary metabolites (sugars, carbohydrates) and phenolic compounds	Lavanya et al. 2017
Lavandula angustifolia Mill.	Total phenolic and flavonoid content	Established optimal K fertilization for enhanced phenolic and flavonoid content	Chrysargyris et al. 2017
Terpenes			
Spilanthes oleracea cv. Jambuarana	*Spilanthes oleracea* cv. Jambuarana	*Spilanthes oleracea* cv. Jambuarana	*Spilanthes oleracea* cv. Jambuarana
Satureja hortensis L.	*Satureja hortensis* L.	*Satureja hortensis* L.	*Satureja hortensis* L.
Melissa officinalis	Citronellal, β-caryophyllene, neral, geranial and geranyl acetate	Vermicompost, biophosphate and chemical fertilizers treatments significantly increased the dominant component content	Mafakheri et al. 2016
Rosmarinus officinalis	Essential oil profile comparison based on 32 components	Vermicompost treatment stimulated phenolic compounds in the essential oil	Ganjali and Kaykhaii 2016
Lavandula angustifolia Mill.	Essential oil profile with main constituents 1,8-cineole, borneol, camphor, α-terpineol, myrtenal	Established optimal K fertilization levels for stimulated root growth and stimulated essential oil yield	Chrysargyris et al. 2017

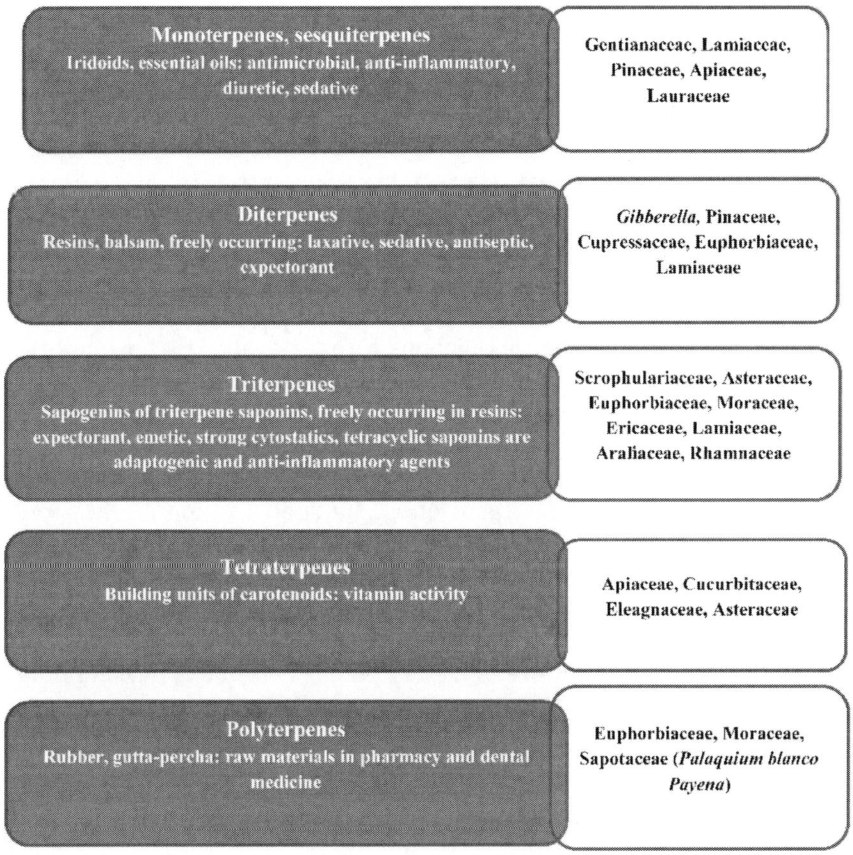

Figure 5. Schematic representation of some biological activities and plant sources of the main groups of terpenes.

- *Phenology state* and the role of growth and development to fulfill the different roles in the plant's life cycle. The phonological stage is decisive on the secondary metabolites produced and has been studied for numerous species of economic importance (Çirak 2007 and references cited within);
- *Genotype and heredity* are also crucial for the production of secondary metabolites. Together with the selection of proper agrotechniques the selection of highly producing genotypes plays a central role for selection starting material for field cultivation (Nejadhabibvash et al. 2012; Kitzberger et al. 2013; Danova 2015; Fang et al. 2017);
- *The role of agrotechnique* for the production of secondary metabolites on the field. Domestication and introducing plant species from their wild habitats to field conditions is related to alteration of many factors which would have been responsible for the production of certain secondary metabolites in their indigenous environments (climatic challenges, soil chemistry, pressure of

specific concrete insects and herbivores). Thus the correct agrotechnique measures are vital for the proper productivity of the crop (Table 2).

3. Application of Plant Derived Secondary Metabolites in Pharmaceutical Practice

Many biologically active compounds of plant origin find medical application as therapeutic molecules either in their indigenous form or as models for the total synthesis and also as matrices for the semi-synthesis of drugs. Examples for terpenes (Figure 5) and phenolics (Figure 6) are illustrative to the wide array of pharmacological activities of secondary metabolites for the human organism (Taiz and Zeiger 2006; Evans 2009).

According to the World Health Organization, about 80% of the World population still largely depends on medicinal plants as a traditional therapeutic method. In addition, up to now, 25% of the prescribed medicines and 50% of the "Over the counter drugs" in industrialized countries are of plant origin (Bajaj et al. 1988; Tripathi and Tripathi, 2003). In addition, about 52% of the anti-inflammatory and 51% of the anticancer drugs are also of plant origin (Arora 2010 and references cited within).

The wide application of secondary metabolites requires the possibility for their abundant delivery for the needs of the practice. Many compounds are currently synthetically produced (for example caffeine, theophylline, theobromine, ephedrine, pseudoephedrine, emetine, papaverine, L-dopa, salicylic acid). Nevertheless, due to their structural complexity, still many therapeutic molecules are being derived from the raw plant material (Bajaj et al. 1988).

Thus, obtaining raw plant material with desirable quality (proper biomass formation and stable and sufficient concentrations of the secondary metabolites) still holds a vital role for the supply of therapeutic molecules, vital for human health and survival.

These qualitative and quantitative characteristics are strongly dependent on the origin of the plant material. Often cultivation in the field of the target crop leads to poor results as compared to the characteristics of the material obtained from the wild habitat, the reason being the dependence of secondary metabolism on the interrelations of the plant organism with its surrounding environment, as discussed above. With its many applications and options, plant cell tissue and organ culture comes as an indispensable supplementary tool to conventional breeding techniques for meeting the needs of medicinal and aromatic plants raw material supply.

Figure 6. Schematic representation of some biological activities and plant sources of the main groups of phenolics.

4. PRINCIPLES OF PLANT CELL TISSUE AND ORGAN CULTURE BREEDING

Plant biotechnology techniques are based on the possibility to cultivate separate cells, tissues, organs or integral plants in culture medium and sterile conditions out of the indigenous natural environments of the plants. This is possible due to the property "totipotency" of the plant cell and its capability to regenerate up to a whole integral organism.

The contemporary stage of development of plant cell tissue and organ cultures techniques has turned the method into a standard procedure technique for modern biotechnology with the following main areas of application (Loyola-Vargas and Vázquez-Flota 2006; Dias et al. 2016):

- Large-scale propagation of plant materials with superior (elite) quality;
- Generation of genetically modified fertile individuals;
- Model system for fundamental plant cell physiology aspects;

- Preservation of endangered species,
- Metabolic engineering of fine chemicals.

4.1. The Plant Cell

The cell is the main building unit of the living plant organism, maintaining the integrity between structure and function (Teiz and Zeiger, 2006). There are several features distinguishing the plant from the animal cell:

- Green plants are the primary producers on the planet, due to their capability to assimilate solar energy and include it into macroergic compounds, during photosynthesis. Thus solar energy is being transformed into chemical energy. Hence are the structural and functional elements unique of the plant cell:
    - Presence of chloroplasts (the chlorophyll containing organelles where photosynthesis occurs);
    - Nucleus of the plant cell (responsible for encoding of enzymes and proteins, realizing the coordination of all processes taking place within the cell);
    - The plant cell membrane (allowing gas and water to pass in and out of the cell while controlling the passage of other molecules);
    - The vacuole (containing cell sap to keep the cell turgid, serving as storage cell compartment where many water soluble polar secondary metabolites are deposited);
    - Cell wall (lignin and pectin incorporated structure which serves for cell strengthening);
    - The cytoplasm (the spatial localization of the enzymes and other proteins used in photosynthesis, primary and secondary metabolites production);
- As already discussed, plants have an attached way of life. Therefore they have evolved complex mechanisms to direct their growth during their whole individual development towards vital resources as light, water and mineral components;
- Terrestrial plant species develop specific supporting structures, in order to accomplish biomass formation towards the sunlight and against the forces of gravity;
- The latter plants are also characterized by the constant loss of water through evaporation, and have therefore established specific mechanisms for prevention of drying;
- Plants are also capable of moving water and minerals from the soil towards the sites where photosynthesis and growth are accomplished, as well of transporting the photosynthetic products towards the non-photosynthesizing tissues and organs;

- The plant cell membrane is surrounded and enforced by a hard cell wall, as neighboring cells are cemented by a pectin middle lamella. As a consequence of this, the development of the plant organism depends entirely on factors as cellular division and growth;
- Plant cell division is realized at specific spatially defined sites called "meristematic centers". They constitute the cell apparatus, in which the processes of cell division, growth and differentiation determine the size, form and structure of the whole plant organism. Neighboring to the sites of cell division are the sites of elongation in which the cells enlarge in width and length. After elongation, the cells undergo differentiation into a specific cell type. Except differentiated cells, however, during its whole life, the plant also produces non-differentiated meristematic cells, which are capable of the initiation of any type of cell, tissue or organ necessary for the plant's survival and development at any time of its individual development. This autonomous feature of the plant cell to store and express at any moment the whole genetic information of the whole organism is called "totipotency".

Plant cell tissue and organ culture is based on cell's totipotency and the possibility to regenerate the whole plant organism from a single cell.

4.2. *In Vitro* Culture Initiation

Sterile working conditions require the sterilization also of starting plant germplasm, which is used for the *in vitro* culture induction. Irrespectively of the starting material origin (collected from the wild or from controlled field/pot cultivation, seeds, above- or underground parts explants) it is vital to provide the maximal possible extent of environmental pathogens elimination without excessive harm being imposed on the vitality of the explants. This is usually achieved by surface sterilization. Initially the explant might be rinsed in up to 70% ethanol which provides on one hand an initial pre-sterilization, but also provides wetting of the explant, due to dissolving of the waxy layer. Then disinfection is performed by the aid of sodium hypochlorite, calcium hypochlorite, mercuric chloride, silver nitrate, hydrogen peroxide, benzalkonium chloride, etc. A detergent, for example Tween 20 or 80 could be added in order to increase the wetting of the explant by the sterilization agent. The choice of the sterilization agent, timing of sterilization, as well as the thorough rinsing with sterile distilled water of its traces are crucial for providing the vitality of the future development and normal morphogenesis of *in vitro* bred plants. After sterilization the explant is placed in medium suitable for either direct morphogenesis of differentiated organs (from existing meristematic centers) or callusogenesis or indirect morphogenesis of differentiated organs occurring through the initially induced callus tissue. The obtained cells or tissues are further on transferred to maintenance medium and the cultivation of the respectively obtained lines takes place.

4.3. Fundamental Aspects of Plant Biotechnology and Requirements for the Maintenance of Plant Cell Tissue and Organ Cultures

The plant cell and tissue cultures are a prospective additional tool and even sometimes an alternative in plant breeding and also in the yield of secondary metabolites in standardized conditions. Due to the long years of fundamental development and achievements, plant biotechnology offers a number of advantages as compared with the conventional plant breeding in the field (Petersen and Alfermann 1993; Murch et al. 2000; Alfermann et al. 2003; Arora 2010):

- Micropropagation and reintroduction into nature of rare, valuable or endangered species. The approach allows for the vegetative reproduction of plants which do not bear seeds, or their germination is difficult or slow;
- Due to the aseptic way of work the obtained plant material is pathogen-free, which makes it possible to quickly reproduce large quantities of clean planting material;
- The limitations on mass propagation of plant material, such as disease, pathogen invasion and unfavorable climatic conditions are being avoided;
- The rapid introduction and testing of new cultivars, selected by the methods of genetic engineering is made possible;
- There is a constant access to cultivated plant material, irrespectively of the fluctuations of the economic, climatic and geographic conditions, characteristic for the natural habitats of the plant species. The losses due to crop destruction by pathogens or herbivores are excluded;
- The standardization of the breeding conditions is related to standardization of the content of the active ingredients, and the yield of toxic/narcotic compounds is accomplished in a strictly controlled environment;
- The yield of secondary metabolites of rare or endangered species is conducted at laboratory conditions, without harming their natural habitats;
- The biotechnological breeding is flexible, and is easily adapted to the market requirements;
- In addition the technique is compact and allows for large scale breeding material preparation at the expense of comparatively small working space requirements as compared with the conventional plant nursery facilities;
- All the above factors contribute for the *in vitro* culture system of medicinal and aromatic plants to become a source of novel natural compounds;

- The *in vitro* plant system is also a source of the isolation of a broad array of enzymes, which can be utilized in the conduction of positional- and stereo-specific chemical reactions, which would otherwise be complicated to be conducted;
- Plant *in vitro* systems are a valuable fundamental model for the elucidation of the biosynthesis of secondary metabolites. They are able to provide plant genetic material unified in its characteristics and in almost unlimited quantities. In addition, the metabolic activity in these conditions is high and for several weeks it is possible to achieve secondary metabolite biosynthesis, which in the natural conditions would require months or even more.

Generally plant cell tissue and organ culture is the cultivation of living plant tissue (with different differentiation degree), at constant environmental conditions and culture medium in which the culture is being developed. *In vitro* growth and development are in complex dependence of numerous factors such as:

4.3.1. Genotypic Prerequisites of the Plant

It is important whether the species is mono- or dicotyledonous, whether it is a gymnosperm or angiosperm, whether it is herbaceous or arboreal, seed setting or not. In medicinal plants cultivation, the susceptibility of the tissues towards the different techniques is also strongly dependent on the type of metabolites produced.

4.3.2. Factors of the Culture Medium. Approaches for Enhancement of Secondary Metabolite Production through Medium Optimization

- *Water constitutes* up to 95% of the culture medium, therefore its quality is of crucial importance. Plant tissue culture technique requires the utilization of distilled water, and protoplast and meristem tissue culture – bi-distilled water;
- *Solidification of the culture medium* is most often performed with agar (linear polygalactoside prepared from red algae (most often of the genera *Gelidium* and *Gracilaria*). Depending on the purposes agar concentration could vary from 0.6 – 0.8%. Agar concentration influences the availability and extent of assimilation of the medium components by the explant. It has been established that agar stimulates shikonin production in *Lithospermum erythror-hizon* suspensions (Fukui et al. 1983). It was further on established that agar interacts with the mineral salts in the medium, thus affecting their availability and up-take capacity by the plant cells (Gribble et al. 2002);
- *Sugar* is exogenously supplemented in the culture medium due to the lower effectiveness and sometimes complete lack of photosynthesis and carbon fixation in *in vitro* conditions. However, *in vitro* photosynthesis has been established for

certain species (Pospišilová et al. 1997). The degree of plant heterotrophy (full dependence on exogenously supplied sugar), mixotrophy (the capability of partial carbon fixation) and even autotrophy (effective *in vitro* photosynthesis and "autonomous nutrition") do not only depend on the photosynthetic capacity of the concrete species but also on the culture medium composition, volume and aperture of the cultivation vessels and on their material and closure (Fujiwara et al. 1989; Kozai 1991; Kozai & Smith 1995; Kozai et al. 2002). According to Lucchesini et al. (2009) at optimal "mixotrophic conditions", *in vitro* shoots could perform photosynthesis simultaneously with the assimilation of exogenously applied sucrose in the culture medium. Therefore the modernization and scale-up of micropropagation techniques often utilize the methods of autotrophic culture in which the plant independently performs photosynthesis. Sugar concentration can vary significantly (1 – 6%), as the modification of this factor is widely utilized in plant biotechnology. For sugars supplementation most often sucrose is used, but some monosaccharides have also been applied (for example glucose and mannitol). Noteworthy, the quantity of the given monosaccharide, equivalent to the respective disaccharide, reduces in double its osmotic potential and hampers water utilization by the explant;

- *The main macro-elements* which are mandatory in the formulation of the culture medium are most often nitrogen, phosphorus, potassium, calcium, magnesium and sulphur;
- *Micro-elements* being supplied are iron, zinc, boron, manganese, copper, cobalt, nickel, aluminum, molybdenum, iodine. The macro- and micro-elements are being supplied in the form of soluble salts, and iron is usually chelated with EDTA in order to avoid its precipitation;
- *Growth regulators (plant hormones)* are vital endogenous factors of growth and development of the plant organism. The best studied groups of hormones are the auxins, gibberellins, cytokinins, ethylene and abscisic acid. The last years led to widening the knowledge on the steroid plant hormones brassinosteroids, which regulate diverse processes of differentiation in the plant organism. A number of other molecules are also related to the herbivore and pathogen resistance, such as jasmonic and salicylic acids (Taiz & Zeiger 2006). With the elaboration of molecular analytical methods in plant physiology, the number of hormones and hormone-like signal molecules is increasing. The exogenous supplementation of plant growth regulators is crucial for targeting, and maintenance of the degree of de-differentiation, micropropagation and rooting, as well as influence of *in vitro* secondary metabolites production. Cytokinins and auxins are usually applied at concentrations 0.01 - 0.5 - 5.0 mg/l. The effect of different plant growth regulators (PGR) on secondary metabolite production has been widely explored in plant cell, tissue and organ culture. A number of literature sources discuss on

the inhibitory activity of 2,4-D on secondary metabolite production in non-differentiated cell cultures such as berberine (Nakagawa et al. 1986), antraquinones (Zenk et al. 1975) and indole alkaloids (Zenk et al. 1977a). In *Morinda citrifolia* cell cultures, low auxin concentrations led to high anthraquinone levels and low growth rate, high auxin concentrations have the reverse effect; 2,4-D was active at approximately a hundred-fold lower concentration than NAA (Hagendoorn et al. 1997). This pattern of inhibitory activity of 2,4-D on several types of secondary metabolites in cell cultures of diverse species, might be attributed to the larger degree of de-differentiation brought by this hormone. Therefore, auxins utilized for callusogenesis are either applied at low concentrations or are being removed from the medium during optimization of secondary metabolites production *in vitro* (Misawa 1994). The alteration of two-stage cultivation has been often applied with the purpose of firstly obtaining intensive biomass formation and then – production of secondary metabolites by reducing the growth rate of the cultures. Examples are the production of shikonin in cell cultures of *Lithospermum erythrorhizon* (Tabata & Fujita 1985), rosmarinic acid by cell culture of *Colleus blumei* (Ulbrich et al. 1985), berberine from cell cultures of *Coptis japonica* (Fujiata & Tabata 1987). The latter examples are illustrative of the stimulation of de-differentiation and inhibitory effect of auxins on secondary metabolites production by non-differentiated cultures. The opposite effect of the same plant growth regulator however has been observed in the differentiated genetically modified "hairy roots" of a number of medicinal plant species. Thus, auxin supplementation (IBA, IAA and NAA) in the culture medium of *Rubia akane* stimulated both growth and alizarin and purpurine production (Park & Lee 2009). Similar results were obtained in lobeline production in *Lobelia inflata* (Yonemitsu et al. 1990), valepotriates in *Centranthus rubber* (Granicher et al. 1995), rutin in *Fagopyrum esculentum* (Lee et al. 2007) and withanolide A in *Withania somnifera* (Murthy et al. 2008). In addition to this finding, in hairy root cultures of *Cichorium intybus* L. cv. Lucknow the combination of even low concentrations of cutokinin (kinetin) in combination with auxin led to de-diffreniation, callus formation and subsequent drop of coumarin content in the cultures (Bais et al. 2001).

- *Amino acids and inositol* – organic compounds such as aminoacids (for example glycine) and the polyol inositol are also exogenously applied to the culture medium;
- *The pH value* of the medium is an important factor for the proper conduction of the physiological processes of the *in vitro* grown plant. Usually pH values vary within the interval 5.5 – 6.0;
- *Other components of the culture medium* – activated charcoal being finely porous carbon material finds wide array of applications in tissue culture. Its effects may

be attributed to establishing a darkened environment for the root; adsorption of undesirable/inhibitory substances; adsorption of growth regulators and other organic compounds, or the slow release of growth promoting substances present in or adsorbed by activated charcoal (Pan and van Staden 1998);

The most broadly used culture medium formulations in plant biotechnology are the ones of Murashige and Skoog (1962), White (1963), Linsmaier and Skoog (1965) and Gamborg (1968).

When discussing the issue of media optimization of secondary metabolite production it is vital to consider that growth and biomass formation are often not related and in many cases have different requirements. As discussed above, secondary metabolites production is related to the interactions of the plant organism with its environment (Harborne 1996). Therefore, the successful application of the methods of plant biotechnology for stimulation of secondary metabolism, the knowledge of the physiological role of the concrete secondary compounds for the plant physiology is of major importance. Thus, in literature there are examples of enhancement of secondary metabolite production achieved by limitation of nutritive resources in the medium (most often decreasing the content of nitrates, potassium, ammonium and phosphates salts results in secondary metabolites stimulation *in vitro*) (Petersen and Alfermann 1993). In this way by providing all other optimal breeding conditions and limiting growth resources, it is possible to stimulate secondary metabolism at the expense of primary metabolism.

4.3.3. Factors of Culture Conditions

- *Sterility* is a major principle for all *in vitro* culture manipulations. This is due to the application in the culture medium of a wide array of components which stimulate biogenic growth and development not only of the plant organism but also of different pathogens like bacteria, fungi and mould. Therefore plant biotechnology is a method for the aseptic cultivation of plant material, which is accomplished by the sterilization of culture media, vessels and instruments. All manipulations are carried out in sterile environment. This is most often achieved by autoclaving at 121°C (1 atm) of consumables, instruments and culture media and work under a constant flow of sterile air in the laminar cabinet. Thermo-degradable components are supplemented to the media by sterile filtration (pore size 0.45μm);
- *Cultivation space* – it is needed to provide a controllable temperature regime. Usually the temperature values are set between 24 - 26°C but this parameter might vary considerably according to the purposes of the cultivation. Thus cold treatment might be applied for the purpose of hardening of explants and higher temperature – for Heat Shock Proteins production and inducing adaptation

reaction towards diverse types of stress. Modification of the temperature regime is a factor in biosynthesis of both primary and secondary metabolites. Thus, the total content of fatty acids is being increased in reducing temperatures in *Catharanthus roseus* suspensions (due to the rise of unsaturated C18 fatty acids), as this effect does not influence indole alkaloids content (Toivonen et al. 1992). Increasing temperatures up to 28°C and 30°C in *Lavandula vera* cell suspensions, has shown the achievement of maximal biomass and rosmarinic acid, respectively (Georgiev et al. 2004). According to the research of the same working group, aeration and agitation in bioreactor conditions stimulate rosmarinic acid production as well (Pavlov et al. 2005);

- *Illumination* is provided by luminescent tubes, whose spectral characteristics, intensity and photoperiod are selected according to the plant species requirements, but also can be a valuable experimental tool according to the culture type and results which are targeted;
- *Humidity, oxygen and carbon dioxide content* depend largely on the microclimate inside of the culture vessels (related to their volume and shape, the aperture diameter and closure type). It is important to also notice that the oxygen which is dissolved in a liquid medium is to a higher extent available for the explant, as compared to the one in agar solidified medium. In addition, gaseous supply is largely controllable in bioreactor systems and is a widely used variable in targeting secondary metabolite production *in vitro.*

4.4. Role of Differentiation and Morphogenesis in Plant Biotechnology. Main Types of *In Vitro* Cultures

The plant organism consists of different organs, which are constituted of different tissues, build up by cells. The vegetative body of vascular plants consists of three main organs: leaves (with its primary vital function of performing photosynthesis), stem (having supportive and translocation functions) and roots (having supportive and assimilation functions). According to the degree of organization of cultured cells *in vitro* cultures can be classified into differentiated and non-differentiated types (De Fossard 1977; Pierik 1987; Razdan 2003).

However, secondary metabolites production most often requires a certain differentiation level of plant tissues and is often lacking in non-differentiated cultures. Therefore secondary metabolite production in non-differentiated cultures often requires additional measures for optimization the production of secondary metabolites which are often empirical. Examples for such approaches will be discussed below.

4.5. Differentiated Cultures. Role of Differentiation in Successful *In Vitro* Yield of Secondary Metabolites

Differentiated cultures are characterized by a diverse degree of organization and communication between the cells, tissues and organs. These are:

- *Cultures of integral sterile plant.* Such cultures can be obtained by directly germinating seeds in sterile conditions, or through morphological differentiation of organs from non-differentiated culture types. They have the highest level of morphological development and communication between the three organ types. In these cultures it is mostly achievable to obtain the anatomical structures needed for production, translocation and accumulation of secondary metabolites;
- *Embryo cultures* are integral whole plants which are obtained by discarding the seed covering, excising of the embryo and growth of the integral plant from it. This approach is applied in cases when for diverse reasons rescue of the fertilized material is required, for example in the case of seeds which have hard covering which hampers the germination process;
- *Organ cultures* are separate plant organs grown in culture medium such as shoot cultures (aerial parts), root cultures, etc. The "hairy roots" are a specific genetically modified root culture type, where genetic material has been transferred to the plant genome from the Ri plasmid of *Agrobacterium rhizogenes* through inoculation of a wounded segment on the plant body and the subsequent infection of the plant with the bacterium. The inserted genes encode proteins which participate in the biosynthesis of hormones stimulating the excessive cell multiplication at the site of the infection. The obtained cell mass produces opines, which serve as nutritive substrate for the plants. This phenomenon is a creation of a biological niche by the bacterium which enables it to develop and reproduce. After the integration and expression of the genetic information, the "hairy root" phenotype can be observed – which is manifested as increased branching and negative gravitropism of the roots. Over 450 species have been reported to be susceptible to the Gr (-) *Agrobacterium rhizogenes* infection. The experimental application of "hairy root" cultures dates from 1980. The main advantage of these cultures is that they are "hormonally autonomous" and can maintain the state of root morphogenesis without formation of aerial parts and exogenous supplementation of growth regulators to the medium (Pistelli et al. 2010).

There are many examples of species being cultivated *in vitro* as differentiated cultures, such as conventional (Becker and Chavadej 1988; Maurmann et al. 2006) and transformed (Gränicher et al. 1992; Gränicher et al. 1995) roots of *Valeriana officinalis*; differentiated shoots of *Hypericum perforatum* (Kirakosyan et al. 2004).

Noteworthy, the development of transformed "hairy roots" often allows for the production of secondary metabolites characteristic for the aerial parts of the respective plants. Thus, artemisinin, isolated of the aerial parts of the field grown *Artemisia annua*, has been successfully yielded from genetically transformed roots of the plant (Weathers et al. 1994; Jaziri et al. 1995; Liu et al. 1999). Another example is the production of the 2-OH-1,4-naphtoquinone lawsone by genetically transformed roots of *Lawsonia inermis* (Bakkali et al., 1997). In addition, important advantages of "hairy roots" are their genetic stability, hormonal autonomy and the relatively constant levels of secondary metabolites produced.

Sometimes, however the successful scale-up of differentiated cultures to bioreactor scale might be difficult due to problems arising from the necessity to supply unified amounts of aeration and nutritive compounds, as well as illumination to a non-homogenous biomass.

4.6. Non-Differentiated Cultures. Main Principles and Approaches to Enhance Secondary Metabolite Production

In this culture type the plant cells are not organized into tissues and organs.

4.6.1. Protoplast Cultures are Obtained by the Enzymatic Degrading of the Cell Wall and the Subsequent Maintenance of Plant Cells Which are Only Surrounded by the Cytoplasmic Membrane

The fusion of protoplasts of different species is a widely utilized approach for obtaining interspecies hybrids.

4.6.2. Callus Cultures are Aggregates of Non-Differentiated Plant Cells Cultivated on Solid Medium

They are obtained through inducing de-differentiation of differentiated plant tissue by means of plant growth regulators treatment of diverse explant types. The balance between cytokinins and auxins is decisive for maintenance of the plant culture in the non-differentiated state. As discussed above, the degree of de-differentiation and "multiplication speed" of the culture are usually inversely related to the secondary metabolites produced by it.

4.6.3. Suspension Cultures are Cells or Actually Cell Aggregates with Different Extent of Aggregation, Cultivated in Liquid Culture Medium

In order to provide a unified nutritive compounds and oxygen supply to the whole cultivated biomass, agitation of culture vessels, usually on an orbital shaker, at 30 – 150 rpm is applied. Depending on the cultivation method plant suspensions can be classified into:

- *Batch cultures* are maintained through a regular sub-culture of a small aliquot part of the generated biomass into fresh culture medium after a defined period of time. Biomass formation has the following sequence: *lag phase* immediately after transfer into the fresh medium – no new growth is occurring yet; *exponential phase* after the adaptation and overcoming the stress of sub-culture manipulations, a period of intensive cell growth is observed; *stationary phase* after 3 – 4 consecutive cell divisions, the nutritional resources are being exhausted and toxic by-products of cellular metabolism are being accumulated, growth slows down and cellular density remains constant. If no sub-culture occurs at this stage, the cell number gradually starts decreasing and the suspension line cannot continue its existence. These suspensions are usually maintained in 100 – 250 ml Erlenmeyer flasks containing 20 – 75 ml of liquid culture medium;
- *Continuous cultures* - this technique provides for the long-term cultivation through the constant supply of fresh medium into the system. Therefore cultures are grown in special vessels, which would allow for the sterile supply of fresh medium and discarding of the old one. These cultures might be of closed type when only the exhausted medium is discarded from the system, but the cells remain and increase their density with a point when saturation is achieved. The open type continuous cultures are characterized with drawing out equal volumes of nutrition medium and cells. The latter line is capable of prolonged maintenance.

4.6.4. Main Practical Problems in Utilization of Non-Differentiated Cultures for Secondary Metabolites Production

Although callus and suspension cultures are characterized by a faster reproduction and biomass formation, as compared with differentiated cultures, practical problems arise which are being solved in dependence of the genotypic peculiarities of each concrete species, being developed.

- *Lack of synchronization of growth and development* - growth and cell division of the many cells/cellular aggregates in the vessel are usually not synchronous in these types of cultures. Therefore a biomass of cell units with different age, instability of enzymatic processes and hence instability of secondary metabolites production is observed. This problem is solved by measures for synchronization through different physical and chemical methods.

- *Genetic instability* – this is another major problem due to the intensively ongoing cell division. Therefore, plant organs or whole plants differentiated and regenerated of non-differentiated plant tissues (indirect morphogenesis) are mostly likely to differ from the parent plants. Desirable or non-desirable traits could be obtained in this way. Selection, identification, isolation and clonal propagation of cultivars with desirable qualities are often applied.
- *Lack of specialized anatomic and morphologic structures* needed for secondary metabolite production, translocation and accumulation. In spite of the many advantages of non-differentiated *in vitro* cultures (such as the rapid growth, the numerous possibilities for optimization of secondary metabolite production), there still remains the main drawback, namely the lack of differentiation. Although the plant cell is characterized by its totipotency, the biosynthesis of many secondary metabolites still requires a certain level of differentiation of the plant tissue. The development of differentiated cultures is a solution when the production of secondary metabolites cannot be realized in the non-differentiated tissues. Thus, it has been established that although the non-differentiated culture displayed a lack of certain secondary metabolites, the induction of somatic embryogenesis, root or aerial parts led to the regaining of the biosynthetic activity of the plant tissues (Charlwood et al. 1990). Another example is the finding that although alkaloids were only present in traces in the suspension of *Leucojum aestivum*, differentiation and subsequent aerial parts induction led to a significant increase of alkaloid content (Pavlov et al. 2007). The lines selected in the latter work produced over 200μg/gDW in liquid medium and displayed stability regarding growth and secondary metabolites production in the liquid system.

Thus plant cell tissue and organ culture systems represent a prospective complementary tool to conventional breeding techniques for the delivery of desirable secondary metabolites. In order to successfully utilize the potential of this valuable technique, however it is necessary to be able to interpret the physiological behavior of the plant as a result of the different optimization measures and relate it to the resulting increased or decreased levels of metabolites produced. There are many examples of the successful delivery of plant secondary metabolites through plant cell tissue and organ cultures, where the productivity has surpassed the levels obtained for the source “mother” plant (Table 3).

Analysis of the plant response to biotic and abiotic stress is an important prerequisite in the development of optimization strategies for secondary metabolites production *in vitro* (Broeckling et al. 2005). Elicitation for example is one of the main approaches as far as phytoalexines production is targeted. Thus, medium optimization, screening and selection could lead to 20-30 times higher levels of the secondary metabolites produced *in vitro* (Verpoorte et al. 2002).

- *Selection of highly productive lines* could be performed even by the naked eye when pigmented compounds are targeted such as the examples with anthocyanins, shikonins or berberine. In other cases the application of analytical methods to monitor the produced raw plant material is necessary. The utilization of different selective methods has led to the achievement of highly productive lines with significant improvement of secondary metabolite production (Table 4).
- *Impact of stress factors* – according to Brodelius (1988) stress factors exogenously applied on the plant cell cultures could be classified in the following categories:
  - *Culture medium parameters* – such factors include carbon, nitrogen, phosphate sources and plant growth regulators;
  - *Physical stress* – extreme light treatment, ultraviolet light, aeration, medium osmolarity and pH. Thus, the application of electric impulses has led to the intercellular accumulation of taxuyunnanine C (Tc) in *Taxus chinensis* suspension (Hong et al. 2004). The phenomenon has been explained by the induction of defense response and possible alteration of the dielectric properties of the cell membrane;
  - *Chemical stress* – heavy metals, abiotic elicitors, etc. Cultivation at Fe^{2+} stress has been reported to induce oxidative stress, lipid peroxidation and cell death related to the increased levels of ethylene and β-thujaplicin production in *Cupressus lusitanica* suspension (Zhao et al. 2005);
  - *Biotic stress* – infection with pathogens/biotic elicitors. In order to be protected from pathogenic microorganisms and herbivores, plants have evolved a complicated inducible chemical defense system. Plants are capable, as a response of external pathogenic stimuli to react by the *de novo* synthesis of certain compounds designated as phytoalexins. The process is called elicitation. Being protective compounds against exogenously arising stimuli, phytoalexins can also be artificially induced by exogenous treatment of *in vitro* tissues by the controlled application of elicitation stress factors. The rapidness of the response can vary from several hours, to several days and often even excretion of the compounds into the medium can be observed (Petersen and Alfermann 1993). Numerous plant species, grown as cell suspensions have shown positive results regarding secondary metabolites elicitation by the exogenous application of methyl jasmonate. Its application initiates the *de novo* transcription of genes, encoding the expression of phenylalanine ammonia lyase, related to the chemical defense protective mechanisms of the plant cell (Gundlach et al. 1992). In literature there are

numerous reports concerning the successful secondary metabolites stimulation *in vitro* as a result of exogenous methyl jasmonate treatments. Such examples are the production of rosmarinic acid in *Lavandula vera* suspension (Georgiev et al. 2007);

- *Application of biosynthetic precursors and biotransformation* – irrespectively of the level of enzyme expression at a given moment, the plant cell possesses the whole genetic information for the full range of all functioning enzymatic systems. Therefore, it is possible to feed cheaper lower molecular precursors of valuable target metabolites, and enhance their yields *in vitro*. One of the first such experiments have been performed by the working group of Zenk who fed tryptophan to *Catharanthus roseus* suspensions (Zenk et al. 1977a) and phenylalanine to *Coleus blumei* suspensions (Zenk et al. 1977 b). The rosmarinic acid production was increased as a result. Plant cell cultures are a valuable source of enzymes, which are readily available in them and much easier to access as compared of fresh plant tissues of different species derived from the field. Extracted enzymes can further on included into various biochemical reactions outside of the plant organism. This makes it possible to complete perform biosynthetic experiments and biotransformation in laboratory conditions (Kutney 1997).

Table 3. Comparative yield of some secondary metabolites of *in vitro* cultures with field samples of the respective plant species

Compound	Plant species/ *In vitro* system	Yield (% of the dry weight)		*In vitro*/ Field content ratio	Reference
		In vitro	Field samples		
Non-differentiated cultures					
Nitrogen containing compounds					
Bisoclaurine	*Stephania cepharantha* Suspension	2.3	0.8	2.9	Akasu et al. 1976
Ajmalicine	*Catharanthus roseus* Suspension	1.0	0.3	3.3	Misawa 1994
Benzylisoquinoline alkaloids	*Coptis japonica* Suspension	11	5 - 10	2.2 - 1.1	Misawa 1994
Berberine	*Thalictrum minor* Suspension	10	0.01	1000	Misawa 1994
Berberine	*Coptis japonica* Suspension	10	2 - 4	5 - 2.5	Misawa 1994
Nicotine	*Nicotiana tabacum* Suspension	3.4	2.0	1.7	Misawa 1994
Caffeine	*Coffea arabica* Suspension	1.6	1.6	1	Govil et al. 2017
Serotonin	*Peganum harmala* Suspension	2	2	1	Govil et al. 2017
Trigonelline	*Trigonella foenum-graecum* Suspension	5	0.4	12.5	Govil et al. 2017
Trigonelline	*Trigonella wilfordii* Suspension	0.2	0.01	20	Govil et al. 2017

Table 3. (Continued)

Compound	Plant species/ *In vitro* system	Yield (% of the dry weight)		*In vitro*/ Field content ratio	Reference
		In vitro	Field samples		
Serpentine	*Catharanthus roseus* suspension	2	0.26	7.7	Govil et al. 2017
Vomilenien	*Rauwolfia serpentine* Suspension	0.214	0.004	53.5	Govil et al. 2017
Phenolics					
Shikonin	*Lithospermum erythrorhizon* Suspension	20	1.5	13.3	Misawa 1994
Antraquinones	*Morinda citrifolia Suspension*	18	0.3	60	Misawa 1994
Antraquinones	*Galium verum* Suspension	5.4	1.2	4.5	Misawa 1994
Antraquinones	*Galium aparine* Suspension	3.8	0.2	19	Misawa 1994
Rosmarinic acid	*Coleus blumeii* Suspension	15	3	5	Misawa 1994
Ubiquinone - 10	*Nicotiana tabacum* Suspension	0.036	0.003	12	Misawa 1994
Anthocyanins	*Vitis* sp. Cell culture	16	10	1.6	Zhong 2001
Anthocyanins	*Euphorbia milli* Cell culture	4	0.3	13.3	Zhong 2001
Anthocyanins	*Perilla frutescens* Cell culture	24	1.5	16	Zhong 2001
Rosmarinic acid	*Coleus blumei* Cell culture	15	3	5	Misawa 1994
Isoflavonoids daidzein, retuzin, genistein and formononetin and the pterocarpans maakiain and medicarpin	*Maackia amurensis* Callus	2.08	0.52	4	Fedoreyev et al. 2004
Shikimic acid	*Galium mollugo* Suspension	10	2.3	4.3	Govil et al. 2017
Terpenoids					
Ginsenosides	*Panax ginseng* Callus	27	4.5	6	Misawa 1994
Triptolide	*Tripteryquium wilfordii* Suspension	0.05	0.001	50	Misawa 1994
Diosgenin	*Dioscorea deltoides* Suspension	7.8	2.4	3.3	Govil et al. 2017
Differentiated cultures					
Camptothecin	*Camptotheca acuminate* roots/hairy roots of *Ophiorrhiza pumila*	0.1	0.02 – 0.5	5 - 0.2	Khani et al. 2012
Podophyllotoxin	*Podophyllum peltatum* Callus derived adventitious roots, transferred onto PGR-free MS medium	1.6	0.2	8	Sakata et al. 1990
Hypericin (Hyp), pseudohypericin (psHyp)	*Hypericum rumeliacum* Shoot cultures	Hyp – 0.03 psHyp – 0.12	Hyp – 0.01 psHyp – 0.01	Hyp – 3 psHyp – 12	Danova et al. 2012a

5. Research Evidence on the Successful Bioreactor Scale-Up of Secondary Metabolites

As discussed above the successful biotechnological yield of bioactive secondary metabolites requires the good knowledge and understanding of the biochemical and physiological aspects of plant growth and development related to secondary metabolism.

Scale-up of secondary metabolites production to bioreactor presumes the stabilization of the laboratory-scale processes and the gradual reaching larger, up to industrial-scale volumes similarly to microbial cultures. Usually scale-up begins in a 250 – 300 ml Erlenmeyer flask in laboratory conditions, goes through volume of up to 1 liter and afterwards the cultured line can to be adapted to 30 – 150 – 1000 l bioreactor. Unlike bacterial cell lines, the plant cells have certain specificities like for example their slower growth, larger cell size, tendency to form cell aggregates, as well as their sensitivity towards agitation and stirring. Therefore special adaptation of microorganism bioreactor processes is needed, as well as search for novel ideas in the tailoring of bioreactor for plant cells. The plant bioreactor processes could be compared to the above described "continued cultures" of suspension lines (Razdan 2003).

5.1. Scale-Up of Non-Differentiated Cell Cultures to Bioreactor

The non-differentiated cell/suspension cultures are considered as the most prospective source for the scaled-up yield of secondary metabolites (Georgiev et al. 2009). One of their biggest advantages is the fast growth and rapid biomass accumulation as compared with the differentiated cultures. As discussed above, the non-differentiated tissue is easier to scale-up to larger volumes and to bioreactor conditions, due to the possibility to provide uniform aeration and nutrition to the cell aggregates than to tissue mass of different differentiation types. As extensively reviewed by Bajaj et al. (1988) great success has been already achieved in the scale-up of secondary metabolite production of non-differentiated plant cell cultures in bioreactors (serpentine production by *Catharanthus roseus* in 85 l; rosmarinic acid – by *Coleus blumei* in 450 l; shikonin derivatives by *Lithospermun erythrorhizon* in 750 l; plant biomass by *Nicotiana tabacum* in 20 000 l; saponins by *Panax ginseng* in 2 000 l; polysaccharides by *Echinacea purpurea* and *Echinacea angustifolia* in 75 000 l).

5.2. Scale-Up of Differentiated Cell Cultures to Bioreactor

The differentiated state of the tissues in these systems is related to the better cell-to-cell communication, presence of secretory and translocation anatomic structures.

Table 4. Selection strategies of highly productive lines based on modification of morphogenesis and growth and development patterns of the *in vitro* cultured plants

Plant species	**Type of *in vitro* culture**	**Optimization measures/ Selection agents**	**Obtained results**	**References**
Selection of explant type as starting material for *in vitro* lines selection				
Rauvolfia tetraphylla L.	Callus	Utilization of different explant types for callus induction	Leaf explants – 90% callus induction response, cotyledon – 64%, hypocotyl – 60%, node and internode – 50%, petiole – 46%, root – 33%; leaf derived callus expressed highest levels of reserpine production	Anitha and Kumari 2013
Utilization of the coloration of the target compounds as selection criteria				
Euphorbia milli	Cell culture	Clonal selection (cell aggregates cloning) and 24-times consecutive subcultures of cells with intensive coloration	Selected line with over seven fold increase of anthocyanins content	Yamamoto et al. 1982
Muscadinia ssp. (muscadine grape), *Vitis aestivalis* var. *Cynthiana*	Cell culture	Choice of variety, selection of appropriate explant, reduction of browning by activated charcoal or Amberlite XAD4, repeated selection and sub-culture of colored clusters	Anthocyanins production, mainly anthocyanidin-3,5-O-diglucosides	Ananga et al. 2013
Induction of morphogenesis and establishment of higher levels of differentiation of the cell lines				
Coptis japonica	Cell culture	Continuous selection of cell aggregates	Lines with rapid growth and enhanced berberine production; 14 l bioreactor scale-up obtained; stability of metabolite levels for over 27 generations	Yamada & Sato 1981
Hypericum perforatum	Suspension	Selection of globular berries-like structured cell aggregates	Hypericin production exceeding the one of the differentiated shoots and callus of the plant	Vardapetyan et al. 2000
Application of chemical selection factors				
Lavandula vera	Suspension	Pimelic acid supple-mentation, gamma ray irradiation	Biotin production	Watanabe et al. 1982
Catharanthus roseus	Suspension	Selection of compact globular aggregates	Alkaloid stimulation	Verpoorte 1996
Rhodiola sachalinensis	Suspension	Selection of compact callus aggregates; higher cytokinin/auxin ratio and sucrose; optimization of inoculum volume	Enhanced salidroside production	Xu et al. 1999

All these factors are prerequisites for the production of higher molecular compounds by the plant organism as its morphological development is closer to the field cultivated plant, as compared with the non-differentiated cultures. Examples are the production of ginsenosides by adventitious roots of *Panax ginseng* C.A. Meyer in a 1 – 10 t two stage bioreactor system (Hahn et al. 2003); saponins by conventional roots of *Panax ginseng* in 5 and 20 l bioreactor (Paek et al. 2009) and ginsenosides (levels of 500 mg/l/day) by conventional roots of *Panax ginseng* in 20 t bioreactor (Charlowood and Charlwood 1991).

6. Commercial Utilization of Plant Biotechnology Tools in Mass Propagation of Planting Material for the Needs of Agriculture. Application of Plant Cell Tissue and Organ Culture Techniques as a Complementary Tool to Conventional Methods of Plant Breeding

The rapid development of plant biotechnology within the last years has led to its mass application for the production of planting material for agricultural needs.

According to the USDA in 2004 "biotechnological crops" constituted up to 46% of corn, 76% of cotton and 85% of soybean plantations, about a decade later these numbers rose up to 90, 90 and about 93%, respectively (Fernandez-Cornejo et al. 2014). Presently in the USA 60 - 70% of the food products on the market contain certain ingredient or oils which are biotechnologically obtained (Biotechnology Innovation Organization 2018).

The biotechnological selection and market realization of new cultivars however require the strict observation and critical application of a number of ethical considerations included into the legislation of the respective countries.

7. Commercial Utilization of Plant Biotechnology Tools for Industrial-Scale Secondary Metabolites Production

7.1. Patented Processes for the Biotechnological Delivery of Secondary Metabolites

The commercial yield of plant secondary metabolites by plant cell tissue and organ cultures is still limited as compared to the wide market application of plant biotechnology in agriculture and ornamental plants breeding. Nevertheless, numerous successful examples show the prospective for the development of this field.

First attempts for the production of secondary metabolites by *in vitro* cultivated plant tissues were performed by the Charles Pfizer Co Company in 1950 (Lombardino 2000; Dias et al. 2016). Production, experimented on *in vitro* cultured water-melon tissue turned unsuccessful. Further on, the first patent obtained in this area was by Routian and Nickell as assignors to Chas. Pizer & Co., Inc., New York, N. Y., a corporation of Delaware (Routian and Nickell 1956). The invention consisted of "a method for cultivating plant tissue under submerged, aerated conditions in liquid culture" and was "also concerned with the production of useful materials by this method". Tissues of several plant species were included, designated as "tissues of those plants classified above the Thallophytes in systematic botany" (including *Rumex acetosa* L., Sweet clover, *Agave toumeyana*, Sunflower stem and petiole, Periwinkle, Tobacco, *Opuntia* cactus, *Datura* and avocado). The delivery of "vitamins, steroids, alkaloids of various types, antimicrobial agents, sugars, enzymes, organic acids, aromatic materials, and so forth" was defined.

Further development after 1978 in the field led to the feasible biotechnological production of secondary metabolites in Germany and Japan (Loyola-Vargas and Vázquez-Flota 2006; Dias et al. 2016). Thus, now-a-days the number of patents, involving plant cell cultures has risen to he impressive number of 28 000, with companies, applying plant cell cultures or hairy roots for the production of ingredient for cosmetics, food and pharmaceutical industry intensively expanding (Ochoa-Villarreal et al. 2016). Some examples of such processes patented by different companies for the industrial and semi-industrial scale delivery of plant secondary metabolites are summarized in Table 5.

The large-scale commercial production of secondary compounds for pharmaceutical industry from *in vitro* culture is recognized for shikonin, ginsenosides, berberine and taxol (Malik et al. 2016). Thus, industrial-scale production of ginsenosides developed by Nitto Denko Corporation, Japan in 1985 is still used till today. The yielded material is produced without the application of genetic engineering and utilizes the indigenous biosynthetic capacity of the wild genotype with only application of optimization methods. The company inserts the material into different drinks.

Initially berberine production by root cultures of *Coptis japonica* and *Thalictrum minus* was introduced by Mitsui Petrochemical Industries. *C. japonica* cell cultures production was firstly developed by Furuya et al. in Kitasato University (1972) and then Yamamoto and Shoyakugaku in Nippon Paint (1981) selected a highly productive line which was adopted by the company.

Table 5. Patented processes for secondary metabolites semi-commercial and commercial production in plant cultures *in vitro* systems

Manufacturer	Biotechnologically produced secondary metabolite	References
A. Nattermann & Cie, Gmbh, Cologne, Germany	*Coleus blumei* – rosmarinic acid - anti-inflammatory	Ochoa-Villarreal et al. 2016
Boehringer (Germany)	*Digitalis lanata* – (cell cultures in 300 l airlift system) digoxin for medical application	Malik et al. 2016
Mitsui petrochemical Ind. (Japan)	*Geranium sp.* – geraniol for flavor and fragrance industries	Misawa 1994; Eibl and Eibl 2002; Petersen and Amstutz 2008; Wilson and Roberts 2012; Malik et al. 2016; Ochoa-Villarreal et al. 2016
	Catharanthus roseus (cell culture) – arbutin for cosmetics – skin lightening and antiseptic agent	
	Carthamus tinctorius – carthamine - natural dye for food and pharmaceutical industries	
	Lythospermum erythrorizon (batch cell cultures 750 l bioreactor) – shikonin – pigment in cosmetics	
	Coptis japonica and *Thalictrum minus* (cell cultures in batch and continuous flow; impeller driven 4000 l bioreactor) – berberins – as anticancer, anti-inflammatory and antibiotic in medicine	
	Rubia akane (cell culture) – purpurin – pigment for dyes production	
	Lithospemum erythrorhizon – shikonin – red pigment for food industries	
Kibun (Japan)	*Carthamus tinctorius* (cell culture) carthamine - natural dye for food and pharmaceutical industries	Malik et al. 2016; Ochoa-Villarreal et al. 2016
Sumitomo Chem. Ind. (Japan)	*Duboisia* sp. (cell cultures) – scopolamine – active ingredient pharmaceutical industry – anticholinergic, anti-muscarinic, used in treatment of motion sickness, nausea and intestinal cramping	Petersen and Amstutz 2008; Malik et al. 2016 Ochoa-Villarreal et al. 2016
Nippon Oil Company (Japan)	*Podophyllum* sp. – podophyllotoxin (cell culture, differentiated organs) – anticancer active ingredient pharmaceutical industry	Petersen and Amstutz 2008; Malik et al. 2016; Ochoa-Villarreal et al. 2016
	Taxus sp. – paclitaxel - active ingredient pharmaceutical industry	
Nippon Paint Company (Japan)	*Euphorbia millii* and *Aralia cordata* (cell cultures in rotary culture system) - dyes for the textile and color for food industries, respectively	Eibl and Eibl 2002; Malik et al. 2016
Osaka (Japan)	*Aralia cordata* – anthocyanins as food coloring agents	Ochoa-Villarreal et al. 2016
ESCA genetics (USA)	*Taxus* sp. – paclitaxel - active ingredient pharmaceutical industry	Petersen and Amstutz 2008; Malik et al. 2016
	Vanilla planifolia (cell cultures in 72 l impeller driven reactor) – vanillin - flavoring agent for food and beverage industry, as well as pharmaceutics	
Phyton Catalytic (USA/Germany)	*Taxus* sp. – paclitaxel - active ingredient pharmaceutical industry; Anticancer World's largest cGMP plant cell culture facility with bioreactors specifically designed to meet the needs of plant cell culture	Petersen and Amstutz 2008; Malik et al. 2016; Ochoa-Villarreal et al. 2016
Nattermann (Germany)	*Coleus blumei* - rosmarinic acid - medicine, food industry as preservative	Petersen and Amstutz 2008
Diversa (Germany)	*Echinacea purpurea* and *E. angustifolia* - polysaccharide complex - active ingredient pharmaceutical industry	Petersen and Amstutz 2008; Malik et al. 2016
Nippon Shinyaki (Japan)	*Beta vulgaris* (cell culture) – betacyanins - natural dye for food and pharmaceutical industries	Petersen and Amstutz 2008; Malik et al. 2016

Table 5. (Continued)

Manufacturer	Biotechnologically produced secondary metabolite	References
Nippon Paint (Japan)	*Euphorbia milli* and *Aralia cordata* – anthocyanins – natural dyes	Petersen and Amstutz 2008
Nitto Denko Corp. (Japan)	*Panax ginseng* (cell and root cultures, 20 000 l) - ginsenosiedes – dietary supplements	Malik et al. 2016; Ochoa-Villarreal et al. 2016
	Rubia akane (cell culture) – purpurin – pigment for dyes production	
CBN Biotech (Republic of Korea)	*Panax ginseng*- ginsenosiedes – pharmaceutical industry	Malik et al. 2016
ROOTec bioactives GmbH (Switzerland)	*Camptotheca acuminata* (hairy roots in mist bioreactor systems) – camptothecin - DNA topoisomerases, Type I inhibitor – pharmaceutical industry	Lehmann et al. 2014; Malik et al. 2016; Ochoa-Villarreal et al. 2016; ROOTee factsheets
	Podophyllum sp. – podophyllotoxin (hairy roots in mist bioreactor systems) - active ingredient pharmaceutical industry	
	Atropa belladonna (hairy roots in mist bioreactor systems) – atropine	
	Carlina acaulis (hairy roots in mist bioreactor systems) flavonoids with	
	atibacterial and diuretic effects, UV protective effects for cosmetics	
	Nicotiana glauca (hairy roots in mist bioreactor systems) – vitamin D3 derivative and alkaloids (anabasine, nicotine,	
	Panax ginseng (hairy roots in mist bioreactor systems) – ginsenosides – food supplements, cosmetics	
Nattermann (Germany)	*Coleus blumei* (cell cultures) – rosmarinic acid – medical application	Malik et al. 2016
Vipont Research Labs (USA)	*Papaver somniferum* (cell cultures in 300 l airlift system) – sanguinarine – cell toxin	Malik et al. 2016
Samyang, Genex Co. Ltd. (Korea)	*Taxus sp.* – paclitaxel - active ingredient pharmaceutical industry	Malik et al. 2016; Ochoa-Villarreal et al. 2016
Bristol-Myers Squibb Co. (Germany)	*Taxus sp.* – paclitaxel - active ingredient pharmaceutical industry	Malik et al. 2016
Unhwa Biotech Corp., (Jeonbuk, South Korea)	*Panax ginseng* (cultivated cambial meristematic cells "stem cell Ttobyeo") - ginsenosiedes – dietary supplements, cosmetic and medical products	Ochoa-Villarreal et al. 2016
Sederma, Le Perray-en-Yvelines (France)	Cell suspension cultures and hairy roots cultures, different species - about fifteen active ingredients for cosmetic use	Ochoa-Villarreal et al. 2016
Lumene Oy (Finland)	*Rubus chamaemorus* (callus cultures) – cosmetic products application - day creams, foundation creams, lipsticks, skin serums, mascaras, products for hair and/or scalp care, washing products for skin or hair, or as products for skin hygiene	Nohynek et al. 2014; EP2817070 (A4) — 2015-12-23
Mibelle (Switzerland)	*Malus domestica Solar vitis* (wave-type bioreactor cultivated suspension) – the whole cells decomposed in high pressure with a lipid fraction, which encapsulated all the cell constituents in liquid nanoparticles to be used in cosmetic products	Schürch et al. 2008; Nohynek et al. 2014; Ochoa-Villarreal et al. 2016
	Argania spinosa (cell suspension) – cosmetic application.	

Purpurin production was firstly developed in cell cultures of *Rubia akane* by Mitsui Petrochemical Ind. Ltd. Mitsui Petrochemical Ind. Ltd. initially developed shikonin production in *Lythospermum erythrorhizon* suspension. CBN Biotech Co, Korea developed ginsenosides production by conventional root cultures of *Panax ginseng* in a research group at the Research Center for the Development of the Advanced Horticultural Technology (HortTech), Chungbuk National University with the financial support of the Ministry of Science and Technology, Engineering Foundation and Agricultural Development institution in Korea. Phyton Biotech, Germany is the international leader for large-scale paclitaxel production by cell cultures of *Taxus* ssp.

8. Short Outline of the Author's Own Experience in the Developmental Patterns and Morphological Features *In Vitro* and the Subsequent Modification of Secondary Metabolite Production of Species of the Genera *Hypericum*, *Sideritis* and *Artemisia*

8.1. *Hypericum* Species *In Vitro* Secondary Metabolites Productivity

Hypericum perforatum L. is a plant species with long years of medicinal application by humankind. It has been used since the 1st century A.D. It is native to Europe, North Africa and Asia, but also naturalized to North America. It possesses a broad spectrum of biological activities, determined by its rich pharmacological spectrum. Its most important biologically active constituents are the polyphenolic compounds, flavonoids, naphthodianthrones and phloroglucinols, terpenes, which have been shown to possess antidepressant, anticancer, antiviral and antibiotic activity.

Throughout the years the species has been an incessant object of scientific interest in terms of secondary metabolites production in plant cell tissue and organ culture (Danova 2015 and references cited within). In our laboratory shoot cultures as a model of elucidation of biotechnological productivity of different *Hypericum* species have been studied.

In a comparative work of the content of hypericin and pseudohypericin, it was established that the Balkan endemic *H. rumeliacum*, representative of the evolutionary more developed *Drosocarpium* section produced both *in situ* and *in vitro* higher hypericin as compared with *H. tetrapterum* (representative of section *Hypericum*, which is evolutionary older as compared with *Drosocarpium*) (Danova et al. 2012a). Interestingly, pseudohypericin level was higher in *H. tetrapterum*, as compared with *H. rumeliacum*, however the total hypericins of *H. rumeliacum* still remained higher. These findings corroborate with the ones of other authors that representatives of the evolutionary more developed *Hypericum* species have higher hypericins producing capacity (Kitanov 2001; Kirakosyan et al. 2003, 2004). Then an *in vitro* experiment on tissue culture development of these species showed that the higher hypericin productivity of *H. rumeliacum* was preserved (Danova et al. 2012a). Moreover, in an experiment of modification of vitamin supplementation of the latter two species, as well as of hypericin non-producing *H. calycinum* showed that the production of hypericins and phenolic and flavonoid compounds can be differentially targeted *in vitro.* It was established that Gamborg vitamins supplementation which increased the index of compactness (number of leaf couples per cm of stem length) and average leaf area also stimulated the polyphenolics levels in all three species, as compared with media with Murashige and Skoog vitamins (Danova 2010; Danova et al. 2012a). When studying the levels of hypericins in this experimental design it was established that they were higher in the medium with lower flavonoid and phenolics levels (Murashige and Skoog vitamins). The morphologic study showed that the number of hypericin dark glands per leaf was a comparatively conservative species-related feature and did not vary significantly within one and the same species, irrespectively of the vitamins modification (when comparing one and the same segment of the stem – basal, medium and apical) (Danova et al. 2012a). However, the morphological differences of obtaining a higher leaf portion within the total biomass in Gamborg treated plants led to a lower dark glands density per mm^2 of leaf area in the latter medium. Thus in Murashige and Skoog medium the total plant biomass contained a larger ratio of dark glands per unit of dry biomass which led to higher hypericins levels per gram dry material. Thus, this simple and reproducible experiment illustrated the importance of growth and development for the biosynthetic capacity of *Hypericum in vitro.*

8.2. *Sideritis scardica* Griesb Productivity *In Vitro*

The Balkan endemic *Sideritis scardica* Griesb. is traditionally utilized as a pulmonary treatment, as well as anti-flu and wound healing remedy (Ivancheva and Stantcheva 2000). It has been established that presently there are less than 2 500 mature individuals in Bulgaria and over 250 in Serbia (IUCN 2016). The plant has been determined with a near threatened status, with a decreasing current population trend (Khela 2013). The low germination rate and collection pressure imposes significant risk on its natural populations. Thus, as a part of wider program for medicinal and aromatic plants germplasm conservation, the species was introduced *in vitro* and experiments on the polyphenolic productivity under different organic and inorganic treatments was experimented. An array of plant growth regulators and activated charcoal (AC) treatments was applied in order to study their effects on the developmental patterns and polyphenolics productivity of the species *in vitro.* Low (0.02 and 0.05 g/l) as well as high (0.2 and 0.5 g/l) concentrations of activated charcoal were applied. The higher AC concentrations (0.2 and 0.5 g/l) led to a profound stimulation of axillary shoot formation, enlargement of leaf area and stimulation of polyphenolic production. In addition, photosystem II quantum yield revealed that the control (untreated) and all carbon-treated variants were physiologically fit. However, a slower apparent electron transport rate for 0.2 and 0.5 g/l treated plants was established, as compared with the control and 0.02 and 0.05 g/l treated plants. Thus, it was hypothesized that the higher leaf mass and leaf area in 0.2 and 0.5 g/l treated *S. scardica* might be a compensatory reaction/mechanism that allows for more efficient light utilization. Thus, the AC efficiency in affecting the biosynthetic capacity and physiological status of the plant, without the addition of organic supplements such as plant growth regulators, seems to be a prospective approach in tissue culture optimization of this specis *in vitro* (Koleva et al. 2016). In order to compare with AC treatments, plant growth regulators (PGR) were also applied to shoot cultures of the plant. The following combinations were experimented: 0.2 mg/l BA + 0.02 mg/l NAA; 0.2 mg/l BA + 0.5 mg/l NAA; 0.2 mg/l BA + 1.0 mg/l NAA; 0.5 mg/l BA + 0.5 mg/l NAA and 0.5 mg/l BA + 1.0 mg/l NAA treatments were compared with the plant growth regulators free control (Danova et al. 2017). The comparison showed that the PGR treatment led to the formation of elongated elyptical leaves with larger areas of the leaf blades, longer petiols and well expressed toothing of the leaf margins. Leaves of AC treated plants were with ovate shape, smaller area of the leaf blade, shorter petiols and less expressed toothing of the leaf margins. The addition of activated charcoal in the medium prolonged the sub-culture period, as compared with the control plants to at least 2 months without significant browning and necrosis of the plantlets. On the contrary, maintenance of *S, scardica* in PGR-free and PRG supplemented media was related to intensive polyphenolics leakage in the medium and required a shorter sub-culture period of maximum 6 weeks. Both treatments increased the polyphenolic levels *in vitro*.

8.3. *Artemisia* Productivity *In Vitro*

The medicinal plant *Artemisia alba* Turra is utilized for various purposes – decoction of the aerial parts is applied as a digestive in traditional medicine; a study on the ethanol extract revealed its anti-inflammatory activity and the essential oils were established to exert strong spasmolytic and antimicrobial properties. However, as summarized above (Table 1) great variability of the essential oil profiles has been reported in literature by many authors. This effect is attributed to the factors of the environment or to the genotype. Shoot cultures of the plant were initiated and supplementations of indole-3-butyric (IBA) acid and benzyl adenine (BA) were experimented: PGR-lacking control, 0.5 mg/l IBA; 1.0 mg/l IBA; 0.2 mg/ BA + 0.5 mg/l IBA and 0.2 mg/l BA + 1.0 mg/l IBA. Two main morphotypes were achieved as a result – the PGR-free control and IBA supplemented plants displayed shoot and root development, and the plants where BA and IBA were combined, displayed lack of root formation and callusogenesis at the explant base (Danova et al. 2012b). Comparison of the essential oils of the aerial parts of the two morphotypes revealed the presence of two main terpenoid profiles – oils with higher monoterpenoid/ sesquiterpenoid ratio (plants with root development, irrespectively of the PGR concentrations) and oils with sesquiterpenoid domination (the group with inhibited rooting). Given the decisive role of roots for cytokinin biogenesis in the plant organism, further research was completed in order to assess the levels of endogenous cytokinins, as well as the chloroplast architecture in the samples of the different treatments. Inhibited rooting also resulted in a significant drop of endogenous isoprenoid CK bioactive-free bases and ribosides as well as CK *N*-glycoconjugates and in decreased *trans*-zeatin (*trans*Z):*cis*-zeatin (*cis*Z) ratio in the aerials. Marked impairment of the structural organization of the photosynthetic apparatus and chloroplast architecture were also observed in samples with suppressed rooting. It is well known that in the plant cell monoterpenoid and *trans*Z type CKs biogenesis are spatially bound to plastids, while sesquiterpenoid and *cis*Z production are compartmented in the cytosol. The observed dependencies suggest an interplay between the biosynthesis of terpenoids and CK bioactive free bases and ribosides in *A. alba in vitro*. Possible moderation of the two terpenoid biosynthetic pathways because of the alteration of chloroplast structure in the experimental model has been hypothesized (Danova et al. 2018). In addition, the dependencies obtained by this experimental design provide clues of the possibilities to target terpenoid production by means affecting morphogenesis in this plant species. In conclusion, secondary metabolism is strongly dependent on growth and development and morphogenetic patterns of the plant organism. Therefore, the knowledge and understanding of these dependencies makes it possible to optimize culture conditions and target secondary metabolite production in both conventionally and biotechnologically cultivated plants and utilize the indigenous biosynthetic capacity of the wild genotype without performing genetic modifications.

Acknowledgments

Partial support of this work by the bilateral agreement between the Czech and Bulgarian Academies of Sciences, as well as grant BG05M2OP001-1.002-0012-C01 for establishment of Center of Competence, Operational Programme "Science and Education for Smart Growth" (SESG) are gratefully acknowledged.

References

Aboukhalid, K., Al Faiz, C., Douaik, A., Bakha, M., Kursa, K., Agacka-Mołdoch, M., Machon, N., Tomi, F. & Lamiri, A. (2017). Influence of Environmental Factors on

Essential Oil Variability in *Origanum compactum* Benth. *Growing Wild in Morocco.*, *14*. https://doi.org/ 10.1002/cbdv.201700158.

Afrin, S., Huang, J. J. & Luo, Z. Y. (2015). JA-mediated transcriptional regulation of secondary metabolism in medicinal plants. *Sci. Bull.*, *60*, 1062–1072.

Agati, G., Brunetti, C., Di Ferdinando, M., Ferrini, F., Pollastri, S. & Tattini, M. (2013). Functional roles of flavonoids in photoprotection: New evidence, lessons from the past. *Plant Physiology and Bio-chemistry.*, *72*, 35-45.

Akasu, M., Itokawa, H. & Fujiata, M. (1976). Bisoclaurine alkaloids in callus cultures of *Stephania cepharantha. Phytochemistry.*, *15*, 471-473.

Alfermann, A. W., Petersen, M. & Fuss, E. (2003). Production of natural products by plant cell biotechnology: results, problems and perspectives. In: Laimer and Rücker W (eds). *Plant tissue culture*: *100 years since Gottlieb Haberlandt*, Springer Wien New York.

Alonso-Amelot, ME., Oliveros-Bastidas, A. & Calcagno-Pisarelli, M. P. (2007). Phenolics and condensed tannins of high altitude *Pteridium arachnoideum* in relation to sunlight exposure, elevation, and rain regime. *Biochemical Systematics and Ecology*, *35*, 1-10.

Ananga, A., Georgiev, V., Ochieng, J., Phills, B. & Tsolova, V. (2013). *Production of Anthocyanins in Grape Cell Cultures: A Potential Source of Raw Material for Pharmaceutical, Food, and Cosmetic Industries*. The Mediterranean Genetic Code Barbara Sladonja, IntechOpen, DOI: 10.5772/54592.

Anitha, S. & Kumari, B. D. R. (2013). *In vitro* callus culture in *Rauvolfia tetraphylla* L.: Indole alkaloid production. *Asian Journal of Plant Sciences.*, *12*, 28-33.

Arora, R. (ed.) (2010). *Medicinal plant biotechnology*, C.A.B. International, Oxfordshire, UK.

Bais, H. P., Sudha, G., George, J. & Ravishankar, G. A. (2001). Influence of exogenous hormones on growth and secondary metabolite production in hairy root cultures of *Cichorium intybus* L. cv. Lucknow local. *In Vitro Cellular & Developmental Biology – Plant.*, *37*, 293–299.

Bajaj, Y. P. S., Furmanowa, M. & Olszowska, O. (1988). Biotechnology of the Micropropagation of Medicinal and Aromatic Plants. In: YPS Bajaj (ed.), *Biotechnology in Agriculture and Forestry*, *Medicinal and Aromatic Plants I*, *4*, pp. 60-103, Springer-Verlag, Berlin Heidelberg.

Bakkali, A. T., Jaziri, M., Foriers, A., Vander Heyden, Y., Vanhaelen, M. & Homès, J. (1997). Lawsone accumulation in normal and transformed cultures of henna, *Lawsonia inermis. Plant Cell Tiss. Organ Cult.*, *51*, 83–87.

Balandrin, M. F. & Klocke, J. A. (1988). I.1 Medicinal, Aromatic and Industrial materials from Plants. In: In: Bajaj YPS (ed) *Biotechnology in agriculture and forestry*, vol. *4*. Springer-Verlag, Berlin, Heidelberg, New York, London, Paris, Tokyo., pp. 3–31.

Barbora, B. C. (2013). Agrotechniques in cultivation of tea in Assam. *Int. J. Tea Sci.*, *9*, 43-54.

Becker, H. & Chavadej, S. (1988). Valepotriates: Production by plant cell cultures. In: Bajaj, Y.P.S. (Ed.), *Biotechnology in Agriculture and Forestry*, Vol. *4*. Medicinal and Aromatic Plants. Springer-Verlag, Berlin.

Becker, C., Klaering, H. P., Kroh, L. W. & Krumbein, A. (2014). Cool-cultivated red leaf lettuce accumulates cyanidin-3-O-(6”-O-malonyl)-glucoside and caffeoylmalic acid. *Food Chemistry*, *146*, 404–411.

Becker, C. & Kläring, H. P. (2016). CO_2 enrichment can produce high red leaf lettuce yield while increasing most flavonoid glycoside and some caffeic acid derivative concentrations. *Food Chemistry*, *199*, 736–745.

Bernáth, J. & Tetenyi, P. (1981). The Effect of environmental factors on growth., development and alkaloid production of Poppy (*Papaver somniferum* L.) II. *Interaction of light and temperature. Riochem. Physiol. Pflanzen.*, *176*, 599-605.

Biotechnology Innovation Organization. (2018). https://www.bio.org/ articles/frequently-asked-questions-agricultural-biotechnology.

Borges, L. S., Vieira, M. A. R., Marques, M. O. M., Vianello, F. & Lima, G. P. P. (2012). Influence of organic and mineral soil fertilization on essential oil of *Spilanthes oleracea* cv. Jambuarana. *Am. J. Plant Physiol.*, *7*, 135-142.

Bouvier, F., Suire, C., d'Harlingue, A., Backhaus, R. A. & Camara, B. (2000). Molecular cloning of geranyl diphosphate synthase and compartmentation of monoterpene synthesis in plant cells. *Plant J.*, *24*, 241–252.

Broeckling, C. D., Huhman, D. V., Farag, M. A., Smith, J. T., May, G. D., Mendes, P., Dixon, R. A. & Sumner, L. W. (2005). Metabolic profiling of *Medicago truncatula* cell cultures reveals the effects of biotic and abiotic elicitors on metabolism, *J. Exp. Bot.*, *56* (410), 323-36.

Brodelius, P. (1988). Stress-induced secondary metabolism in plant cell cultures, in: Plant Cell Biotechnology, NATO ASI. (Pais, M.S, Mavituna, F., Novais, J.M., Eds.) *Ser H*, *18*, 195-209.

Caldwell, M., Robberecht, R. & Flint, S. (1983). Internal filters: Prospects for UV-acclimation in higher plants. *Physiol. Plant.*, *58*, 445-450.

Carbonell-Bejerano, P., Diago, M. P., Martínez-Abaigar, J., Martínez-Zapater, J. M., Tardáguila, J. & Núñez-Olivera, E. (2014). Solar ultraviolet radiation is necessary to enhance grapevine fruit ripening transcriptional and phenolic responses. *BMC Plant Biology.*, *14*, 183.

Castells, E., Berhow, M. A., Vaughn, S. F. & Berenbaum, M. R. (2005). Geographic variation in alkaloid production in *Conium maculatum* populations experiencing differential herbivory by *Agonopterix alstroemeriana*. *Journal of Chemical Ecology.*, *31*, 1693-1709.

Chan, K. X., Crisp, P. A., Estavillo, G. M. & Pogson, B. J. (2011). Chloroplast-to-nucleus communication. Current knowledge, experi-mental strategies and relationship to drought stress signaling. *Plant Signaling & Behavior.*, *5*, 12, 1575-1582.

Charlwood, B. V., Charlwood, K. A. & Molinatorres, J. (1990). Accumulation of secondary compounds by organized plant culture, in: *Secondary Products from Plant Tissue Culture* (Charlwood, B.V., Rhodes M.J.C., Eds.), 167-200.

Charlwood, B. V. & Charlwood, K. A. (1991). Terpenoid production in plant cell culture. In: Harborne JB, Tomas-Barberan FA (eds.) *Ecological chemistry and biochemistry of plant terpenoids*. Oxford: Clarendon Press., *95*, 132.

Chrysargyris, A., Drouza, C. & Tzortzakis, N. (2017). Optimization of potassium fertilization/nutrition for growth, physiological development, essential oil composition and antioxidant activity of *Lavandula angustifolia* Mill. *Journal of Soil Science and Plant Nutrition.*, *17*, 291-306.

Çirak, C., Radušienė, J., Janulis, V. & Ivanauskas, L. (2007). Secondary metabolites in *Hypericum perfoliatum*: variation among plant parts and phenological stages. *Botanica Helvetica.*, *1*, 29–36.

Danova, K. (2010). Production of polyphenolic compounds in shoot cultures of *Hypericum* species characteristic for the Balkan flora. *Botanica Serbica*, *34*(1), 29–36.

Danova, K. (2015). Potential of the Balkan Flora as a Source of Prospective *Hypericum* Genotypes for the Conventional and Biotechnological Delivery of Phytopharmaceuticals. Chapter 2 in: *Hypericum: Botanical Sources, Medical Properties and Health Effects*. Howard R. Davis (Ed), Series Plant Science Research and Practices, Nova Science Publishers, USA, ISBN: 978-1-63482-701-0, pages 19-52.

Danova, K., Nikolova-Damianova, B., Denev, R. & Dimitrov, D. (2012a). Influence of vitamins on polyphenolic content, morphological development, and stress response in shoot cultures of Hypericum spp. *Plant Cell Tiss Organ Cult*, *110*, 383–393.

Danova, K., Todorova, M., Trendafilova, A. & Evstatieva, L. (2012b). Cytokinin and auxin effect on the terpenoid profile of the essential oil and morphological characteristics of shoot cultures of *Artemisia alba*. *Natural Product Communications.*, *7*, 1075 – 1076.

Danova, K., Koleva, P., Aneva, I. & Evstatieva, L. (2017). Multiplication and polyphenolics production of *Sideritis scardica* through different tissue culture approaches. *Proceedings of Seminar of Ecology*, 2016, Union of Scientists, Bulgaria, 21-22 April 2016, ISBN: 979-853-476-132-4, pp. 100-104.

Danova, K., Motyka, V., Todorova, M., Trendafilova, A., Krumova, S., Dobrev, P., Andreeva, T., Oreshkova, Ts., Taneva, St. & Evstatieva, L. (2018). Effect of cytokinin and auxin treatments on morphogenesis, terpenoid biosynthesis, photosystem structural organization, and endogenous isoprenoid cytokinin profile in *Artemisia alba* Turra *in vitro. J Plant Growth Regul.*, *37*, 403–418.

De La Cruz Medina, J., Rodriguez Jiménes, G. C., García, H. S. & co-authors: Zarrabal, T. L. R., Alvarado, M. Á. G., José, V. & Olvera, R. (2009). Vanilla: Post-harvest Operations. *FAO Compendium.* http:// www.fao.org/.

Dias, M. I., Sousa, M. J., Alves, R. C. & Ferreira, I. C. F. R. (2016). Exploring plant tissue culture to improve the production of phenolic compounds: A review. *Industrial Crops and Products*, *82*, 9–22.

Eibl, R. & Eibl, D. (2002). In: Oksman-Caldentey KM, Barz WH (eds.) *Plant Biotechnology and Transgenic Plants*. CRC Press, 2002.

EP2817070, (A4) — 2015-12-23 – European Patent COSMETIC *Compositions Containing Cloudberry Cell Culture Preparation*. patent link.

Evans, W. C. (2009). *Trease and Evans Pharmacognosy*. 16th Edition, Elsevier.

Fang, R., Redfern, S. P., Kirkup, D., Porter, E. A., Kite, G. C., Terry, L. A., Berry, M. J. & Simmonds, M. S. J. (2017). Variation of theanine, phenolic, and methylxanthine compounds in 21 cultivars of *Camellia sinensis* harvested in different seasons. *Food Chemistry.*, *220*, 517–526.

Fedoreyev, S. A., Kulesh, N. I., Glebko, L. I., Pokushalova, T. V., Veselova, M. V., Saratikov, A. S., Vengerovskii, A. I. & Chuchalin, V. S. (2004). Maksar: a preparation based on Amur Maackia. *Pharm. Chem. J.*, *38*, 605–610.

Fernandez-Cornejo, J., Wechsler, S., Livingston, M. & Mitchell, L. (2014). Genetically Engineered Crops in the United States, United States *Department of Agriculture, Economic Research Report Number*, 162, https://www.ers.usda.gov/webdocs/ publications/45179/43668_ err162.pdf.

de, Fossard. (1977). The horizons of tissue culture propagation. *Seminar NSW ASS.* Univ. Sydney, 1-147.

Fujiata, Y. & Tabata, M. (1987). Secondary metabolites from plant cells – pharmaceutical application and progress in commercial production, in *Plant tissue and Cell Culture* (Green, C.E. et al.), New York: Alan R. Liss., 169-185.

Fujiwara, K., Kozai, T., Nakajo, J. & Watanabe. (1989). Effect of closures on light intensities in plant tissue culture vessels. *J Agric Meteorol.*, *45*, 143-149.

Fukui H., Yoshikawa N. & Tabata, M. (1983). Induction of shikonin formation by agar in Lithospermum erythrorhizon cell suspension, *Phytochem.*, *22*, 241-2453.

Furuya, T., Shono, K. & Ikuta, A. (1972). Isolation of berberine from callus tissue of *Coptis japonica*. *Phytochemistry.*, *11*, 175.

Fürstenberg-Hägg, J., Zagrobelny, M. & Bak, S. (2013). Plant Defense against Insect Herbivores. *Int. J. Mol. Sci.*, *14*, 10242-10297.

Gamborg, O. L., Miller, R. A. & Ojima, K. (1968). Nutreint requerments of suspension culture of soybean root cells. *Exp Cell Res.*, *50*, 151-158.

Ganjali, A. & Kaykhaii, M. (2016). Investigating the essential oil composition of *Rosmarinus officinalis* before and after fertilizing with vermicompost. *Journal of Essential Oil Bearing Plants.*, *20*, 1413-1417.

Georgiev, M., Pavlov, A. & Ilieva, M. (2004). Rosmarinic acid production by *Lavandula vera* MM cell suspension: the effect of temperature. *Biotech. Lett.*, *26*, 855-856.

Georgiev, M., Kuzeva, S., Pavlov, A., Kovacheva, E. & Ilieva, M. (2007). *Elicitation of rosmarinic acid by Lavandula vera MM cell suspension culture with abiotic elicitors.*, *23*, 301–304.

Georgiev, M., Weber, J. & Maciuk, A. (2009). Bioprocessing of plant cell cultures for mass production of targeted compounds. *Applied microbiology and biotechnology.*, *83*, 809-23.

Gersbach, P. V. (2002). The essential oil secretory structures of *Prostanthera ovalifolia* (Lamiaceae). *Annals of Botany.*, *89*, 255-260.

Gerson, E. A. & Kelsey, R. G. (1998). Variation of piperidine alkaloids in ponderosa (*Pinus ponderosa*) and lodgepole pine (*P. contorta*) foliage from central Oregon. *Journal of Chemical Ecology.*, *24*, 815-827.

Gorunovic, M. S., Bogavac, P. M., Chalchat, J. C. & Garry, R. P. (1992). Analysis of the essential oil of *Artemisia lobelii All. Pharmazie.*, *47*, 803-804.

Govil, C. M., Aggrawal, A. & Sharma, J. (2017). *Plant Biotechnology and Genetic Engineering*. PHI Learning, Private Limited, Delhi-110092.

Gränicher, F., Christen, P. & KapétanidisI, M. (1995). Production of valepotriates by hairy root cultures of *Centranthus rubber* DC. *Plant Cell Rep.*, *14*, 294-298.

Gränicher, F., Christen, P. H. & Kapétanidis, I. (1992). High-yield production of valepotriates by hairy root cultures of *Valeriana officinalis* L. var. *sambucifolia* Mikan. *Plant Cell Rep.*, *11*, 339-342.

Gribble, K., Conroy, J. P., Holford, P. & Milham, P. J. (2002). *In vitro* uptake of minerals by *Gypsophila paniculata* and hybrid eucalypts, and relevance to media mineral formation, *Australian Journ. of Botany*, *50*, 713-723.

Gundlach, H., Muller, M. J., Kutchan, M. J. & Zenk, M. H. (1992). Jasmonic acid is a signal transducer in elicitor-induced plant cell cultures, In Plant Biology, *Proc. Natl. Acad. Sci. USA*, *89*, 2389-2393. Heidelberg New York, pp. 294-309.

Gupta, R. & Chakrabarty, S. K. (2013). Gibberellic acid in plant. Still a mystery unresolved. *Plant Signaling & Behavior*, *8*, 9, e25504.

Hagendoorn, M. J. M., Jamar, D. C. L., Meykamp, B. & van der Plas, L. H. W. (1997). *Cell division versus secondary metabolite production in Morinda citrifolia cell suspensions.*, *150*, 325-330.

Hahn, E. J., Kim, Y. S., Yu, K. W., Jeong, C. S. & Paek, K. Y. (2003). Adventitious root cultures of *Panax ginseng* c.v. *meyer* and ginsedoside production through large-scale bioreactor system. *J Plant Biotechnol.*, *5*, 1-6.

Hanosová, H., Koprna, R., Valík, J., Knoppová, L., Frébort, I., Dzurová, L. & Galuszka, P. (2015). Improving field production of ergot alkaloids by application of gametocide on rye host plants. *New Biotechnology.*, *32*, 739-746.

Harbaum-Piayda, B., Walter, B., Bengtsson, G. B., Hubbermann, E. M., Bilger, W. & Schwarz, K. (2010). *Influence of pre-harvest UV-B irradiation and normal or controlled atmosphere storage on flavonoid and hydroxycinnamic acid contents of pak choi (Brassica campestris L. ssp. chinensis var. communis).*, *56*, 202-208.

Harborne, J. B. (1996). Recent advances in chemical ecology, *Nat. Prod. Rep.*, *12*, 83–98.

Hassan, S. & Mathesius, U. (2012). The role of flavonoids in root–rhizosphere signaling: opportunities and challenges for improving plant–microbe interactions. *Journal of Experimental Botany*, Vol. *63*, No. 9, pp. 3429–3444.

Hassiotis, C. N., Ntana, F., Lazari, D. M., Poulios, S. & Vlachonasios, K. E. (2014). Environmental and developmental factors affect essential oil production and quality of *Lavandula angustifolia* during flowering period. *Industrial Crops and Products*, *62*, 359–366.

Hazlett, D. L. & Sawyer, N. W. (2008). Distribution of alkaloid-rich plant species in shortgrass steppe vegetation. *Conservation Biology.*, *12*, 1260-1268.

Hiyama, H., Ozawa, A., Kumazawa, H. & Takeda, O. (2017). Stabilization of ephedrine alkaloid content in *Ephedra sinica* by selective breeding and stolon propagation. *Biol. Pharm. Bull.*, *40*, 43–48.

Hong, Ye., Lin-Ling, Huang., Shu-De, Chen. & Jian-Jiang, Zhong. (2004). *Pulsed electric field stimulates plant secondary metabolism in suspension cultures of Taxus chinensis*, *Biotechnol. and Bioengineering*, *88* (6), 788-795.

Honório, I. C. G., Bonfim, F. P. G., Montoya, S. G., Casali, V. W. D., Leite, J. P. V. & Cecon, P. R. (2016). Growth., development and content of flavonoids in calendula (*Calendula officinalis* L.). *Acta Scientiarum. Agronomy.*, *38*, 69-75.

Hounsome, N., Hounsome, B., Tomos, D. & Edwards-Jones. (2008). Plant metabolites and nutritional quality of vegetables. *Journal of Food Science.*, *73*, 48-65.

Hunt, M. G., Rasmussen, S., Newton, P. C. D., Parsons, A. J. & Newman, J. A. (2005). Near-term impacts of elevated CO_2, nitrogen and fungal endophyte-infection on *Lolium perenne* L. growth, chemical composition and alkaloid production. *Plant, Cell and Environment.*, *28*, 1345–1354.

ICUN. (2016). *The IUCN Red List of Threatened Species*, http://www.iucnredlist.org/details/203271/.

Ivancheva, S. & Stantcheva, B. (2000). Ethnobotanical inventory of medicinal plants in Bulgaria. *J Ethnopharmacol.*, *69*, 165–172.

Jansen, G., Jürgens, H. U., Schliephake, E., Seddig, S. & Ordon, F. (2015). Effects of growing system and season on the alkaloid content and yield of different sweet *L. angustifolius* genotypes. *Journal of Applied Botany and Food Quality.*, *88*, 1 – 4.

Jayanth, S. G., Anilkumar, A. S. & Mallikarjunagowda, A. P. (2015). Spike yield and alkaloid production in long pepper as influenced by precision farming techniques under protected cultivation. *Int. Journ. Sci. Nat.*, *6*, 293-295.

Jaziri, M., Shimomura, K., Yoshimatsu, K., Fauconnier, M. L., Marlier, M. & Homes, J. (1995). Establishment of normal and transformed root cultures of *Artemisia annua* L. for artemisinin production. *J. Plant Physiol.*, *145*, 175–177.

Jović, D., Đoković, D. D., Tešević, V., Vajs, V., Tasić, S. & Milosavljević, S. (2001). Essential oil of the aerial parts of *Artemisia lobelia*. *Lekovite sirovine (Medicinal raw materials).*, *21*, 45-50.

Kasahara, H., Takei, K., Ueda, N., Hishiyama, S., Yamaya, T., Kamiya, Y., Yamaguchi, S. & Sakakibara, H. (2004). Distinct isoprenoid origins of *cis*- and *trans*-Zeatin biosyntheses in *Arabidopsis*. *J Biol Chem*, *79*, 14049–14054.

Khani, S., Barar, J., Movafeghi, A. & Omidi, Y. (2012). Production of anticancer secondary metabolites: impacts of bioprocess engineering. In: Orhan I (Ed.) *Biotechnological Production of Plant Secondary Metabolites*. Bentham Science Publishers.

Khela, S. (2013). *Sideritis scardica. The IUCN Red List of Threatened Species*, e.T203271A2762714. http://dx.doi.org/10.2305/ IUCN.UK. 2013-2.RLTS.T203271A2762714.en.

Kitanov, G. M. (2001). Hypericin and pseudohypericin in some *Hypericum* species. *Biochem Syst Ecol*, *29*, 171–178.

Kirakosyan, A., Gibson, D. M. & Sirvent, T. (2003). A comparative study of *Hypericum perforatum* plants as sources of hypericins and hyperforins. *J Herbs Spi Med Plants*, *10*, 73–88.

Kirakosyan, A., Sirvent, T. M., Gibson, D. M. & Kaufman, P. B. (2004). The production of hypericins and hyperforin by *in vitro* cultures of St. John's wort (*Hypericum perforatum*). *Biotechnol Appl Biochem*, *39*, 71–81.

Kitzberger, C. S. G., dos Santos Scholz, M. B., Pereira, L. F. P., Vieira, L. G. E., Sera, T., Silva, J. B. G. D. & de Toledo Benassi, M. (2013). Diterpenes in green and roasted coffee of *Coffea arabica* cultivars growing in the same edapho-climatic conditions. *Journal of Food Composition and Analysis*, *30*, 52–57.

Koleva, P., Petrova, N., Krumova, S., Velikova, V., Aneva, I., Evstatieva, L., Wolfram, E. & Danova, K. (2016). Effect of activated charcoal on the developmental patterns, polyphenolics productivity and photosynthetic activity of *Sideritis scardica in vitro*. *Planta Med.*, *82*(S 01): S1-S381, DOI: 10.1055/s-0036-1596817.

Kozai, T. (1991). Micropropagation under photoautotrophic conditions. – In: Debergh, P.C., Zimmerman, R.H. (ed.): *Micropropagation. Technology and Application.*, pp. 447-469. Kluwer Academic Publishers, Dordrecht - Boston - London.

Kozai, T. & Smith, M. A. L. (1995). Environmental control in plant tissue culture - general introduction and overview. – In: Aitken-Christie, J., Kozai, T., Smith, M.L.

(ed.): *Automation and Environmental Control in Plant Tissue Culture*, pp. 301-318. Kluwer Academic Publishers, Dordrecht - Boston - London.

Kozai, T., Koyama, Y. & Watanabe, I. (2002). Multiplication of potato plantlets *in vitro* with sugar-free medium under high photosynthetic photon flux. *Acta. Hort.*, pp. 230.

Kougan, G. B., Tabopda, T., Kuete, V. & Verpoorte, R. (2013). Chapter 6 – Simple Phenols, Phenolic Acids, and Related Esters from the Medicinal Plants of Africa. In: Victor Kuete (Ed.) *Medicinal Plant Research in Africa. Pharmacology and Chemistry*. Elsevier.

Kutney, J. P. (1997). Plant cell culture combined with chemistry – routes to clinically important compounds, *Pure & Appl. Chem.*, *69*, 431-436.

Ky, C. L., Louarn, J., Dussert, S., Guyot, B., Hamon, S. & Noirot, M. (2001). Caffeine, trigonelline, chlorogenic acid and sucrose diversity in wild *Coffea arabica* L. and *C. canephora P. accessions. Food Chemistry*, *75*, 223–230.

Laughlin, J. C. (1980). The effect of time of harvest on the yield components of poppies (*Papaver somniferum* L.). *The Journal of Agricultural Science.*, *95*, 667-676.

Lavanya, C., Ashoka, J., Sreenivasa, A. G., Sushila, N. & Beladhadi, B. V. (2017). Effect of elevated carbon dioxide and temperature on growth, yield and quality parameters of mulberry. *Entomology, Ornithology & Herpetology: Current Research.*, *6*, 196.

Lee, S. Y., Cho, S. I., Park, M. H., Kim, Y. K., Choi, J. E. & Park, S. U. (2007). Growth and rutin production in hairy root cultures of buckwheat (*Fagopyrum esculentum* M.). *Prep. Biochem. Biotechnol.*, *37*, 239-246.

Lehmann, N., Dittler, I., Lämsä, M., Ritala, A., Risher, H., Eibl, D., Oksman-Caldentey, K. M. & Eibl, R. (2014). Disposable bioreactors for cultivation of plant cell cultures. In: Paek KY, Murthy HN, Zhong JJ (Eds.) (2014) *Production of Biomass and Bioactive Compounds Using Bioreactor Technology*. Springer Dordrecht Heidelberg New York London.

Li, S. H., Nagy, N., Hammerbacher, A., Krokene, P., Niu, X. M., Gershenzon, J. & Schneider, B. (2012). Localization of phenolics in phloem parenchyma cells of Norway spruce (*Picea abies*). *Chembiochem: a European journal of chemical biology.*, *13*, 2707-13.

Linsmaier, E. F. & Skoog, F. (1965). Organic growth factor requirements of tobacco tissue cultures, *Physiol. Plant.*, *18*, 100-127.

Liu, C. Z., Wang, Y. C., Zhao, B., Guo, C., Ouyang, F., Ye, H. C. & Li, G. F. (1999). Development of a nutrient mist bioreactor for growth of hairy roots. *In Vitro Cell. Dev. Biol. Plant.*, *35*, 271–274.

Lokar, L. C., Maurich, V., Mellerio, G., Moneghini, M. & Poldini, L. (1987). *Variation in terpene composition of Artemisia alba in relation to environmental conditions.*, *15*, 327-333.

Lombardino, J. G. (2000). A brief history of Pfizer Central Research. *Bull. Hist. Chem.*, *25*, 10.

Lošák, T. & Richter, R. (2004). Split nitrogen doses and their efficiency in poppy (*Papaver somniferum* L.) nutrition. *Plant Soil Environ.*, *50*, 484–488.

Loyola-Vargas, V. M. & Vázquez-Flota, F. (Eds.), (2006). *Plant Cell Culture Protocols*, vol. *318*. Humana Press, Totowa, New Jersey.

Lucchesini, M., Bertoli, A., Mensuali-Sodi, A. & Pistelli, L. (2009). Establishment of *in vitro* tissue cultures from *Echinacea angustifolia* D.C. adult plants for the production of phytochemical compounds. *Scientia Horticulturae*, *122*, 484–490.

Luthria, D. L., Mukhopadhyay, S. & Krizek. (2006). Content of total phenolics and phenolic acids in tomato (*Lycopersicon esculentum* Mill.) fruits as influenced by cultivar and solar UV radiation. *Journal of Food Composition and Analysis.*, *19*, 771–777.

Lydon, J., Casale, J. F., Kong, H., Sullivan, J. H., Daughtry, C. S. T. & Bailey, B. (2009). The effects of ambient solar UV radiation on alkaloid production by *Erythroxylum novogranatense* var. *novo-granatense. Photochemistry and Photobiology.*, *85*, 1156–1161.

Mafakheri, S., Hajivand, Sh., Zarrabi, M. M. & Arvane, A. (2016). Effect of bio and chemical fertilizers on the essential oil content and constituents of *Melissa officinalis* (Lemon balm). *Journal of Essential Oil Bearing Plants.*, *19*, 1277-1285.

Maggio, A., Rosselli, S., Bruno, M., Spadaro, V., Raimondo, F. M. & Senatore, F. (2012). Chemical Composition of essential oil from Italian populations of *Artemisia alba* Turra (Asteraceae). *Molecules.*, 17, 10232-10241.

Mahdavi-Damghania, A., Kamkarb, B., Al-Ahmadib, M. J., Testi, L., Munoz-Ledesma, F. J. & Villalobos, F. J. (2010). Water stress effects on growth, development and yield of opium poppy (*Papaver somniferum* L.). *Agricultural Water Management.*, *97*, 1582–1590.

Malik, S., Bhushan, S., Sharma, M. & Ahuja, P. S. (2016). Biotechnological approaches to the production of shikonins: a critical review with recent updates. *Crit Rev Biotechnol.*, *36*, 327-40.

Maurmann, N., De Carvalho, C. M. B., Silva, A. I. L., Fett-Neto, A. G., Von Poser, G. L. & Rech, S. B. (2006). Valepotriates accumulation in callus., suspended cells and untransformed root cultures of *Valeriana glechomifolia. In Vitro Cell Dev Biol Plant.*, *42*, 50-53.

Misawa, M. (1994). Plant tissue culture: an alternative for production of useful metabolites. In: *FAO Corporate Document Repository*, ch. 7, Products of interest in Industry.

Murashige, T. & Skoog, F. (1962). A revised medium for rapid growth and bioassays with tobacco tissue cultures. *Physiol Plantarum*, *15*, 473-497.

Murch, S. J., KrishnaRaj, S. & Saxena P. K. (2000). Tryptophan is a precursor for melatonin and serotonin biosynthesis in *in vitro* regenerated St. John's wort (*Hypericum perforatum* L. cv. Anthos) plants. *Plant Cell Reports*, *19*, 698-704.

Murthy, H. N., Dijkstra, C., Anthony, P., White, D. A., Davey, M. R., Power, J. B., Hahn, E. J. & Paek, K. Y. (2008). Establishment of *Withania somnifera* hairy root cultures for the production of withanolide A. *J Integr Plant Biol.*, *50*, 975-981.

Najafian, S. & Zahedifar, M. (2014). Antioxidant activity and essential oil composition of *Satureja hortensis* L. as influenced by sulfur fertilizer. *J Sci Food Agric.*, *95*, 2404-8.

Nakagawa, K., Fukui, H. & Tabata, M. (1986). Hormonal regulation of berberine production in cell suspension cultures of Thalictrum minus, *Plant Cell. Rep.*, *5*, 69-71.

Nejadhabibvash, F., Rahmani, F., Heidari, R., Jamei, R. & Azimi, F. (2002). Study of inheritance and environment on tropane alkaloids within *Hyoscyamus* species. *Austr. Journ. Crop Sci.*, *6*, 1428-1434.

Nilsamranchit, S., Ogaki, K., Ashida, K. & Sugino, M. (1996). Effects of application of N-P-K fertilizer on the growth and yield of a medicinal plant *Geranium thunbergii* Sieb. Et Zucc. *Mem. Fac. Agr. Kinki Univ.*, *29*, 1-10.

Ochoa-Villarreal, M., Howat, S., Hong, S. M., Jang, M. O., Jin, Y. W., Lee, E. K. & Loake, G. J. (2016). Plant cell culture strategies for the production of natural products. *BMB Rep.*, *49*, 149-158.

Paek, K. J., Murthy, H. N., Hahn, E. J. & Zhong, J. J. (2009). Large scale culture of ginseng adventitious roots for the production of ginsenosides. *Adv Biochem Engin/Biotechnol.*, *113*, 151-176.

Pan, M. J. & van Staden, J. (1998). The use of charcoal in *in vitro* culture – A review. *Plant Growth Regulation.*, *26*, 155-163.

Pandey, G., Khatoon, S., Pandey, M. M. & Rawat, A. K. S. (2017). Altitudinal variation of berberine, total phenolics and flavonoid content in *Thalictrum foliolosum* and their correlation with antimicrobial and antioxidant activities. *Journal of Ayurveda and Integrative Medicine* (Article in press) http://dx.doi.org/10.1016/j.jaim.2017.02.010.

Park, S. U. & Lee, S. Y. (2009). Anthraquinone production by hairy root culture of *Rubia akane* Nakai: Influence of media and auxin treatment. *Scientific Research and Essay*, *4*, 690-693.

Patra, B., Schluttenhofer, C., Wu, Y., Pattanaik, S. & Yuan, L. (2013). Transcriptional regulation of secondary metabolite biosynthesis in plants. *Biochimica et Biophysica Acta (BBA) - Gene Regulatory Mechanisms.*, *1829*, 1236-1247.

Pavlov, A., Georgiev, M., Panchev, I. & Ilieva, M. (2005). Optimization of rosmarinic acid production by Lavandula vera MM plant cell suspension in a laboratory bioreactor. *Biotechnol Prog.*, *21*(2), 394-6.

Pavlov, A., Berkov, S., Courot, E., Gocheva, T., Tuneva, D., Pandova, B., Georgiev, M., Georgiev, V., Yanev, S., Burrus, M. & Ilieva, M. (2007). Galanthamine production by *Leucojum aestivum in vitro* systems. *Process Biochemistry*, *42*, 734–739.

Petersen, M. & Alfermann, A. W. (1993). Plant cell culture, Ch. 17: in *Biotechnology* (eds) H. J. Rehm and G. Reed, in coop. With A. Püchler and P. Stadtler, *1*, 578-633.

Petersen, F. & Amstutz, R. (2008). *Natural compounds as drugs*, vol. *1*, Birkhäuser, Basel, Switzerland.

Pierik, R. L. M. (1987). *In vitro culture of higher plants*. Martinus Nijhoff Publishers, Dordrecht/Boston/Lankaster.

Pistelli, L., Giovannini, A., Ruffoni, B., Bertoli, A. & Pistelli, L. (2010). Hairy root cultures for secondary metabolites production. *Adv Exp Med Biol.*, *698*, 167-84.

Pošpišilová, J. Čatský. & Šesták, Z. (1997). Photosynthesis in plants cultivated *in vitro*. In: (Pessarakli M ed.) *Handbook of photosynthesis*, Marcel Dekker Inc., New York, p. 525-530.

Radulović, N. & Blagojević, P. (2010). *Volatile profiles of Artemisia alba from contrasting serpentine and calcareous habitats.*, *5*, 1117 – 1122.

Ram, M., Ram, B. & Kumar, S. (2010). Plant spacing and the yield of leaves and alkaloids in *Duboisia myoporoides*. *Journal of Herbs, Spices & Medicinal Plants.*, *3*, 77-80

Razdan, M. K. (2003). *Introduction to plant tissue culture*. 3rd ed. Scientific publishers Inc., New Hampshire, USA.

ROOTec factsheets. www.rootec.com.

Routian, J. B. & Nickell, L. G. (1956). US Patent, *2*, 747, 334.

Ryu, H. W., Yuk, a. H. J., An, J. H., Kim, D. Y., Song, H. H. & Oh, S. R. (2017). Comparison of secondary metabolite changes in *Camellia sinensis* leaves depending on the growth stage. *Food Control*, *73*, 916-921.

Sakata, K., Morita, E. & Takezono, T. (1990). Production of podophyllotoxin by organ cultures presented at the *Ann. Meeting Soc. Agric. Chem. Fukuoka*, Japan, March 30 – April 2 1990.

Schürch, C., Blum, P. & Zülli, F. (2008). Potential of plant cells in culture for cosmetic application. *Phytochem. Rev.*, *7*, 599–605.

Sedaghathoor, S., Torkashvand, A. M., Hashemabadi, D. & Kaviani, B. (2009). Yield and quality response of tea plant to fertilizers. *African Journal of Agricultural Research*, Vol. *4* (6), pp. 568-570.

Shukla, H. Y. & Shukla, P. K. (2012). Effect of stand geometry and plant growth regulators on root yield and alkaloid content of Ashwagandha (*Withania somnifera* Dunal.) *Int. J. Med. Arom. Plants.*, *2*, 390-395.

Singh, B. & Sharma, R. A. (2015). Plant terpenes: defense responses, phylogenetic analysis, regulation and clinical applications. *3 Biotech.*, *5*, 129–151.

Song, R., Kelman, D., Johns, K. L. & Wright, A. D. (2012). Correlation between leaf age, shade levels, and characteristic beneficial natural constituents of tea (*Camellia sinensis*) grown in Hawaii. *Food Chemistry.*, *133*, 707–714.

Stojanovic, G., Palic, R. & Mitrovic, J. (2000). Chemical composition and antimicrobial activity of the essential oil of *Artemisia lobelii* All. *Journal of Essential Oil Research.*, *12*, 621-624.

Tabata, M. & Fujita, Y. (1985). Production of shikonin by plant cell cultures, in: *Biotechnology in Plant Science* (Day, P., Zaitlin, M., Holländer, A. Eds.), Orlando: Academic press, 207-218.

Taiz, L. & Zeiger, E. (2006). Secondary metabolites In: Plant Physiology 4th ed, Sinauer Associeates Inc. Publishers, *Biology, Neuroscience & psychology.*, 283-308.

Toivonen, L., Laakso, S. & Rosenguist, H. (1992). The effect of temperature on growth, indole alkaloid accumulation and lipid composition of *Catharanthus roseus* cell suspension cultures. *Plant cell reports.*, *11*, 390-394.

Tripathi, L. & Tripathi, J. N. (2003). Role of biotechnology in medicinal plants. *Tropical Journal of Pharmaceutical Research*, *2*, 243-253.

Ulbrich, B., Wiesner, W. & Arens, H. (1985). Large scale production of rosmarinic acid from plant cell cultures of *Coleus blumei*, in: *Primary and secondary metabolism of Plant Cell Cultures* (Neumann, K.H., Barz, W., Reinhard, E., Eds.), Berlin: Springer 293-303.

Vardapetyan, H. R., Kirakosyan, A. B. & Charchoglyan, A. G. (2000). The kinetics regularities of the globular structures growth in cell cultures of *Hypericum perforatum* L. *Biotechnologia*, *4*, 53–58.

Verpoorte, R. (1996). Plant cell biotechnological research in the Netherlands. In: *Plant Cell Culture Secondary Metabolism Toward Industrial Application* (DiCosmo, F. and Misawa, M., eds.), CRC Press, Boca Raton, FL, 203–229.

Verpoorte, R., Contin, A. & Memelnik, J. (2002). Biotechnology for the production of plant secondary metabolites, *Phytochem. Rev.*, *1*, 13-25.

Wang, Y. S., Gao, L. P., Shana, Y., Liua, Y. J., Tian, Y. W. & Xia, T. (2012). Influence of shade on flavonoid biosynthesis in tea (*Camellia sinensis*, (L.) O. Kuntze). *Scientia Horticulturae*, *141*, 7–16.

Watanabe, K., Jano, S. I. & Yamada, Y. (1982). The selection of cultured plant cells, producing high levels of biotin. *Phytochemistry.*, *21*, 513.

Weathers, P. J., Cheetham, R. D., Follansbee, E. & Teoh, T. (1994). Artemisinin production by transformed roots of *Artemisia annua*. *Biotechnol. Lett.*, *16*, 1281–1286.

White, P. R. (1963). *The Cultivation of Animal and Plant Cells*, 2nd ed, New York: Ronald Press.

Wilson, S. A. & Roberts, S. C. (2012). Recent advances towards development and commercialization of plant cell culture processes for synthesis of biomolecules. *Plant Biotechnol J.*, *10*, 249–268.

Yamada, Y. & Sato, F. (1981). Production of berberine in cultured cells of *Coptis japonica*. *Phytochemistry*, *20*, 545– 547.

Yamamato, H. (1981). *Shoyakugaku (In Japanese).*, *35*, 15.

Yamamoto, Y., Mizuguchi, R. & Yamada, Y. (1982). Selection of high and stable pigment-producing strain in cultured *Euphorbia millii* cells. *Theor. Appl. Genet.*, *61*, 113.

Yonemitsu, H., Shimomura, K., Satake, M., Mochida, S., Tanaka, M., Endo, T. & Kaji, A. A. (1990). Lobeline production by hairy root culture of *Lobelia inflata* L. *Plant Cell Rep.*, *9*, 307-310.

Yu, J., Yuan, S., Pang, H., Zhang, X., Jia, X., Tang, Z. & Zu, Y. (2010). Distribution and accumulation of vindoline, catharanthine and vinblastine in *Catharanthus roseus* cultivated in China. *Zhongguo Zhong Yao Za Zhi.*, *35*, 3093-6 (article in Chinese).

Xu, J. F., Ying, P. Q., Han, A. M. & Su, Z. G. (1999). Enhanced salidroside production in liquid-cultivated compact callus aggregates of *Rhodiola sachalinensis*: manipulation of plant growth regulators and sucrose. *Plant Cell Tissue Organ Cult.*, *55*, 53-58.

Zenk, M. H., El-Shagi, H., Schulte, U., et al. (1975). Antraquinone production by cell suspension cultures of *Morinda citrifolia*, *Planta Med. Suppl.*, 79-101.

Zenk, M. H., El-Shagi, H., Arens, H., Stöckigt, J., Weiler, E. W. & Deus, B. (1977a). Formation of indole alkaloids serpentine and ajmalicine in cell suspension cultures of Catharanthus roseus, in: *Plant Tiss. Cult. and its Bio-Technol. Applicat.* (Barz, W., Reinhard, E., Zenk, M.H., Eds.), Berlin-Heidelberg, New York: Springer.

Zenk, M. H., El-Shagi, H. & Ulbrich, B. (1977b). Production of rosmarinic acid by cell suspension cultures of *Coleus blumei*, *Naturwissen-schaften*, *64*, 585-586.

Zhao, J., Fujita, K. & Sakai, K. (2005). Oxidative stress in plant cell culture: A role in production of β-thujaplicin by *Cupressis lusitanica* suspension culture, *Biotechnol and Bioengineering*, Volume *90*, Issue 5, Pages 621 – 631.

Zhong, J. J. (2001). Biochemical Engineering of the Production of Plant-Specific Secondary Metabolites by Cell Suspension Cultures. *Advances in Biochemical Engineering/Biotechnology.*, *72*, 1-26.

In: Recent Advances in Plant Research
Editors: B. Welch and M. Wilkerson
ISBN: 978-1-53614-170-2

Chapter 2

FABACEAE LINDL.: BIODIVERSITY, TAXONOMY AND NEW RECORDS FOR NORTHEAST REGION OF BRAZIL

Gustavo S. Gomes*[1,*]*, Guilherme S. Silva*[2]*,
Gonçalo M. Conceição*[3]*, Regigláucia R. Oliveira*[3]*,
***Domingos L. S. Silva*[3] *and Paula R. P. Martins*[3]**
[1]Department of Chemistry and Biology,
State University of Maranhão, Caxias, Maranhão, Brazil
[2]Postgraduate Program in Botany of the National Research
Institute of the Amazon-INPA, Manaus, Amazonas, Brazil
[3]Postgraduate Program in Biodiversity, Environment and Health/
PPGBAS State University of Maranhão/UEMA,
Caxias, Maranhão, Brazil

ABSTRACT

Fabaceae is one of the largest families among the angiosperms, with representatives widely distributed throughout the globe, where 770 sorts and 19,500 species are recognized and divided into six subfamilies (*Caesalpinioideae*, *Cercidoideae*, Detioideae, Dialioideae, Duparquet-ioideae, and *Papilionoideae*). Floristic and taxonomic works indicate *Fabaceae* as the largest species family in Brazil and most representative in forest formations. The Cerrado has a significant representation of the family, where the state of Maranhão covers a broad area. In this way, the research herein meant to map out the diversity of *Fabaceae* in vegetational fragments of the Cerrado in the state of Maranhão. Monthly expeditions were carried out in the period between September 2016 and June 2017 for observation, collection, and identification of botanical material. The Cerrado area sampled belongs to the municipality of São João do Sóter, in the eastern part of Maranhense. After the collections, the specimens were herborized, analyzed with the help of taxonomic keys and specialized literature for identification, and determined by area

* Corresponding Author Email: gustavogomes@aluno.uema.br.

experts. A total of 68 specimens, 31 genera, and 45 species were cataloged. Of all the specimens collected, the subfamilies *Papilionoideae* and *Caesalpinioideae* were the most representative of 21 species each. The prevailing growth habit was the bush type (20) in terms of other species. Concerning the physiognomies of the Cerrado, predominance of the species in the gallery forest environment (28) was observed. Three new ledgers to the Maranhão were obtained: Desmodium subsecundum Vogel. *Aeschynomene* viscidula Michx and *Vigna lasiocarpa* (Mart. ex Benth.), with V. lasiocarpa acting as a new record for the northeast region of Brazil. Taxonomic keys, descriptions and photos accompany all the studied species, helping in composing a taxonomic treatment. In this way, it can be established that the Cerrado of Maranhão possesses a diversity of species for the *Fabaceae* family and that the research carried out has provided a basis for later studies, since there are few for Maranhão.

Keywords: Cerrado, floristic, *Papilionoideae*, shrubs, *Vigna lasiocarpa*

INTRODUCTION

Within the large diversity of angiosperms, *Fabaceae* is one of the largest botanical families, with 770 genera and 19.500 species (Lewis et al., 2005, 2013, LPWG, 2017) that are divided into six subfamilies (*Caesalpinioideae*, *Cercidoideae*, *Detarioideae*, Dialioideae, Duparquet-ioideae, and *Papilionoideae*) (LPWG, 2017). In Brazil, there are 2.837 species (1.535 endemics) grouped in 222 genera (Flora do Brasil 2020). *Fabaceae*, formerly known only as Leguminosae, was traditionally handled by Bentham (1865) as a single family subordinated into three subfamilies (*Caesalpinioideae*, Mimosoideae, and *Papilionoideae*).

Later Hutchinson (1964) and Cronquist (1981) refuted this division proposing that these subfamilies were considered as independent families. According to APG IV (2016), the subfamilies are currently considered members of a single family (*Fabaceae*), indexed in the Fabales order, of the Eurosids clade.

Fabaceae is a family of great economic importance, its plants acting as exemplary sources of protein and micronutrients that would present many health benefits, especially to individuals living in developing countries. (Graham & Vance 2003; Yahara et al., 2013). This particular family of plants are ecologically important because they have adapted to the rapid urbanization of their endemic environment, particularly to the changes in nitrogen levels and the number of fixative bacteria they present.

Bacterium of the genus *Rhizobium*, located in the root nodules found in many species, convert atmospheric nitrogen into ammonia, a soluble form that can be used by 20 other plants, resulting in extremely valuable species as suppliers of natural fertilizers, in addition to their considerable importance in agriculture, their capacity to occupy different habitats and diverse forms of life (Lewis et al., 2005; Sprent et al., 2013 Yahara et al., 2013).

The Cerrado has more than 4.800 species of endemic plants and vertebrates, being considered as a global biodiversity hotspot. This phytogeographical domain covers three of the largest hydrographic basins in South America, accounting for 43% of Brazil's surface waters outside the Amazon (Strassburg et al., 2017).

Due to its considerable biodiversity, this phytogeographical domain has been the focus of several botanical studies, and therefore, work on the *Fabaceae* family in the Cerrado is necessary due to its floristic richness (Batalha, 2011), in addition to the fact that this biome is in danger of extinction (Strassburg et al., 2017). In the state of Maranhão, there is an ecotone with the Amazon rainforest and Caatinga (Maranhão, 2011). In Maranhão, the municipality of São João do Sóter is in this phytogeographic domain, possessing fragments untouched by humans and conserving its biodiversity.

The state of Maranhão has large extensions of the Cerrado biome, with forest formations of significant extensions (deciduous rainforests and, to a lesser extent, dense ombrophilous forests), as well as a large area of ecological tension with the Amazonian biome (Maranhão, 2011).

Thus, the objective of this chapter and the research found within is to understand the diversity of *Fabaceae* in vegetative fragments of Cerrado in the state of Maranhão in order to: Demonstrate the morphological aspects of the family *Fabaceae*; to elaborate on taxonomic keys for the species studied; to determine the types of growth habits, phytogeographic domains and environments of occurrence of the species, and to contribute to the knowledge concerning the flora of the Cerrado from Maranhão.

Material and Methods

Location and Characterization of the Study Area

The municipality of São João do Sóter, located in Maranhão, carry over 1.438.1 km² and has 17,238 inhabitants in the last census. The population density is 11.9 inhabitants per km² in the municipality territory. Located at 108 meters altitude, at the following geographical coordinates: 5° 6 '28' 'S. and 43° 48' 34 "W (Figure 1). In the Municipality, three Cerrado areas were sampled: Pedras, Redondo, and Serra do Cajuí Village (IBGE, 2010).

The vegetational fragments sampled are composed of physiognomies, coconut and dirty cerrado of herbaceous and perennial plants. The floristic survey comprised three stages: fieldwork, preparation and analysis, and identification of collected material. The collection of the botanical material was carried out during September 2016 to June 2017, through the active search method, with a monthly visit to the study area during the dry and rainy months.

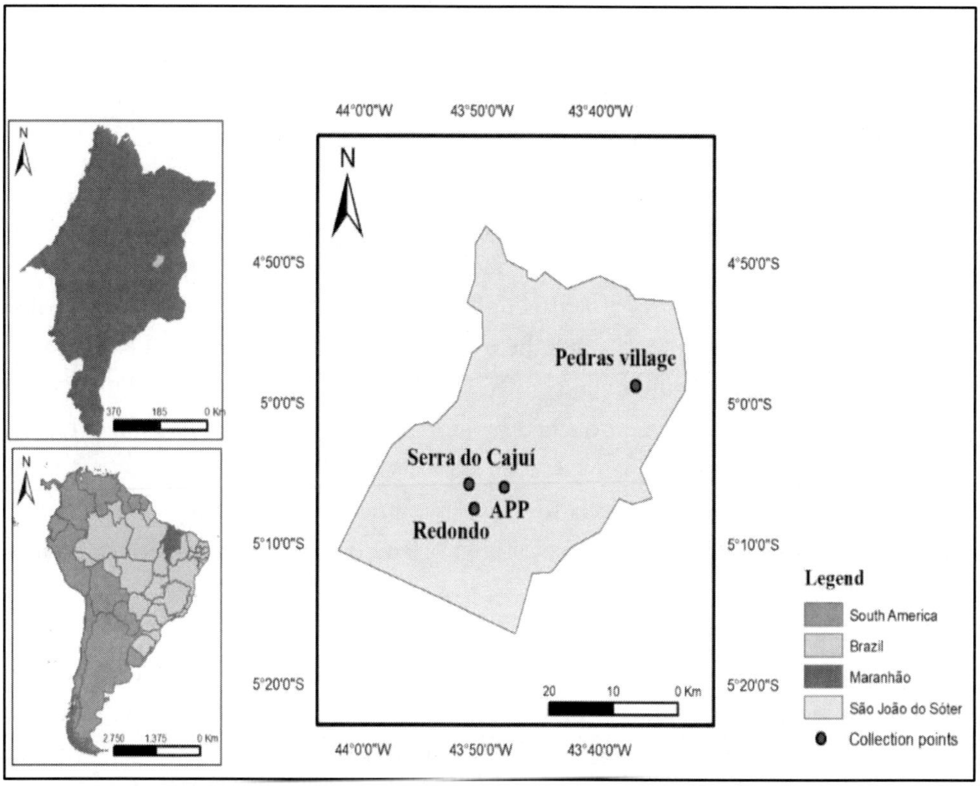

Figure 1. Map of the Municipality of São João do Sóter, Maranhão/Brazil, Collection area of the especies of *Fabaceae*. Source: IBGE; 2006, Google Earth 2014.

In the fieldwork, the specimens were collected in the fertile state, following the protocol of Bononi; Fidalgo (1989). Several collections of notes were made, such as: date and location of collection (locality, municipality, state and geographic coordinates), the name of the collector, and the physical characteristics of the environment (such as slope, soil type, of water courses, etc.), concomitant with notes on the specimens: plant habit, height, bark characteristic, color and texture of leaves, flowers and fruits and other striking features that aided in identification.

In the preparation of the botanical material, it was herborized and processed as well as conditioned as exsiccates in Herbarium Prof. Aluízio Bittencourt (HABIT), Center for Higher Studies of Caxias - CESC/UEMA. In the identification stage, the specimens were recognized at the species level taking into account the comparison with the specialized material, specific bibliographies, and taxonomic keys, and/or determined and confirmed by a family taxonomist. For the taxonomic treatment, all information and morphological characteristics of each species were observed and recorded. These characteristics formed a description of the plant along with the information of the author of the species, the primary work, geographical distribution and phytogeographical domains where the species occurs in the Brazilian territory. Concerning the data of the author for all species, the tropic data network (2018) was used. With the botanical description, dichotomous taxonomic keys of the described species were elaborated, using the taxonomic criteria already presented in the botanical description that better identifies the studied species.

RESULTS AND DISCUSSION

Taking into account floristic aspects, *Fabaceae* is one of the families among the angiosperms that has the richest species in the different phytogeographic domains of Brazil according to Queiroz (2009). 68 specimens were listed from the main data, distributed among 31 genera and 45 species. The representation of species of the *Fabaceae* subfamilies sampled in this research shows that of the 68 specimens studied, 21 species belong to the subfamily *Papilionoideae*, 21 belong to the subfamily *Caesalpinioideae*, 2 species belong to the subfamily *Cercidoideae* and 1 species belongs to the subfamily *Detarioideae*, as shown in Table 1.

As for species richness, the subfamilies *Papilionoideae* and *Caesalpinioideae* were the most representative, sampled 21 and 19 species respectively (table. 1). *Papilionoideae* is a monophyletic subfamily, with high reliability of phylogenetic reconstruction. The subfamily includes many species of economic importance (Tozzi, 2016). *Caesalpinioideae* in its present circumscription contains 148 genera and 4.400 species. Pantropical distribution is common in dewy and dry regions, with a handful of species that extend to the temperate zone (LPWG, 2017). Brazilian distribution is represented by about 52 genera and 810 species (Flora do Brasil 2020).

In the physiognomy of occurrence, the specimens were collected, mainly in gallery forests, with 28 species (Fig. 2). Among the plant formations of the Cerrado, the gallery forest, also known as the riparian forest or riparian forest by some researchers, is characterized by being associated with watercourses, and this formation (although small) possesses richness, genetic diversity, and acts in the protection of water resources (Ribeiro, 1998). The two most representative genera in the survey were *Aeschynomene* and *Mimosa* with four species each (Fig. 3).

The genus *Aeschynomene* has a pantropical distribution, with about 180 species (Lewis et al., 2005). *Mimosa* L. comprises 540 species, being the second largest genus of the *Mimosoide clade* (Simon et al., 2011). It is distributed mainly in the neotropical region, counting on approximately 496 endemic taxon of the neotropics and 40 native species of the old world (Simon et al., 2011; Barneby, 1991).

Table 1. Representation of *Fabaceae* species within each subfamily sampled

N°	Subfamily	Species
1.	*Caesalpinioideae*	*Caesalpinia pulcherrima* (L.) Sw.
2.		*Cenostigma macrophyllum* Tul.
3.		*Chamaecrista flexuosa* (L.) Greene.
4.		*Chamaecrista nictitans* (L.) Moench.
5.		*Chamaecrista rotundifolia* (Pers.) Greene.
6.		*Delonix regia* (Bojer ex Hook.) Raf.
7.		*Dimorphandra gardneriana* Tul.
8.		*Inga edulis* Mart.
9.		*Inga thibaudiana* DC.
10.		*Mimosa caesalpinifolia* Benth.
11.		*Mimosa pudica* L.
12.		*Mimosa sensitiva* L.
13.		*Mimosa xanthocentra* Mart.
14.		*Parkia platycephala* Benth.
15.		*Stryphnodendron adstringens* (Mart.) Coville.
16.		*Libidibia ferrea* (Mart. Ex Tul.) L.P Queiroz
17.		*Plathymenia reticulata* Benth.
18.		*Senna multijuga* (Rich.) H.S. Irwin & Barneby.
19.		*Senna reticulata* (Willd.) H.S. Irwin & Barneby.
20.	*Cercidoideae*	*Bauhinia dubia* G. Don.
21.		*Bauhinia pulchella* Benth.
22.		*Phanera variegata* (L.) Benth.
23.	*Detarioideae*	*Tamarindus indica* L.
24.		*Hymenaea stigonocarpa* Mart. ex Hayne.
25.	*Papilionoideae*	*Abrus fruticulosus* Wight & Arn.
26.		*Aeschynomene brasiliana* (Poir.) DC.
27.		*Aeschymoneme histrix* Poir.
28.		*Aeschynomene paniculata* Willd. ex Vogel
29.		***Aeschynomene viscidula* Michx.
30.		*Centrosema brasiliana* (L.) Benth.
31.		*Clitoria guianensis* (Aubl.) Benth.
32.		*Crotalaria retusa* L.
33.		*Crotalaria stipularia* Desv.
34.		*Desmodium barbatum* (L.) Benth.
35.		*Desmodium incanum* (Sw.) DC.
36.		***Desmodium subsecundum* Vogel.
37.		Dioclea bicolor Benth.
	Papilionoideae	*Galactia jussiaeana* Kunth.
38.		*Indigofera suffruticosa* Mill.
39.		*Macroptilium atropurpureum* (Sessé & Moc. ex DC.) Urb.
40.		*Macroptilium lathyroides* (L.) Urb.
41.		*Periandra heterophyla* Benth.
42.		*Phaseolus vulgaris* L.
43.		*Stylosanthes viscosa* (L.) Sw.
44.		**Vigna lasiocarpa* (Mart. ex Benth.) Verdc.

(*) indicates a new record for the Northeast; (**) indicates a new record for Maranhão

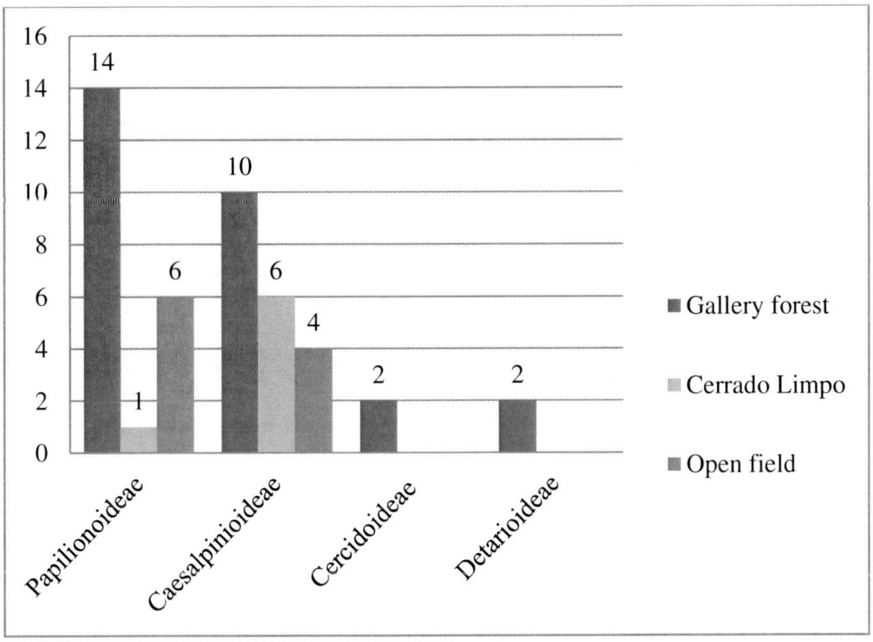

Figure 2. Representation of the environments of collection of *Fabaceae* subfamilies in the collection environments.

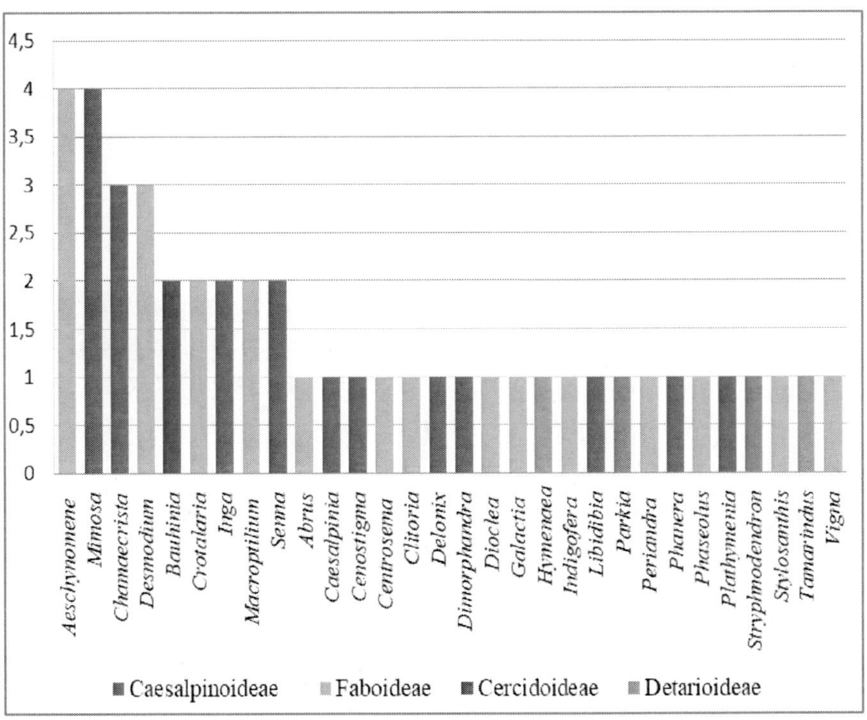

Figure 3. Representation of the number of species of each genus studied.

Regarding this lifeform, the predominant habit of growth was the bush type with 18 species (Fig. 4).

The habit of this family is assorted, from trees, shrubs and sub-bushes to upright herbs, which creep or even climb (JUDD et al., 1999). Shrubs are characterized by woody structures of varying size, but not more than 6 m in height, and possess stems with branches much closer to the ground (Ormond, 2006). The bushes are out of reach of woody plants, in which the support and stem tissues form layers that are added year after year (Bononi; Fidalgo, 1989).

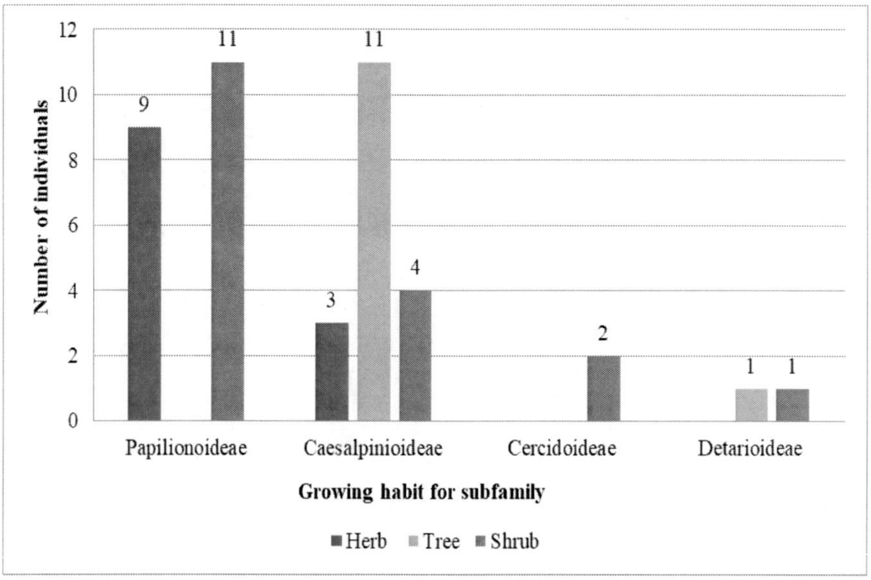

Figure 4. Growth habits of *Fabaceae* species sampled in the study.

As for the number of specimens of each species, the species *Phanera variegata* (L.) Benth., *Aeschymoneme histrix* Poir., *Libidibia ferrea* (Mart. Ex Tul.) L.P Queiroz and *Chamaecrista flexuosa* (L.) Greene. were the most representative in the survey, with P. variegata being the most representative with 5 specimens, and the other 3 specimens each. P. variegata is widely planted in the tropics and hot regions of the world (Isely, 1990). P. variegata is distributed in the Amazonian phytogeographical domain and is found in Colombia, French Guiana and Suriname. In Brazil, it is native throughout the North Region, in the states of Maranhão and Mata Atlântica, and in the state of São Paulo (Domingos, Carpelari Jr, 2016).

Three new occurrences were sampled at Maranhão: *Desmodium subsecundum* Vogel., *Aeschynomene viscidula* Michx. and *Vigna lasiocarpa* (Mart. ex Benth.) Verdc., with *V. lasiocarpa* being a new record for the Brazilian Northeast. *Vigna lasiocarpa* can be recognized by its habit, stipules elliptic to lanceolate, inflorescence multiflora, leguminous oblong and marmoreadas seeds, and black and beige colors. It is associated with humid environments and flooded areas (Maréchal et al., 1978).

Desmodium subsecundum is endemic in Brazil, where it occurs in the Federal District and in the states of Bahia, Espírito Santo, Goiás, Mato Grosso do Sul, Minas Gerais, Pará, Paraná, Rio de Janeiro, Rio Grande do Sul, Santa Catarina, São Paulo and Tocantins (Lima et al., 2014). *Aeschynomene* viscidula in Brazil occurs in the states of BA and CE (Fernandes, 1996). It has also been quoted in the state of MS (Silva et al., 2007) and in the states of PE, AL, and SE (Lima; Oliveira, 2010).

KEY OF SUBFAMILY CAESALPINOIDEAE

1. Grass, shrub or trees with leaves that are always bipinnates with inflorescence of the glomerulus or raceme types
2. Trees that are more than 10m in height, always with a woody stem
    - *Stryphnodendron adstrigens*: Puberulent rusty branches, lower crown, 30 to 45 leaflets per leaf, apex of rounded leaflet, leaflet with 0.4cm of length and 0.1cm wide
    - *Parkia platycephala*: Grayish, not pubertal branches, high crown, 35 to 50 leaflets per leaf, apex of acute leaflet, leaflet with 0.6 cm of length. and 0.2 cm long
3. Herbs or subshrub less than 10 m high, woody or herbaceous stems
    - Individuals less than 1m in height, small leaflets less than 1cm in length and less than 0.5cm in width:
    - *Mimosa pudica:* Symmetrical lanceolate leaflets, dry fruit type crusty, 1.5 cm long and 0.3 cm wide.
    - *Mimosa xanthocentra* Leaflets oblong-lanceolates, fruit legume dry with indumentum, with 9 cm of comp. and 0.3cm width
    - Individuals over 1m in height, large leaflets larger than 1cm in length and more than 0.5cm wide.
    - *Mimosa sensitiva:* Leaflets tetrafoliolates, asymmetric, 2 pairs of leaflets for leaf, dry fruit of the craspedium type, with approximately 2,5cm of length. and 0.3 cm long
    - *Mimosa caesalpiniifolia:* Leaflets symmetrical, with 6 to 12 pairs of leaflets per leaf, fruit, legume, with 5cm of length and 0.3cm wide.
4. Grass, shrub, or trees with leaves, bipinnates or imparipinnates, with inflorescences type raceme, never glomerulus.
5. Tree or shrub, with woody stem.
    - Trees with pubertals branches.

- *Cenostigma macrophyllum:* Stipules rudimentary or without the presence of nectary, fruit of the legume type with falcate apex.
- Sticks not rudimentary or with the presence of nectary, legume fruit without falcate apex.
  - Winged petiole with petiole nectary.
- *Inga edulis:* Nectary greater than 2mm, rounded longitudinal at the base of the rachis petioles.

  Inga thibaudiana: Nectary less than 2mm, rounded not longitudinal at the base of the rakers petiole.
  - Absence of winged petiole, absence of nectary petiole: 12
- *Senna reticulata:* Presence of long stipules in the shape of an orange sickle, with leaflets 10 cm long and 4.2 cm long.
- *Senna multijuga:* Presence of small elongated stipules, leaflets 6.7 cm long, and 2.6 cm long.

6. Trees with glabrous branches that are a little puberulent.
   - Inflorescence raceme with red or yellow flowers.
   - *Delonix regia:* Flowers with red petals, lush banner, legume 10 cm.
   - *Libidibia ferrea:* Flowers with yellow petals with reddish banner.
   - Inflorescence not raceme with green or white flowers.
   - *Dimorphnadra gardneriana:* Inflorescence with more than 10 racemes, green with presence of stamens and staminoids.
   - *Plathymenia reticulata*: Inflorescence with less than 10 racemes, white without the presence of stamens and staminoids.
7. Subshrub or herb.
   - *Caesalpinia pulcherrima:* Presence of aculeus in the branches, showy red flowers, legume fruit with long apex.
     - Absence of aculeus in the branches, luxuriant flowers, legume fruit without long apex.
   - *Chamaecrista nictitans:* Prostrate growth, bipinnates leaves, legume fruit with 2 to 3 items.
     - Growth creep or erect decumbent, bifoliolates leaves, legume fruit with more than 3 items.
   - *Chamaecrista rotundifolia:* Creeping growth, leaflets 3.2 cm in length and cm long, linear type fruit with 2.2 cm long and fruit petiole extended with 1.5 cm of comp
   - *Chamaecrista flexuosa*: Decumbent erect growth, leaflets with 4cm of length and 1.2cm long, a linear legume fruit type of 4.5cm length with fruit petiole not very extended with 0.5cm of length.

- *Chamaecrista flexuosa*: Decumbent erect growth, leaflets with 4cm of length and 1.2cm long, a linear legume fruit type of 4.5cm length with fruit petiole not very extended with 0.5cm of length

DESCRIPTION OF SUBFAMILY *CAESALPINIOIDEAE*

Stryphnodendron adstringens

(Mart.) Coville (Century Dict.) 11: 111, 1910.

Common Name

Barbatimão

Species Description

Tree with woody trunk fissured with approximately 10m of height, sympodial growth, branches blackish parts, absence of stipules, inerms, cylindrical, glabrescent branches, glandular trichomes in the branches, with 0.3cm of length. 0.4 inches long composites bipinnates leaves, alternating between 5.5 cm of length and 0.3 of long deciduous, presents of 30 to 45 pairs of leaflets, leaflets with 0.4 cm of length and 0.1 cm wide, rounded apex, smooth margin, paralelinnerved vein, absent flower, absent fruit.

Material Examined

Brazil. Maranhão: São João do Sóter, Pedras Village, 23/VI/17, G. S. Gomes; G. M. Conceição, 43 (HABIT).

Geographical Distribution

North (Tocantins); Northeast (Bahia); Center-West (Distrito Federal, Goiás, Mato Grosso do Sul, Mato Grosso); Southeast (Minas Gerais, São Paulo); South (Paraná) (Flora do Brasil 2020).

Phytogeographical Domains

Caatinga, Cerrado (Flora do Brasil 2020).

Parkia platycephala

Benth. J. Bot. (Hooker) 4 (30): 329, 1841.

Common Name

Faveira-de-bolota, Visgueiro.

Species Description

Tree with woody stem, sympodial growth of tall crown with 15m of height, grayish branches, abstract stipules, composite bipinnates leaves, presence of 35 to 50 pairs of leaflets, spiral alternate phyllotaxy, diminutive lanceolate leaflets, 6cm of length. and 0.2 cm wide, petiole with 7 cm, long leaves with 19.5 cm showing pulvin, whole margin, acute leaf apex, penninerved vein, cuminous inflorescence, dry fruit type legume, with approximately 10 cm of length and 2.5 cm wide.

Material Examined

Brazil. Maranhão: São João do Sóter, Permanent Protection Area, 08/X/16, G. S. Gomes; G. M. Conceição, 4 (HABIT).

Geographic Distribution

North (Pará, Tocantins); Northeast (Bahia, Ceará, Maranhão, Paraíba, Pernambuco, Piauí, Rio Grande do Norte); Center-West (Distrito Federal, Goiás, Mato Grosso) (Flora do Brasil 2020).

Phytogeographical Domain

Amazon forest, Caatinga, Cerrado (Flora do Brasil 2020).

Mimosa pudica

L. Gard. Dict. (Ed. 8) no. 4, 1768.

Common Name

Dormideira, Sensitiva, Dorme-dorme, Mulher-fecha-porta, Malícia-de-mulher.

Species Description

Shrub with woody stem, it presents 20 cm of height, aculeous in the form of sickle with 0,5cm of length, and 0.2 mm long, bipinnates composites leaves with 6 to 10 pairs of leaflets, 3,2 cm long pinnas, phyllotaxy alternate distical, lanceolate leaflet, with 0.9 cm of length, and 0.2 cm wide, 2.3 cm petiole, 3.9 cm leaf, whole margin, lanceolate leaf apex, paralelinnerved vein, glomerule cimose inflorescence, dichla-mydeous, heterochlamydeous, zygomorphic symmetry, polystemone free stamens, pink color, fruit legume craspedium type, 1.5 cm long and 0.3 cm wide.

Material Examined

Brazil. Maranhão: São João do Sóter, Pedras Village, 23/VI/17, G. S. Gomes; G. M. Conceição 42 (HABIT).

Geographic Distribution

North (Amazonas, Pará, Rondônia, Roraima); Northeast (Bahia, Maranhão, Pernambuco); Center-West (Distrito Federal, Mato Grosso do Sul, Mato Grosso); Southeast (Espírito Santo, Minas Gerais, Rio de Janeiro, São Paulo); South (Paraná, Rio Grande do Sul, Santa Catarina) (Flora Do Brasil 2020).

Phytogeographical Domains

Amazon forest, Caatinga, Cerrado, Atlantic forest (Flora do Brasil 2020).

Mimosa xanthocentra

Mart. Flora 21 (2, Biebl.): 50.1838.

Common Name

Dorme-dorme

Species Description

Subshrub with approximately 60 cm of height, woody stem, prostrate sympodal growth, inermous, cylindrical branches, aculeous of 0,2cm length, and 0.2cm wide, on the stem; stiples with 0.3cm of length, and 0.2cm wide, triangular, alternate composite sheets with 2 cm long. and 0.4 cm wide, bipinnates, with 10 to 15 pairs of leaflets, petioles 0.2 cm in length, absent extraphleic nectaries; leaflets 0.7 cm in length and 0.4 mm long, oblong-lanceolate, flower absent, legume fruit with indumentum, indecent, with 9 cm of comp. and 0.3 cm wide.

Material Examined

Brazil. Maranhão: São João do Sóter, Pedras Village, 23/VI/17, G. S. Gomes; G. M. Conceição, 41 (HABIT).

Geographic Distribution

North (Pará, Rondônia, Tocantins); Northeast (Bahia, Ceará, Maranhão, Piauí); Center-West (Distrito Federal, Goiás, Mato Grosso do Sul, Mato Grosso); Southeast (Minas Gerais, Rio de Janeiro, São Paulo); South (Paraná, Santa Catarina) (Flora do Brasil 2020).

Phytogeographical Domains

Amazon forest, Cerrado, Atlantic forest (Flora do Brasil 2020).

Mimosa Sensitiva

L. Sp. Pl. 1: 518, 1753.

Common Name

Malícia.

Species Description

Shrub with 2 m length, branched sympodial growth, woody stem, rudimentary aculeous branches present in every branch of the shrub with 0.2 cm of length, and 0.1 cm wide, greenish branches, tetrafoliolate composites leaves, alternate phyllotaxy, asymmetrical leaflets, unbalanced, 2.6 x 1.0 cm, petiole with 3.0 cm of length, leaf 5.6 cm, whole leaf margin, acute leaf apex, penninerved vein, cimose, spiky, white, flower absent, fruit craspedium type, with approximately 2,5 cm of length. and 0.3 cm wide.

Material Examined

Brazil. Maranhão: São João do Sóter, Pedras Village, 23/VI/17, G. S. Gomes; G. M. Conceição, 15, 39 (HABIT).

Geographic Distribution

North (Amazonas, Pará, Rondônia, Tocantins); Northeast (Alagoas, Bahia, Ceará, Maranhão, Pernambuco, Piauí, Rio Grande do Norte, Sergipe); Center-West (Distrito Federal, Mato Grosso do Sul, Mato Grosso); Southeast (Espírito Santo, Minas Gerais, Rio de Janeiro, São Paulo) (Flora do Brasil 2020).

Phytogeographical Domain

Amazon forest, Caatinga, Cerrado, Atlantic forest (Flora do Brasil 2020).

Mimosa caesalpiniifolia

Benth. (J. Bot. (Hooker) 4 (31): 392, 1841.

Common Name

Unha de gato.

Species Description

Shrub with woody stem, with 3 m height, branched sympodial growth, prostrate, grayish branches to blackened form rope, aculeous in sickle were present with 0.8cm of length and 0.2cm wide. bipinnates composite leaves with 6 to 12 pairs of pinnas, alternate phyllotaxy, ovate leaflet, 2.6 cm long and 1.6 cm wide, petiole with 0.2 cm of length, leaf approximately 8.0 cm, whole margin, rounded leaf apex, penninerved grove, cumin-shaped, spiky inflorescence, white coloring, absent flower, legume fruit, with 5cm of length and 0.3 cm wide.

Material Examined

Brazil. Maranhão: São João do Sóter, Permanent Protection Area, 08/X/16, G. S. Gomes; G. M. Conceição, 13, 14 (HABIT).

Geographical Distribution

North (Amazonas, Pará, Rondônia); Northeast (Alagoas, Bahia, Ceará, Maranhão, Paraíba, Pernambuco, Piauí, Rio Grande do Norte); Center-West (Distrito Federal, Goiás, Mato Grosso do Sul); Southeast (Espírito Santo, Minas Gerais, Rio de Janeiro, São Paulo); South (Paraná, Santa Catarina) (Flora do Brasil 2020).

Phytogeographical Domain

Amazon forest, Caatinga, Cerrado, Atlantic forest (Flora do Brasil 2020).

Cenostigma macrophyllum

Tul. Ann. Sci. Nat., Bot., Ser. 2, 20: 141, pl. 3, 1843.

Common Name

Caneleiro, Canela de Velho, Maraximbé or Fava do Campo.

Species Description

Tree with woody stem with approximately 10 m of height, sympodial growth; rudimentary stipules present with 0.3 cm of length, with puberulent garments; petiole with approximately 3.0 cm of length, and 0.4cm wide; bipinnates composite leaves with 5 to 8 pairs of leaflets, alternate phyllotaxy, absence of nectaries, 13 cm length; ovate leaflets, 9.7 cm long, 2.6 cm long, whole leaf margin, mucronate leaf apex; flower absent and cimose inflorescence with floral buds with 0.9 cm of length; fruit of the fallow legume type, 12.7 cm long and 1.2 cm long.

Material Examined

Brazil. Maranhão: São João do Sóter, Pedras Village, 23/VI/17, G. S. Gomes; G. M. Conceição, 27 (HABIT).

Geographical Distribution

North (Tocantins); Northeast (Bahia, Ceará, Maranhão, Piauí); Center-West (Distrito Federal, Goiás, Mato Grosso do Sul, Mato Grosso); Southeast (Minas Gerais); South (Paraná) (Flora do Brasil 2020).

Phytogeographical Domain

Amazon forest, Caatinga, Cerrado (Flora do Brasil 2020).

Inga edulis

Mart. Flora 20 (2, Beibl.): 113-114, 1837.

Common Name

Ingá de metro

Species Description

Trees of 7 m high, sympodial growth of low branch, woody stem, branches with attached structures, tomentose stipules, cylindrical pebbles of 0.2 cm of length and 0.2 of width, winged deciduous, bipinnates composite leaves, with 3 to 8 pairs of leaflets, smooth margin, acute apex, penninerved vein, petioles with 0.4 cm of length and 0.2cm wide, petiolate with 13.5cm of length, and 8 cm in length, winged leaf 4.6 cm., nectary leaves of 0.1cm in length and 0.1 in width, broad, concave, sessile in the circular shaped stem, absent flowers and absent fruits.

Material Examined

Brazil. Maranhão: São João do Sóter, Pedras Village, 23 / VI / 17, G. S. Gomes; G. M. Conceição, 37, 38 (HABIT).

Geographical Distribution

North (Acre, Amazonas, Amapá, Pará, Rondônia, Roraima); Northeast (Bahia, Paraíba, Pernambuco); Center-West (Mato Grosso); Southeast (Espírito Santo, Minas Gerais, Rio de Janeiro, São Paulo); South (Paraná, Santa Catarina) (Flora do Brasil 2020).

Phytogeographical Domains

Amazon forest, Caatinga, Cerrado, Atlantic forest (Flora do Brasil 2020).

Senna reticulata

(Willd.) H. S. Irwin & Barneby. Mem. New York Bot. Gard. 35: 458, 1982.

Common Name

Maria-mole, Mata-pasto.

Species Description

Tree with woody stem, approximately 5m high, sympodial growth, puberulent blackish branches, sickle type stipules, orange, present with 0.8 cm of length, and 0.3cm wide, composite leaves with 15 to 20 pairs of pinnas, spiral alternating phyllotaxy, oblong leaflets, 10 cm in length and 4.2 cm wide, petiole with 4.5 cm of length, leaf with approximately 18 cm, whole margin with emarginated leaf apex, penninerved vein, raceme inflorescence with several racemes produced, dioecious, complete, heterochlamydeous, dialystemonous, free stamens, fruit dehiscent legume, with 14 cm of length and 2.2 cm wide.

Material Examined

Brazil. Maranhão: São João do Sóter, Serra do Cajuí Village, 14/II/17, G. S. Gomes; G. M. Conceição, 22 (HABIT).

Geographic Distribution

North (Acre, Amazonas, Amapá, Pará, Rondônia); Northeast (Bahia, Ceará, Maranhão, Pernambuco, Piauí); Center-west (Goiás, Mato Grosso); Southeast (Minas Gerais, São Paulo) (Flora do Brasil 2020).

Phytogeographical Domains

Amazon forest, Caatinga, Cerrado (Flora do Brasil 2020).

Senna multijuga

(Rich.) H.S. Irwin & Barneby Mem. New York Bot. Gard. 35: 492, 1982.

Common Name

Canafístula, Aleluia

Species Description

Tree with woody stem, 7m high with sympodial growth, with grayish parts about 10m, grayish branches, composite leaves, 8 to 12 pairs of pinnas, petiole with up to 2 cm of length, and 0.9 cm wide, leaflets of 6.7 cm in length and 2.6 cm of length. petiolate, oblong, glabrous or puberulent tops, nectary 0.2 cm in length, the raceme spinal cord at the base of the petiole in the first jug; canaliculate petiole with 0.2 cm of length and 0,1 of width., yellow dioecious flower, complete (sepals and petals), dichlamydeous, heterochlamydeous, dialystemonous, free stamens, absent fruit.

Material Examined

Brazil. Maranhão: São João do Sóter, Pedras Village, 23/VI/17, G. S. Gomes; G. M. Conceição, 25, 26 (HABIT).

Geographical Distribution

North (Acre, Amazonas, Amapá, Pará, Rondônia, Roraima, Tocantins); Northeast (Alagoas, Bahia, Maranhão, Rio Grande do Norte); Center-West (Distrito Federal, Goiás, Mato Grosso do Sul, Mato Grosso); Southeast (Espírito Santo, Minas Gerais, Rio de Janeiro, São Paulo); South (Paraná, Rio Grande do Sul, Santa Catarina) (Flora do Brasil 2020).

Phytogeographical Domains

Amazon forest, Caatinga, Cerrado, Atlantic forest (Flora do Brasil 2020).

Dimorphandra gardneriana

Tul. Arch. Mus. Hist. Nat., 4: 185, 1844.

Common Name

Fava d'anta.

Species Description

Tree with stem trunk with 12 m long, sympodial growth, woody stem, shape rope, abstipate stipules, composite leaves with 30 cm of length and 14 cm long, bipinnates, opposite phyllotaxy, oblong leaflet shape, with 1.8 cm of length and 1.0 cm wide, presents about 15 to 25 pairs of leaflets, whole leaf margin, rounded leaf apex, truncated base, diminutive petiole with approximately 0.2 cm of length, raceme inflorescence, with approximately 10 racemes, with approximately 10 cm of length and 3 cm wide, shows a light greenish color with a crescent, showing stamens and tiny staminoids, absent fruit.

Material Examined

BRAZIL. Maranhão: São João do Sóter, Pedras Village, 23 / VI / 17, G. S. Gomes; G. M. Conceição, 32, 33 (HABIT).

Geographic Distribution

North (Pará, Tocantins); Northeast (Bahia, Ceará, Maranhão, Pernambuco, Piauí); Center-west (Goiás, Mato Grosso); Southeast (Minas Gerais) (Flora do Brasil, 2017).

Phytogeographical Domain

Caatinga, Cerrado (Flora do Brasil 2020).

Plathymenia reticulata

Benth. J. Bot. (Hooker) 4 (30): 334, 1841.

Common Name

Amarelo acende-candeia, Amarelinho, Candeia, Oiteira, Paricazinho, Pau-amarelo, Pau-de-candeia, Vinhático-branco, Vinhático-castanho, Vinhático-da-mata, Vinhático-do-campo, Vinhático-do-mato, Vinhático-rajado e Vinhático-testa-de-boi.

Species Description

Tree with woody stem, sympodial growth, height of 12 m, form embira, abstact stipules, absent nectary, leaves composed of approximately 20 to 30 pairs of leaflets, bipinnates with pulvinus, alternate phyllotaxy, oblong leaflet, 2 cm long and 0.6 cm long, petiole with 3.0 cm, long leaves with 14 cm approximately, whole margin, emarginated leaf apex, penninerved vein, raceme inflorescence approximately 4 cm long and 0.3 wide, single racemes, miniature white flowers, absent fruit.

Material Examined

Brazil. Maranhão: São João do Sóter, Permanent Protection Area; Redondo Village; Populated Stones, 08/X/16; 13/II/2016; 23/VI/2017, G. S. Gomes; G. M. Conceição, 16, 24, 68 (HABIT).

Geographic Distribution

North (Pará); Northeast (Bahia, Ceará, Maranhão, Piauí); Center-West (Distrito Federal, Goiás, Mato Grosso do Sul, Mato Grosso); Southeast (Espírito Santo, Minas Gerais, Rio de Janeiro, São Paulo); South (Paraná) (Flora do Brasil 2020).

Phytogeographical Domains

Amazon forest, Caatinga, Cerrado, Atlantic forest (Flora do Brasil 2020).

Delonix regia

(Bojer ex Hook.) Raf. Fl. Tellur. 2: 92, 1836 [1837].

Common Name

Flamboyant

Species Description

Tree with sympodial growth, woody stem, approximately 15 m high, shape embira, abstipate stipules, composite bipinnates leaves with 30 cm of length and 15 cm wide, alternate Phyllotaxy, oblong leaflet shape, with 30 to 50 pairs of leaflets, with 7 mm of length and 0.2 cm long, rounded apex, smooth margin, symmetrical base, petiole with 2,5 cm of length and 0.4 cm wide, petiole with 0.3 cm. and 0.1 cm long, absent flower, fruit of the dehiscent type, with 15 cm of length and 3cm wide

Material Examined

Brazil. Maranhão: São João do Sóter, Pedras Village, 23/VI/17, G. S. Gomes; G. M. Conceição, 40 (HABIT).

Geographical Distribution

North (Acre, Amazonas, Amapá, Pará, Rondônia, Roraima, Tocantins); Northeast (Alagoas, Bahia, Ceará, Maranhão, Paraíba, Pernambuco, Piauí, Rio Grande do Norte, Sergipe); Center-west (Distrito Federal, Goiás, Mato Grosso do Sul, Mato Grosso); Southeast (Espírito Santo, Minas Gerais, Rio de Janeiro, São Paulo); South (Paraná, Rio Grande do Sul, Santa Catarina) (Flora do Brasil 2020).

Phytogeographical Domains

Amazon forest, Caatinga, Cerrado, Atlantic forest, Pantanal (Flora do Brasil 2020).

Libidibia ferrea

(Mart. Ex Tul.) LP Queiroz Legum. Caatinga 130, 2009.

Common Name

Pau-Ferro.

Species Description

Tree with woody stem, 5m high, sympodial growth, absent stipules, puberulent branches rusty, bipinnates composite leaves with approximately 10 to 15 pairs of leaflets, alternate phyllotaxy, leaflets oblong, with 3.6 cm of length., and 1.0 cm wide, petiole with 1.1 cm of length and 0.3 cm wide, leaf with approximately 6.5 cm, whole leaf margin, rounded leaf apex, penninerved vein, glabrous leaflets, indeterminate inflorescence, flower with approximately 1.0 cm of length and 0.5 cm wide, yellow flowers, complete, dichlamydeous, heterochlamydeous, showy standard, dialysemons, legume fruit dry type, with 7.3 cm of length and 2.0 cm wide.

Material Examined

Brazil. Maranhão: São João do Sóter, Pedras Village, 23/VI/17, G. S. Gomes; G. M. Conceição, 28, 29, 30 (HABIT).

Geographic Distribution

Northeast (Alagoas, Bahia, Ceará, Maranhão, Paraíba, Pernambuco, Piauí, Rio Grande do Norte, Sergipe); Southeast (Espírito Santo, Minas Gerais, Rio de Janeiro) (Flora do Brasil 2020).

Phytogeographical Domains

Caatinga, Cerrado, Atlantic forest (Flora do Brasil 2020).

Caesalpinia pulcherrima

(L.) Sw. Observ. Bot. 166, 1791.

Common Name

Flamboyanzinho.

Species Description

Woody trunk tree, 3 to 4m high, shows sympodial growth, presence of aculeous in the trunk with 0,4 cm of length, and 0.2 cm wide, composite leaves bipinnates with 5 pairs of pinnas, 30 cm in length, phyllotaxy alternates, without stipules, petiole with 0,5 cm of length and 0.3 cm wide, green leaflets of 7 to 11 pairs, with 6.1 cm of length and 3.4 cm wide, red flower with long stamens of 4 cm of length. zygomorph symmetry legume fruit dry type with 9,3 cm of length and 2.5 cm wide.

Material Examined

Brazil. Maranhão: São João do Sóter, Pedras Village, 13/II/17, G. S. Gomes; G. M. Conceição, 31 (Habit).

Geographical Distribution

North (Acre, Amazonas, Amapá, Pará, Rondônia, Roraima, Tocantins); Northeast (Alagoas, Bahia, Ceará, Maranhão, Paraíba, Pernambuco, Piauí, Rio Grande do Norte, Sergipe); Center-West (Distrito Federal, Goiás, Mato Grosso do Sul, Mato Grosso); Southeast (Espírito Santo, Minas Gerais, Rio de Janeiro, São Paulo); South (Paraná, Rio Grande do Sul, Santa Catarina) (Flora do Brasil 2020).

Phytogeographical Domains

Amazon forest, Cerrado, Atlantic forest (Flora do Brasil 2020).

Inga thibaudiana

DC. Prodr. 2: 434-435, 1825.

Species Description

Shrub with 3 m height, woody stem, simpodial growth, with cylindrical root, presents indumentum in the branches, form rope, compound leaves with 11,1 cm of length and 9.5 cm wide, with 3 to 8 pairs of leaflets, alternating leaflets with 4.9 cm of comp. and 3.8 cm wide. Coriaceous in the abaxial surface, with a smooth margin, penninerved vein, petiole with tiny extraphalic nectaries, at the base of the leaflets, with a rounded nectary, fruit absent.

Material Examined

Brazil. Maranhão: São João do Sóter, Pedras Village, 23/VI/17, G. S. Gomes; G. M. Conceição, 34, 35 (HABIT).

Geographical Distribution

North (Acre, Amazonas, Amapá, Pará, Rondônia, Roraima); Northeast (Alagoas, Bahia, Ceará, Maranhão, Paraíba, Pernambuco); Center-West (Distrito Federal, Goiás, Mato Grosso do Sul, Mato Grosso); Southeast (Espírito Santo, Minas Gerais, Rio de Janeiro, São Paulo) (Flora do Brasil 2020).

Phytogeographical Domains

Amazon forest, Caatinga, Cerrado, Atlantic forest (Flora do Brasil 2020)

Chamaecrista nictitans

(L.) Moench. Methodus 272, 1794.

Species Description

Herb with approximately 40cm of height, herbaceous stem, prostrate growth, branches of 28 cm of length and 16.5 inches wide, with stipules on the petiole with approximately 0.4 cm of length and 1 cm long, composite leaves, bipinnates, oblong leaflets of approximately 0.1 cm in length, and 0.1cm wide, rachis with 0.3 cm of length, absent flower and legume fruit with 3.5 cm of length and 0.3 cm wide.

Material Examined

Brazil. Maranhão: São João do Sóter, Pedras Village, 23/VI/17, G. S. Gomes; G. M. Conceição, 58 (HABIT).

Geographical Distribution

North (Acre, Amazonas, Amapá, Pará, Rondônia, Roraima, Tocantins); Northeast (Alagoas, Bahia, Ceará, Maranhão, Paraíba, Pernam-buco, Piauí, Sergipe); Center-West (Distrito Federal, Goiás, Mato Grosso do Sul, Mato Grosso); Southeast (Espírito Santo, Minas Gerais, Rio de Janeiro, São Paulo); South (Paraná, Rio Grande do Sul, Santa Catarina) (Flora do Brasil 2020).

Phytogeographical Domains

Amazon forest, Caatinga, Cerrado, Atlantic forest, Pantanal (Flora do Brasil 2020).

Chamaecrista rotundifolia

(Pers.) Greene. Pittonia 4 (20D): 31, 1899.

Common Name

Erva-de-Coração, Pasto-Rasteiro, Coração, Alfafa-Nativa, Mata-Pasto, Acácia-Rasteira, Fedegoso.

Species Description

Herb stem, triangular stipules present with 0.1 cm of length and 0.2 cm wide. Bifoliate composite leaves, alternate phyllotaxy, 3.2 cm obovate leaf, and 3 cm wide, whole leaf margin, obtuse apex, yellow zygomorphic flower, aclamydeous, legume fruit dry type with 2.2 cm in length and fruit petiole extended with 1.5 cm of length.

Material Examined

Brazil. Maranhão: São João do Sóter, Pedras Village, 23/V /17, G. S. Gomes; G. M. Conceição, 66 (HABIT).

Geographic Distribution

North (Rondônia, Tocantins); Northeast (Alagoas, Bahia, Ceará, Maranhão, Paraíba, Pernambuco, Piauí, Rio Grande do Norte, Sergipe); Center-West (Distrito Federal, Goiás, Mato Grosso do Sul, Mato Grosso); Southeast (Espírito Santo, Minas Gerais, Rio de Janeiro, São Paulo); South (Paraná, Rio Grande do Sul) (Flora do Brasil, 2017).

Phytogeographical Domains

Amazon forest, Caatinga, Cerrado, Atlantic forest, Pampa, Pantanal (Flora do Brasil 2020).

Chamaecrista flexuosa

(L.) Greene Pittonia 4 (20D): 27, 1899.

Common Name

Peninha

Species Description

Shrubs with woody stem, erect to decumbent growth, approximately 20 cm in length, perennial, erect, with triangular stipules of 5 cm of length and 3.2 cm wide, bilobate leaves with 4cm of length, leaflets, with 5 to 10 pairs, with 4cm of length and 1.2 cm long, extrafollicular nectary leaflets linear-lanceolate to linear oblong or closely oblong-elliptic, straight to slightly distally falcate, persistent, heteromorphic and asymmetric stipules, lanceolate acuminate or ovate acuminate, absent flower, legume fruit with 4.5 cm in length and 0.5 cm wide.

Material Examined

Brazil. Maranhão: São João do Sóter, Pedras Village, 23/VI/17, G. S. Gomes; G. M. Conceição, 48, 49, 50 (HABIT).

Geographical Distribution

North (Acre, Amazonas, Amapá, Pará, Rondônia, Roraima, Tocantins); Northeast (Alagoas, Bahia, Ceará, Maranhão, Paraíba, Pernambuco, Piauí, Rio Grande do Norte, Sergipe); Center-West (Distrito Federal, Goiás, Mato Grosso do Sul, Mato Grosso); Southeast (Espírito Santo, Minas Gerais, Rio de Janeiro, São Paulo); South (Paraná, Rio Grande do Sul, Santa Catarina) (Flora do Brasil 2020).

Phytogeographical Domains

Amazon forest, Caatinga, Cerrado, Atlantic forest, Pampa, Pantanal (Flora do Brasil 2020)

KEY OF SUBFAMILY *CERCIDOIDEAE*

- *Phanera variegata*: Herb or shrubs with grampiform branches scandante, single leaves with split limb
- Herb, shrub or tree without grampiform branches, single leaves with or without split limb.
- *Bauhinia pulchella:* Bilobed simple leaf, small 3.5 cm., 2.6 cm wide, legume fruit type, 6 cm long, 2 cm long.
- *Bauhinia dubia:* Single codiform leaf, uncut in limbus, large 7.5 cm long, and 4.3 cm wide; legume fruit with 14,5cm of length, and it's 1,5cm wide.

DESCRIPTION OF SUBFAMILY *CERCIDOIDEAE*

Phanera variegata

(L.) Benth. Pl. Jungh. 2: 262, 1852.

Common Name

Pata-de-vaca

Species Description

Shrub with approximately 2 m, sympodial growth, woody stem, blackened branches, leaves with tiny stipules to absent, single leaves with 7.0 cm of length and 4.6 cm long, bilobate simple leaflets, rounded apex, petiole with 4.3 cm and length and 0.8 cm wide phyllotaxy alternates, vein cunninerved, absence of nectaries, presence of grampiform branches scandants with approximately 6 cm of length and 0.3 cm wide, absent fruits and absent flower.

Material Examined

Brazil. Maranhão: São João do Sóter, Permanent Protection Area; Redondo Village; Pedras Village, 08/X/16; 13/II/2016; 23/VI/2017, G. S. Gomes; G. M. Conceição, 2, 3, 20, 21, 36 (HABIT).

Geographical Distribution

North (Acre, Amazonas, Amapá, Pará, Rondônia, Roraima, Tocantins); Northeast (Ceará, Maranhão, Piauí); Center-West (Goiás, Mato Grosso do Sul, Mato Grosso) (Flora do Brasil 2020).

Phytogeographical Domains

Amazon forest, Cerrado (Flora do Brasil 2020).

Bauhinia pulchella

Benth. Fl. Bras. 15 (2): 190, 1870.

Common Name

Mororó-de-bode.

Species Description

Shrub with woody stem, approximately 3m high, with sympodial growth, presents rudimentary stipules with 0.3 cm of length, branches glabrous; petiole with 0.6 cm of length and 0.2 cm wide; single sheets, 6.0 cm long and 4.3 cm wide, spiral alternating phyllotaxy, absence of nectary; leaflet with bilobate limbum, 3.5 cm., 2.6 cm wide, whole margin, rounded leaf apex; inflorescence cyanotic, monoicious, dichlamideous, heterochlamydeous, zygomorph, chalice gamossepalous, with 1.0cm of length glabrous, linear peduncle; corolla linear dialypetalous with 1.5-2.5 cm of length; free stamens; dried legume fruit type, 6 cm long 2 cm long.

Material Examined

Brazil. Maranhão: São João do Sóter, Redondo Village, 13/II/17, G. S. Gomes; G. M. Conceição, 17 (HABIT).

Geographic Distribution

North (Pará, Rondônia, Tocantins); Northeast (Bahia, Ceará, Maranhão, Pernambuco, Piauí, Rio Grande do Norte); Center-West (Goiás, Mato Grosso do Sul, Mato Grosso); Southeast (Minas Gerais) (Flora do Brasil 2020).

Phytogeographical Domain

Amazon forest, Caatinga, Cerrado (Flora do Brasil 2020).

Bauhinia dubia

G. Don Gen. Hist. 2: 463, 1832.

Common Name

Monoró, pata-de-vaca.

Species Description

Shrub with woody stem of sympodial growth with approximately 3m in length, rudimentary stipules present with 0.2 cm in length; glabrous branch; petiole of 4 cm long and 0.2 cm wide; sheet 7.5 cm long, and 4.3 cm wide width, alternate phyllotaxy, absence of nectaries; simple leaflets cordiform to sagittate, 3.4 cm long, and 3.6 cm wide, cm, whole leaf margin, rounded apex; inflorescence raceme; floral bud with 0.8 - 2.5 cm length, dried fruit, dehiscent legume type with 14.5 cm of length and 1.5cm wide.

Material Examined

Brazil. Maranhão: São João do Sóter, Village Redondo 13/II/17, G. S. Gomes; G. M. Conceição, 23 (HABIT).

Geographic Distribution

North (Amazonas, Pará, Tocantins); Northeast (Ceará, Maranhão, Piauí) (Flora do Brasil 2020).

Phytogeographical Domains

Amazon forest, Cerrado (Flora do Brasil 2020).

KEY OF SUBFAMILY *DETARIOIDEAE*

- *Tamarindus indica:* Composite bipinnates leaves, presents 28 to 40 pairs of leaflets, alternating phyllotaxy.
- *Hymenaea phyllotaxy:* Composite bifoliolate leaves, not dystonic alternate phyllotaxy.

Tamarindus indica

L. (Sp. Pl.)1: 34, 1753.

Common Name

Tamarindo.

Species Description

Tree with 12 m length, woody stem, symmetrical growth, high crown, brownish branches, rudimentary stipules present with 0.2 cm of length, composite leaves bipinnates, presents 28 to 40 pairs of leaflets, oblong leaflet with 1.0 cm of length and 0.3 cm wide, petiole with 3.0 cm of length and leaf with approximately 8.0 cm of length, whole leaf margin, rounded leaf apex, paralelinerved vein, flower absent, absent fruit.

Material Examined

Brazil. Maranhão: São João do Sóter, Permanent Protection Area, 08/X/16, G. S. Gomes; G. M. Conceição, 5 (HABIT).

Geographical Distribution

North (Acre, Amazonas, Amapá, Pará, Rondônia, Roraima, Tocantins); Northeast (Alagoas, Bahia, Ceará, Maranhão, Paraíba, Pernambuco, Piauí, Rio Grande do Norte, Sergipe); Center-West (Distrito Federal, Goiás, Mato Grosso do Sul, Mato Grosso); Southeast (Espírito Santo, Minas Gerais, Rio de Janeiro, São Paulo); South (Paraná, Rio Grande do Sul, Santa Catarina) (Flora do Brasil 2020).

Phytogeographical Domain

Amazon forest, Caatinga, Cerrado, Atlantic forest (Flora do Brasil 2020).

Hymenaea stigonocarpa

Mart. Getreue Darstell. Gew. 11: sub pl. 13, 1830.

Common Name

Jatobá do Cerrado.

Species Description

Tree with 15 m height, woody stem, elongated sympodial growth, glabrous branches, bifoliolate composite leaves, alternate phillotaxy, asymmetrical leaflet form ovate, without indumentum, leaflet with 11 cm of length and 5 cm wide, petiole with 2.0 cm, leaf with 8.0 cm, presence of leaf blade limbs, whole margin, emarginated leaf apex, penninerved veins to cunnineverd, absent flower, legume fruit dry type, with approximately 7.0 cm long and 4 cm wide.

Material Examined

Brazil. Maranhão: São João do Sóter, Redondo Village, 23/VI/17, G. S. Gomes; G. M. Conceição, 19 (HABIT).

Geographic Distribution

North (Pará); Northeast (Bahia, Ceará, Maranhão); Center-West (Distrito Federal, Goiás, Mato Grosso do Sul, Mato Grosso); Southeast (Minas Gerais, São Paulo) (Flora do Brasil 2020).

Phytogeographical Domain

Amazon forest, Caatinga, Cerrado, Pantanal (Flora do Brasil 2020).

KEY OF SUBFAMILY *PAPILIONOIDEAE*

1. *Dioclea bicolor*: Liane, fruit type vegetable with 3 seeds, without forming of articles, with 4 cm in length, margin revolute of legume
2. Herb or Subshrub, legume fruit with 3 more seeds or with formation of articles, without revolute margin.
3. Subshrub with height equal to or greater than 1m, only erect growth
   - *Indigofera suffruticosa:* Composite leaves imparipinnates, with 8 to 12 pairs of leaflets, fruit legume type curved, forming in bunches in the main branch, with 8,0 cm of length.
4. Leaves simple or trifoliolate, fruit legume lomentum type, not form clusters in the main branch.
   - *Desmodium subsecundum:* Leaves simple large ovate, with 20 cm of length and 10 cm long, long fruit forming branches, arranged alternately.
     - Small trifoliolate leaves, small fruit forming branches, arranged not alternate leaves form.
   - *Desmodium incanum:* Elliptic leaflets, 2.4 cm long and 1.1 cm wide, fruit type lomentum, 1.1 cm long, with 3 to 5 articles.
   - *Desmodium barbatum* Elliptic leaflets, 7 cm long and 3.6 cm wide, fruit lomentum type, 3.3 cm long and 0.8 cm long, has 3 to 6 items dressed.
5. Climbing or herbaceous herb with height equal to or less than 1,5m, erect, decumbent or prostrate growth.
   - Decoding growth climbing herbs.
   - *Abrus fruticulosus:* Leaves composed of 8 to 15 pairs of leaflets, bipinnates with 2.3 cm of length. and 0.9 cm wide, pseudo-raceme inflorescence
   - Leaves trifoliolate, inflorescence raceme or cimose.
6. Long lanceolate leaflet.
   - *Centrosema brasiliana* Leaflet with 5,2 cm of length and 3 cm long, papilionaceous flower, lilac dark to purple petals.

- *Clitoria guianensis* Leaflets with 5 cm of length and 1 cm wide, papilionaceous flower, petals light to whitish lilac.

7. Leaflets ovate or obovate.
   - *Periandra heterophyllia:* Leaflets ovate, with 10 cm of length and 6.1 cm wide, mucronate leaf apex, marked penninerved vein, red papilionaceous flower.
   - *Vigna lasiocarpa:* Leaflets obovate 4.5 cm long and 3 cm wide, acute leaf apex, unmarked penninerved vein, yellow papilionaceous flower.
8. Herbaceous herb of erect or prostrate growth.
   - Simple leaves, legume fruit capsule type.
   - *Crotalaria stipularia:* Leaflet with 4.7 cm and 2.0 cm long, winged petiole with 0.3 cm, fruit with 2.3 cm
     in length.
   - *Crotalaria retusa:* Leaflet with 4,2 cm of length and 1.5 cm wide, not winged petiole with 0.3 cm, fruit with 3.0 cm in length.
9. Composite leaves, typical legume fruit or lomentum.
   - *Aeschynomene viscidula:* Compound leaves bipinnates, legume lomentum14. Leaflets with up to 6 pairs of pinnas, equal to or less than 0, 5cm in length, lomentum with up to 3 articles.
10. Leaflets with more than 6 pairs of pinnas, greater than 0.5 cm long, legume lomentum with more than 3 articles.
    - *Aeschynomene brasiliana*: Leaflets of the base of the pinna bigger than the apex, obovate, with the maximum of 12 pairs of pinnas.
11. Leaflets of the base never greater than the one of the apex, oblong, with the maximum of more than 12 pairs of pinnas.
    - *Aeschynomene histrix:* Leaflets 0.7 cm long, 0.2 mm wide, acuminate leaf apex, legume with up to 7 articles.
    - *Aeschynomene paniculata:* Leaflets 1.0 cm long, 0.3 mm wide, rounded leaf apex, legume with up to 5 articles.
12. Composite leaves trifoliolate, fruit legume type.
13. Leaf lobes lateral, leaflet equal to or greater than 4cm in length.
    - *Phaseolus vulgaris*: Leaflets with 8cm in length, truncated raquer com 8cm de length, rachis truncated.
    - Leaflets, 4 cm long, not truncated raquer.
    - *Macroptilium atropurpureum:* Lateral leaflets with diminutive petiole to sessile, fruit legume type with 6 cm of length indehiscent.
    - *Macroptilium lathyroides:* Side leaflets with elongated petiole, fruit legume type with 5.6cm of length.
      - Not lobulated lateral leaflets, leaflet smaller than 4cm in length.

- *Galactia jussiaeana* Leaf apex emarginate, leaflet with 3cm of length and 1.7 cm wide, oblong.
- *Stylosanthes viscosa* Acute leaf apex, leaflet with 1.9 cm of length and 0.6 cm wide, elliptical.

Dioclea bicolor

Benth. (Comm. Legum. Gen.) 69, 1837.

Common Name

Mucuna.

Species Description

Liane with woody stem, presents approximately 2 m high, erect growth, shape rope, present linear stipules with 0.6 cm of length., trifoliolate composite leaves, distical alternate phyllotaxy, large oblong lanceolate leaflet with 8,4 cm long and 5.0 cm wide, petiole with 4.6 cm, and leaf with approximately 13.1 cm, whole leaf margin, rounded leaf apex, penninerved vein, raceme inflorescence, fruit legume type, with 4 cm long, few pilose, with 2 to 5 seeds.

Material Examined

Brazil. Maranhão: São João do Sóter, Permanent Protection Area, 08/X/16, G. S. Gomes; G. M. Conceição, 10 (HABIT).

Geographical Distribution

North (Acre, Amazonas, Pará); Northeast (Ceará, Maranhão); Center-West (Goiás, Mato Grosso) (Flora do Brasil 2020).

Phytogeographical Domain

Amazon forest, Caatinga, Cerrado (Flora do Brasil 2020).

Indigofera suffruticosa

Mill. Gard. Dict. (Ed. 8) Indigofera no. 2,1768.

Common Name

Anil, Indigo, Anileira

Species Description

Shrub with woody stem, 2 m long, symmetrical erect growth, present linear stipules with 0.5 cm of length, imparipinnates composite leaves, with 8 to 12 pairs of leaflets, alternate phyllotaxy, lanceolate oblong leaflet, with 3.0 cm in length and 1.2 cm wide, petiole with 2.0 cm, leaf 9.3 cm long, crenellated leaf margin, mucronate leaf apex, penninerved vein, symmetric leaflet, raceme Inflorescence, curved legume type dry fruit, forming in bunches in the main branch, with 8.0 cm of length, with fruit with 1.5 cm of length approximately.

Material Examined

Brazil. Maranhão: São João do Sóter, Pedras Village, 23/VI/17, G. S. Gomes; G. M. Conceição 56 (HABIT).

Geographical Distribution

North (Acre, Amazonas, Amapá, Pará, Rondônia, Roraima, Tocantins); Northeast (Alagoas, Bahia, Ceará, Maranhão, Paraíba, Pernambuco, Piauí, Rio Grande do Norte, Sergipe); Center-West (Distrito Federal), Goiás, Mato Grosso do Sul, Mato Grosso); Southeast (Espírito Santo, Minas Gerais, Rio de Janeiro, São Paulo); South (Paraná, Rio Grande do Sul, Santa Catarina) (Flora do Brasil 2020).

Phytogeographical Domains

Amazon forest, Caatinga, Cerrado, Atlantic forest (Flora do Brasil 2020).

Desmodium subsecundum

Vogel. (Linnaea) 12: 99, 1838.

Species Description

Subshrub with approximately 1 m of height, erect growth, woody stem, triangular stipules with approximately 0.4 cm of comp. and 0.1 cm wide, single large ovate leaves with some 3.4 cm in length and 2 cm long. others with 20 cm of length and 10 cm wide, trichomes on the stem of approximately 0.1 cm long, and 0.1 long, rather, 3 cm long petiole and 0.4 cm wide rachis measuring 0.4 cm wide whole margin, rounded apex, ovate leaflet, symmetrical, penninerved vein, absent flower, fruit legume type of approximately 0.1 cm of length and 0.1 cm wide arranged alternately no petiole, having 3 to 5 articles with adherent trichomes.

Material Examined

Brazil. Maranhão: São João do Sóter, Pedras Village, 23/VI/17, G. S. Gomes; G. M. Conceição, 65 (HABIT).

Geographical Distribution

North (Acre, Pará, Tocantins); Northeast (Bahia); Center-West (Distrito Federal, Goiás, Mato Grosso do Sul); Southeast (Espírito Santo, Minas Gerais, Rio de Janeiro, São Paulo); South (Paraná, Rio Grande do Sul, Santa Catarina) (Flora do Brasil, 2017). This species is a new occurrence for the State of Maranhão.

Phytogeographical Domains

Amazon forest, Cerrado, Atlantic forest (Flora do Brasil, 2020).

Desmodium incanum

(Sw.) DC. Prodr. 2: 332,1825.

Common Name

Pega-pega, mata-pasto, amor-do-campo

Species Description

Herbaceous subshrub, decumbent erect growth, is approximately 1m high, triangular stipules present with 0.6 cm of compost. at the base of the petiole, composite trifoliolate leaf, alternating phyllotaxy, elliptic leaflets, 2.4 cm long and 1.1 cm wide, 1.5 cm petiole, approximately 4.0 cm long leaf, whole leaf margin, apex rounded, symmetrical, penninerved vein, presence of trichomes, medium pilosity, inflorescence raceme, absent flower, dry fruit lomentum type, with 1,1 cm in length. Possessing 3 to 5 articles with adherent trichomes.

Material Examined

Brazil. Maranhão: São João do Sóter, Pedras Village, 23/VI/17, G. S. Gomes; G. M. Conceição, 51, 52 (HABIT).

Geographical Distribution

North (Acre, Amazonas, Amapá, Pará, Rondônia, Roraima, Tocantins); Northeast (Alagoas, Bahia, Ceará, Maranhão, Paraíba, Pernambuco, Piauí, Rio Grande do Norte, Sergipe); Center-West (Distrito Federal, Goiás, Mato Grosso do Sul, Mato Grosso); South (Paraná, Rio Grande do Sul, Santa Catarina) (Flora do Brasil 2020).

Phytogeographical Domains

Amazon forest, Caatinga, Cerrado, Atlantic forest, Pampa, Pantanal (Flora do Brasil 2020).

Desmodium barbatum

(L.) Benth. Vidensk. Meddel. Dansk Naturhist. Foren. Kjøbenhavn 1853 (1-2): 18,1853.

Common Name

Pega-pega.

Species Description

Subshrub with 80 cm of height, woody stem, erect growth, triangular stipules present with 0.6 cm of length., composite leaves trifoliolates, phyllotaxy distical alternate, leaflets elliptic symmetrical, with 7 cm of length and 3,6 cm of width, petiole with 1.5 cm length, leaf with approximately 4.0 cm of length, whole leaf margin, rounded apex, pilosity with whitish trichomes, penninerved vein, inflorescence raceme, absent flower, dried fruit lomentum type, 3 cm of length, and 0.8 cm long, has 3 to 6 items, clingy, clingy.

Material Examined

Brazil. Maranhão: São João do Sóter, Pedras Village, 23/VI/17, G. S. Gomes; G. M. Conceição, 67 (HABIT).

Geographical Distribution

North (Acre, Amazonas, Amapá, Pará, Rondônia, Roraima, Tocantins); Northeast (Alagoas, Bahia, Ceará, Maranhão, Paraíba, Pernambuco, Piauí, Rio Grande do Norte, Sergipe); Center-West (Distrito Federal, Goiás, Mato Grosso do Sul, Mato Grosso); Southeast (Espírito Santo, Minas Gerais, Rio de Janeiro, São Paulo); South (Paraná, Rio Grande do Sul, Santa Catarina) (Flora do Brasil 2020).

Phytogeographical Domains

Amazon forest, Caatinga, Cerrado, Atlantic forest, Pampa, Pantanal (Flora do Brasil 2020).

Abrus fruticulosus

Wight & Arn. (Prod. Fl. Ind. Orient.) 1: 236, 1834.

Species Description

Herb with approximately 1m of length, with herbaceous stem, indumentum in the present branch. Persistent, tiny linear stipules at the base of the petiole, petiole with 3.5 cm of length and 2 cm long; leaves composite of 8 to 15 pairs of leaflets, bipinnates with 2.3 cm of length and 0.9 cm wide obovate, linear leaflets; legume, rounded apex, symmetrical limbus, Inflorescence: position of the pseudo- raceme inflorescence, flower absent, typical fruit with 3.5 cm of length and 1 cm wide apex sharp to curved.

Material Examined

Brazil. Maranhão: São João do Sóter, Pedras Village, 23/VI/17, G. S. Gomes; G. M. Conceição, 57 (HABIT).

Geographical Distribution

North (Amazonas, Pará, Rondônia); Northeast (Maranhão); Center-West (Mato Grosso do Sul, Mato Grosso) (Flora do Brasil 2020).

Phytogeographical Domains

Amazon forest, Cerrado (Flora do Brasil 2020).

Centrosema brasiliana

(L.) Benth Comm. Legum. Gen. 54,1837.

Common Name

Feijão bravo

Species Description

Herb, herbaceous stem, approximately 1.5 m in length, presents triangular ochrea stipules at the base of the petiole of 0.3 cm in length and 0.2 mm long, trifoliolate composite leaves, with an opposite pair at the base and one elongated by the rachis, 7 cm long leaf, alternate phyllotaxy, leaflets 5.2 cm long and 3 cm long lanceolate to linear, smooth margin, acute apex, penninerved vein, little evident trichomes, papilionaceous flower, purple petals, with heterochlamydeous axillary inflorescence, dichlamydeous, dialystemonous, complete, typical legume fruit 7,0 cm length and 0.3-0.4 cm wide, linear.

Material Examined

Brazil. Maranhão: São João do Sóter, Permanent Protection Area; Populated Stones, 08/X/16; 23/VI/2017, G. S. Gomes; G. M. Conceição, 9, 44, 45 (Habit).

Geographic Distribution

North (Amazonas, Amapá, Pará, Roraima); Northeast (Alagoas, Bahia, Ceará, Maranhão, Paraíba, Pernambuco, Piauí, Rio Grande do Norte, Sergipe); Center-West (Distrito Federal, Goiás, Mato Grosso do Sul, Mato Grosso); Southeast (Espírito Santo, Minas Gerais, Rio de Janeiro, São Paulo); South (Paraná, Santa Catarina) (Flora do Brasil 2020).

Phytogeographical Domains

Amazon forest, Caatinga, Cerrado, Atlantic forest, Pantanal (Flora do Brasil 2020).

Clitoria guianensis

(Aubl.) Benth. J. Proc. Linn. Soc., Bot. 2: 40,1858.

Common Name

Feijão do Campo

Species Description

Herb, approximately 50 cm tall, herbaceous stem, ovate stipules, striated, at the base of the petiole with 0.3 cm of length and 0.1 mm long, composite leaf trifoliolate with 5 cm of length. and 1 cm wide, lanceolate leaflet, alternates phyllotaxy, acute apex, penninerved veins, symmetric, present trichomes, axillary inflorescence, flower, monoicious, subsessile; long tubular calyx whitish, striated, papilionaceous corolla, unguiculated petals, lilac, orbicular banner, lato; free wings, obovados, keel falcates, unguiculates petals, heterochlamydeous, dichlamydeous, dialystemonous, absent fruit.

Material Examined

Brazil. Maranhão: São João do Sóter, Pedras Village, 23/VI/17, G. S. Gomes; G. M. Conceição, 60 (HABIT).

Geographic Distribution

North (Amazonas, Amapá, Pará, Roraima, Tocantins) Northeast (Bahia, Ceará, Maranhão, Pernambuco, Piauí) Southeast (Espírito Santo, Minas Gerais, São Paulo) South (Paraná) (Flora do Brasil 2020).

Phytogeographical Domain

Amazon forest, Caatinga, Cerrado, Atlantic forest, Pantanal (Flora do Brasil 2020).

Periandra heterophylla

Benth. (Comm. Legum. Gen.) 57, 1837.

Species Description

Herbaceous stem-type climbing vine, 1.20 m long, decumbent erect growth, presence of linear stiples with 0.4 cm of length, absence of nectary, composite leaves, trifoliolate, alternate, leaflet ovate-lanceolate, 10 cm long and 6.1 cm wide, petiole with 5.0 cm, leaf with approximately 8.0 cm, whole margin, mucronate leaf apex, marked penninerved vein, symmetrical leaflet, presence of trichomes in leaflets, homogeneous pilosity, cimose inflorescence with 0.8 cm of length, flower absent, fruit absent.

Material Examined

Brazil. Maranhão: São João do Sóter, Pedras Village, 23/VI/17, G. S. Gomes; G. M. Conceição, 53 (HABIT).

Geographic Distribution

North (Pará, Tocantins); Northeast (Bahia, Maranhão, Piauí); Center-West (Goiás, Mato Grosso do Sul, Mato Grosso); Southeast (Minas Gerais, São Paulo) (Flora do Brasil 2020).

Phytogeographical Domain

Amazon forest, Cerrado (Flora do Brasil, 2017).

Vigna lasiocarpa

(Mart.ex Benth.) Verdc. (Kew Bull.) 24 (3): 539.1970.

Common Name

Feijãozinho-do-brejo.

Species Description

Herb with approximately 30 cm in length, erect decumbent growth, herbaceous stem, greenish branches, triangular stipules of 3 mm of length and 1 mm wide at the base of the petiole. Leaves composite trifoliolate, alternate, adaxial structure with secondary penninerved vein marked in leaflets, leaflets obovate with 4.5 cm of length and 3 cm long, acute leaf apex, petiole with 4 cm of length. 1 mm wide, full margin, symmetrical, presence of trichomes in leaflets. Inflorescence raceme, axial, yellow flower, zygomorph, achlamydeous, papilionaceous, dry fruit of the long legume type with 3 cm of length and 0.2 cm wide, with trichomes.

Material Examined

Brazil. Maranhão: São João do Sóter, Permanent Protection Area, 08/X/16, G. S. Gomes; G. M. Conceição, 63 (HABIT).

Geographic Distribution

North (Amazonas, Amapá, Pará, Roraima); Center-West (Mato Grosso do Sul, Mato Grosso); South (Paraná, Rio Grande do Sul, Santa Catarina) (Flora do Brasil 2020). This species is a new occurrence for the Northeast region and the State of Maranhão.

Phytogeographical Domain

Amazon forest, Cerrado, Atlantic forest, Pantanal (Flora do Brasil 2020).

Crotalaria stipularia

Desv. J. Bot. Agric. 3: 76, 1814.

Common Name

Xique-xique

Species Description

Herb stem-like stem herb, decumbent erect growth, 65 cm long, linear stipules present with 0.3 cm of length and 0.1 cm wide nectaries absent, single leaves, alternate phyllotaxy, leaflets oblong lanceolate, with 4,7 cm of length. and 2,0 cm wide, winged petiole with 0.3 cm, leaf approximately 5.1 cm and 3 cm wide, leaf margin whole, apex emargined, penninerved vein, symmetric, presence of white trichomes, pilose, yellow papilionaceous flower with 0.3 cm of length and 0.3 inches with zygomorphic symmetry inflorescence cimose, heterochlamydeous, dichlamydeous, dialystemonous, dry fruit, capsule, with approximately 2.3 cm in length.

Material Examined

Brazil. Maranhão: São João do Sóter, Pedras Village, 23/VI/17, G. S. Gomes; G. M. Conceição, 47 (HABIT).

Geographical Distribution

North (Acre, Pará, Roraima); Northeast (Alagoas, Bahia, Ceará, Maranhão, Paraíba, Pernambuco, Piauí, Sergipe); Center-West (Distrito Federal, Goiás, Mato Grosso do Sul, Mato Grosso); Southeast (Espírito Santo, Minas Gerais, Rio de Janeiro, São Paulo) (Flora do Brasil 2020).

Phytogeographical Domains

Amazon forest, Caatinga, Cerrado, Atlantic forest, Pantanal (Flora do Brasil 2020).

Crotalaria retusa

L. Sp. Pl. 2: 715, 1753.

Common Name

Chocalho de Cascavel, Guizo de cascavel, xique-xique.

Species Description

Herb stem type herb, 50 cm high, decumbent erect growth, linear stipules present with 0.4 cm of length, absent nectaries, single leaves, alternate phyllotaxy, obovate leaflets, 4.2 cm I bought 1.5 cm wide, 0.3 cm petiole, 4.6 cm long leaf and 4 cm wide, whole leaf margin, emarginated apex, penninerved vein, cuminous inflorescence, absent flower, dried fruit, capsule, with approximately 3.0 cm of length, and 0.8 cm wide with legume acute terminal.

Material Examined

Brazil. Maranhão: São João do Sóter, Permanent Protection Area, 08/X/16, G. S. Gomes; G. M. Conceição, 12 (Habit).

Geographic Distribution

North (Pará); Northeast (Bahia, Maranhão, Piauí); Southeast (Minas Gerais, Rio de Janeiro, São Paulo); South (Paraná, Rio Grande do Sul, Santa Catarina) (Flora do Brasil 2020).

Phytogeographical Domain

Amazon forest, Caatinga, Cerrado, Atlantic forest, Pampa (Flora do Brasil 2020).

Aeschynomene viscidula

Mich. Fl. Bor.Amer. 2: 74-75, 1803.

Species Description

Herb with approximately 50 cm of height, herbaceous stem, prostrate growth, branches with 32 cm of length and 19 cm wide, triangular stipules with 0.4 mm of length. and 0.2 mm wide, are found at the base of the petiole, composites bipinnates leaves, with 3 to 6 pairs of leaflets subopposed, leaflet with 1.7 cm of length. and 1 cm wide, petiole of 0.5 cm of length, and 0.1 mm wide, obovate leaflets 0.5 cm long, and 0.6 mm wide, trichomes, flower with campanulate calyx, papilionceous, yellow cream, showy standard, heterochlamydeous, dialytemones, dialypetalous, Fruit lomentum legume type, with articles joined by isthmus with 1 cm of length and 0.3 mm in length, features 3 hairpieces.

Material Examined

Brazil. Maranhão: São João do Sóter, Pedras Village, 23/VI/17, G. S. Gomes; G. M. Conceição, 59 (HABIT).

Geographical Distribution

North (Tocantins); Northeast (Alagoas, Bahia, Ceará, Paraíba, Pernambuco, Piauí, Rio Grande do Norte, Sergipe); Center-West (Mato Grosso do Sul); Southeast (Espírito Santo, Minas Gerais) (Flora do Brasil 2020). This species is a new occurrence for the State of Maranhão.

Phytogeographical Domains

Caatinga, Cerrado, Atlantic forest (Flora do Brasil 2020).

Aeschynomene brasiliana

(Poir.) DC. Prodr. 2: 322, 1825.

Species Description

Herb with approximately 50 cm of height, woody stem, erect growth to decumbent, branches with gray parts, triangular stipules with 1 cm of length. and 0.3 cm wide, composite bipinnates leaves have 6 to 12 pairs of leaflets, with 6 cm of length, and 1.4 cm wide, leaflets obovates with 0.8 cm of length, and 0.4 cm wide, leaflets of the base larger than the leaflets of the apex of the pinnas, petiole with 1 cm of length, and 0.2 cm wide, presence of trichomes with 1 mm of length, smooth margin, apex of rounded leaflet, symmetrical, with penninerved veins, inflorescence raceme, flower absent, fruit legume with 0.3 cm of length and 0.2 inches long, arranged in the petiole with 0.3 cm of length has 3 to 6 articles.

Material Examined

Brazil. Maranhão: São João do Sóter, Pedras Village, 23/VI/17, G. S. Gomes; G. M. Conceição, 64 (HABIT).

Geographic Distribution

North (Amazonas, Amapá, Pará, Rondônia, Roraima, Tocantins); Northeast (Alagoas, Bahia, Ceará, Maranhão, Paraíba, Pernambuco, Piauí); Center-West (Distrito Federal, Goiás, Mato Grosso do Sul, Mato Grosso); Southeast (Espírito Santo, Minas Gerais, Rio de Janeiro, São Paulo); South (Paraná) (Flora do Brasil, 2017).

Phytogeographical Domains

Amazon forest, Caatinga, Cerrado, Atlantic forest, Pantanal (Flora do Brasil 2020).

Aeschynomene histrix

Poir. Encycl., Suppl. 4 (1): 77-78, 1816.

Common Name

Lentilha do Campo; sensitiva modesta.

Species Description

Herb, presents on average 45 cm in height, erect decumbent growth, triangular stipules present with 0.3cm of length to 0.1 cm wide. it presents at the base of the petiole, striated, composite leaves bipinnates with 10 to 18 pairs of leaflets, alternate phyllotaxy; petiole 4-7 cm long and 0.4 cm wide; leaflets 0.7 cm. and 0.2 cm wide, opposite, oblong, whole leaf margin, apex acuminate, with whitish trichomes; flower absent; dry fruit, lomentum legume, with about 3cm length presents from 4 to 7 items dressed.

Material Examined

Brazil. Maranhão: São João do Sóter, Permanent Protection Area, 08/X/16, G. S. Gomes; G. M. Conceição, 6, 7, 8 (Habit).

Geographic Distribution

North (Acre, Amazonas, Amapá, Pará, Roraima, Tocantins); Northeast (Alagoas, Bahia, Ceará, Maranhão, Paraíba, Pernambuco, Piauí, Sergipe); Center-West (Distrito Federal, Goiás, Mato Grosso do Sul, Mato Grosso); Southeast (Minas Gerais, São Paulo); South (Paraná) (Flora do Brasil 2020).

Phytogeographical Domain

Amazon forest, Caatinga, Cerrado, Atlantic forest, Pantanal (Flora do Brasil 2020).

Aeschynomene paniculata

Willd. ex Vogel (Linnaea) 12: 95-96, 1838.

Common Name

Cortiça.

Species Description

Herb, with 60 cm height, decumbent erect growth, grayish branches, presence of triangular stipules with 0.1 cm of length. and 0.1 cm wide; garment with few whitish trichomes; 0.6cm petiole; leaves approximately 6,5 cm long and 3 cm long composed of 12 to 25 pairs of leaflets, bipinnates, alternating phyllotaxy, absence of nectaries; oblong leaflet, 1.0 cm long, 0.3 wide, whole margin, rounded leaf apex; median leaflets larger than the base and apex, cimose inflorescence with 0.3 cm of length, absent flower, fruit legume lomentum with 5 articles.

Material Examined

Brazil. Maranhão: São João do Sóter, Pedras Village, 23/VI/17, G. S. Gomes; G. M. Conceição 55 (HABIT).

Geographical Distribution

North (Acre, Amazonas, Amapá, Pará, Rondônia, Roraima, Tocantins); Northeast (Bahia, Ceará, Maranhão, Piauí, Sergipe); Center-West (Distrito Federal, Goiás, Mato Grosso do Sul, Mato Grosso); Southeast (Minas Gerais, Rio de Janeiro, São Paulo); South (Paraná) (Flora do Brasil 2020).

Phytogeographical Domain

Amazon forest, Caatinga, Cerrado, Atlantic forest, Pantanal (Flora do Brasil 2020).

Phaseolus vulgaris

L. Sp. Pl. 2: 723,1753.

Common Name

Feijão-base, feijão-comum, feijoeiro

Species Description

Herb stem climbing herb, prostrate erect growth, 20 cm long, greenish branches, absent stipules, trifoliolate composite leaves, alternating, leaflets cordate to sagittate, 8 cm in length, and 4.3 cm wide, petiole with 0.2 cm of length and 0.1 cm in length, leaves 8.2 cm in length, entire ondulate margin, penninerved vein, acute leaf apex, symmetrical, reduced pulvin, truncated raquer, absent fruit and flower absent.

Material Examined

Brazil. Maranhão: São João do Sóter, Permanent Protection Area, 08/X/16, G. S. Gomes; G. M. Conceição, 01 (HABIT).

Geographic Distribution

North (Pará); Northeast (Alagoas, Bahia, Ceará, Maranhão, Paraíba, Pernambuco, Piauí, Sergipe); Center-West (Distrito Federal, Goiás, Mato Grosso do Sul, Mato Grosso); Southeast (Espírito Santo, Minas Gerais, Rio de Janeiro, São Paulo); South (Paraná, Rio Grande do Sul, Santa Catarina) (Flora do Brasil 2020).

Phytogeographical Domains

Amazon forest, Caatinga, Cerrado, Atlantic forest (Flora do Brasil 2020).

Macroptilium atropurpureum

(Sessé & Moc. ex DC.) Urb. Symb. Antill. 9 (4): 457, 1928.

Common Name

Leucena, Siratro

Species Description

Herb with approximately 60 cm of length, Herbaceous stem, erect decumbent growth, green branches with 21,5 cm of length, and 16 cm wide triangular stipules of approximately 0.2 cm in length, and 0.1 cm wide, composite leaves, alternate, trifoliolate with 6 cm of length and 4 cm long leaflets oblong-lanceolate to sagittate of 4 cm of length and 1.8 cm wide ondulate margin, rounded apex, penninerved vein, symmetrical, slightly pilose, lobed lateral leaflets sessile and main leaflet extended by rachis, absent flower, dry fruit legume type with 6 cm of length, and 0.4 cm wide, indehiscent.

Material Examined

Brazil. Maranhão: São João do Sóter, Pedras Village, 23/VI/17, G. S. Gomes; G. M. Conceição, 61, 62 (HABIT).

Geographic Distribution

North (Amazonas, Pará, Roraima); Northeast (Alagoas, Bahia, Maranhão, Paraíba, Pernambuco, Rio Grande do Norte, Sergipe); Center-West (Distrito Federal, Goiás, Mato Grosso do Sul, Mato Grosso); Southeast (Espírito Santo, Minas Gerais, Rio de Janeiro, São Paulo); South (Paraná, Rio Grande do Sul, Santa Catarina) (Flora do Brasil 2020).

Phytogeographical Domain

Amazon forest, Caatinga, Cerrado, Atlantic forest, Pantanal (Flora do Brasil 2020).

Macroptilium lathyroides

(L.) Urb. Symb. Antill. 9 (4): 457, 1928.

Common Name

Feijão-de-rola.

Species Description

Herb, decumbent erect growth, approximately 30 cm in length, greenish branches, triangular stipules present with 0.5 cm of length and 0.2 cm wide, presence of nectary rounded in the petiole, composite leaves trifoliolate, phyllotaxy distical alternate, leaflet lanceolate to linear, with 4 cm of length and 2.6 cm wide, 5 cm petiole, 7.0 cm long leaf, whole leaf margin, acute leaf apex, symmetrical, leaf apex obtuse to codiform, presence of scattered trichomes, absent flower, dry fruit type legume long, approximately 5.6 cm.

Material Examined

Brazil. Maranhão: São João do Sóter, Pedras Village, 23/VI/17, G. S. Gomes; G. M. Conceição, 54 (HABIT).

Geographic Distribution

North (Amazonas, Pará, Roraima); Northeast (Alagoas, Bahia, Ceará, Maranhão, Paraíba, Pernambuco, Piauí); Center-West (Distrito Federal, Goiás, Mato Grosso do Sul, Mato Grosso); Southeast (Espírito Santo, Minas Gerais, Rio de Janeiro, São Paulo); South (Paraná) (Flora do Brasil 2020).

Phytogeographical Domain

Amazon forest, Cerrado, Atlantic forest, Pantanal (Flora do Brasil 2020).

Galactia jussiaeana

Kunth. Mimos. 196-200, pl. 55, 1824.

Species Description

Herb, decumbent erect growth, 1 m long, presence of triangular stiples with 0.4 cm, trifoliate composite leaves, alternate phyllotaxy, oblong leaflets with trichomes, leaflet with 3 cm of length. and 1.7 cm wide, petiole 3.0 cm, leaf approximately 6,0 cm, whole leaf margin, emarginated leaf apex, penninerved vein, trichomes on leaflets and branches, pilose, symmetrical leaflet, absent flower, fruit dry legume type with 6 cm of length and 0.3 cm wide, terminal apex of falcate fruit.

Material Examined

Brazil. Maranhão: São João do Sóter, Serra do Cajuí Village, 14/II/17, G. S. Gomes; G. M. Conceição, 18 (HABIT).

Geographic Distribution

North (Amazonas, Amapá, Pará, Rondônia, Roraima); Northeast (Bahia, Maranhão, Piauí); Center-West (Goiás) (Flora do Brasil 2020).

Phytogeographical Domain

Amazon forest, Caatinga, Cerrado (Flora do Brasil 2020).

Stylosanthes viscosa

(L.) Sw. Prodr. 108,1788.

Species Description

Herb, 30 cm long, decumbent erect growth, greenish branches, with triangular stipules of 0.5 cm length in the base of petiole, composite leaves trifoliolate, phyllotaxy distical alternate, leaflets elliptical, with 1.9 cm in length. and 0.6 cm wide, petiole with 0.5 cm of length, and 0.2 cm wide, leaf with approximately 2.5 cm, whole margin, acute leaf apex, penninerved vein, presence of glandular trichomes in all branches, homogeneous pilosity, symmetrical leaflets, cimous inflorescence with 4,5 cm of length, dichlamydeous, heterochlamydeous, gamossepals with 0.4 cm of comp., gamopetal with zygomorph symmetry, papilionaceous yellow flowers with, striated standard marked with red coloration. Fruit absent.

Material Examined

Brazil. Maranhão: São João do Sóter, Permanent Protection Area, 08/X/16, G. S. Gomes; G. M. Conceição, 46 (HABIT).

Geographical Distribution

North (Acre, Amazonas, Amapá, Pará, Rondônia, Roraima, Tocantins); Northeast (Alagoas, Bahia, Ceará, Maranhão, Paraíba, Pernambuco, Piauí, Rio Grande do Norte, Sergipe); Center-West (Distrito Federal, Goiás, Mato Grosso do Sul, Mato Grosso); Southeast (Espírito Santo, Minas Gerais, Rio de Janeiro, São Paulo); South (Paraná, Rio Grande do Sul, Santa Catarina) (Flora do Brasil 2020).

Phytogeographical Domain

Amazon forest, Caatinga, Cerrado, Atlantic forest, Pampa, Pantanal (Flora do Brasil 2020).

CONCLUSION

In this research, three new occurrences were obtained for Maranhão: Desmodium subsecundum (*Papilionoideae*), *Aeschynomene* viscidula (*Papilionoideae*) and *Vigna lasiocarpa* (*Papilionoideae*), with V. lasiocarpa being a new ledger to the northeastern region. This research is relevant for the northeastern flora through the increase of knowledge of the *Fabaceae* family characteristics and the expansion of the geographic distribution of the species in the region. *Fabaceae* proved to be very diverse in the Cerrado of Maranhense, making this the largest survey in the family's geographical area for the state through a pioneering research for the area, building an important knowledge base in the field of botanical studies, thus offering taxonomic keys that will serve as recognition and identification of the species for the Cerrado of the state.

Producing taxonomic works on *Fabaceae* is of extreme importance, so it is possible to understand several characteristics which are still unknown. The morphological and taxonomic diversity of *Fabaceae* is expressed in its high number of taxon that is distributed in almost all regions of the world.

Fabaceae proved to be very diverse in the Cerrado of the Maranhense, making the work the largest survey in the family's geographical area for the state through a pioneering research for the area, building an important knowledge in the field of botanical study, where the taxonomic keys produced will serve as recognition and identification of the species for the Cerrado.

The importance of the research is also evidenced by providing information for the conservation and management of these new occurrences from taxonomic data. It can be mentioned that the species indexed as a new occurrence will contribute to the construction of Brazilian flora, through the REFLORA 2020 Project, which intends to index all Brazilian plant species until the year 2020, fortifying the data of Maranhão and the northeast region, thus guaranteeing greater representation of Brazilian biodiversity.

REFERENCES

APG IV. (2016). An update of the Angiosperm Phylogeny Group classification for the orders and families of flowering plants: APG IV. *Botanical Journal of the Linnean Society*. v.181, p. 1–20.

Barneby, R. C. (1991). *Sensitiva ecensitae: a description of the genus Mimosa Linnaeus (Mimosaceae) in the new world.* Bronx: The New York Botanical Garden, p. 835.

Batalha, M. A. (2011) O cerrado não é um bioma [The cerrado is not a biome]. *Biota Neotropical.* v. 11, n. 1, p. 21-24.

Bentham, G. Leguminosae. In: Bentham, G.; Hooker, J. D. (1865) *Sistens dicotyledonum polypetalarum ordines XI: Leguminosas-Myrtaceaeas* [*Scientific dicotyledonum polypetalarum orders 11*]. Genera Plantarum. London: Lovell Reeve e Co, v. l, n. 2, p. 434-600.

Cronquist, A. (1981). *An integrated system of classification of flowering plants.* Columbia University Press, New York. p. 1262.

Domingos, A. H.; Carpelari J. R. L. (2016). *Plantas medicinais patas-de-vaca* [*Medicinal plants, cow legs*]. Piracicaba: ESALQ - Divisão de Biblioteca, p. 32.

Fernandes, A. (1996). *O taxon Aeschynomene no Brasil* [*The taxon Aeschynomene in Brazil*]. Fortaleza, CE: Editora da UFC. 128 p.

Fidalgo, O.; Bononi, V. L. R. (1989). *Técnicas de coleta, preservação e herborização do material botânico* [*Techniques of collection, preservation and herborization of the botanical material*]. Instituto de Botânica de São Paulo [Institute of Botony of São Paulo]. p. 62.

Flora Do Brasil. *Lista de Espécies da Flora do Brasil* [*List of Species of Flora of Brazil*]. Jardim Botânico do Rio de Janeiro. Disponível em: <http://floradobrasil.jbrj.gov.br/>. Acesso em: 27/07/2015.

Graham, P. H.; Vance, C. P. (2003). Legumes: Importance and constraints to greater use. *Plant Physiol.* 131: 872–877.

Hutchinson, J. (1964). *The genera of flowering plants.* Oxford University Press, Oxford. v. 1. p. 516.

IBGE. *Indicadores sociais municipais: uma análise dos resultados do universo do censo demográfico 2010* [*Municipal social indicators: an analysis of the results of the 2010 demographic census universe*]. Rio de Janeiro. IBGE. p. 151, 2010. Disponível em: <http://www.ibge.gov.

br/home/estatistica/populacao/censo2010/indicadores_sociais_municipais/indicadores_sociais_municipais.pdf>. Acesso em: 21/06/2017.

Isely, D. Orbexilum. (1990). In: *Vascular Flora Southeastern United States.* Chapel Hill: University of North Carolina Press. v. 2, p. 99–102.

Lewis, G. P. (1987). *Legumes of Bahia.* Royal Botanic Gardens, Kew. p. 369.

Lewis, G. P.; Schrire, B. D.; Mackinder, B. A.; Lock, J. M. (2005). *Legumes of the World.* Royal Botanic Gardens, Kew. p. 577.

Lima, L, C. P.; Queiroz, L. P.; Tozzi, A. M. G. A.; Lewis, G. P. A. (2014). Taxonomic Revision of *Desmodium* (Leguminosae, *Papilionoideae*) in Brazil. *Phytotaxa.* v. 169, p. 001–119.

Lima, L. C. P.; Oliveira, M. L. A. A. Aeschynomene. In: R. C. Forzza et al. *Lista de Espécies da Flora do Brasil* [*List of Species of Flora of Brazil*]. Jardim Botânico do Rio de Janeiro. 2010. Disponível em: http://floradobrasil.jbrj.gov.br/2010/FB022778 Acesso: em 20/07/ 2017.

LPWG. Legume Phylogeny Working Group. (2017) A new subfamily classification of the Leguminosae based on a taxonomically comprehensive phylogeny. *Taxon.* v. 66, n.1, p. 44–77. Doi: https://doi.org/10.12705/661.3.

______. (2013) Legume Phylogeny Working Group. Legume phylogeny and classification in the 21st century: Progress, prospects and lessons for other species-rich clades. *Taxon.* v. 62, p. 217–248. Doi: https://doi.org/10.5167/uzh-78167.

Maranhão. (2011). *Plano de Ação Para Prevenção e Controle do Desmatamento e das Queimadas no Estado do Maranhão* [*Plan of Action for the Prevention and Control of Deforestation and Burning in the State of Maranhão*]. Governo do Estado do Maranhão. Secretaria de Estado do Meio Ambiente e Recursos Naturais. p. 110.

Maréchal, R.; Mascherpa, J. M.; Stainier, F. (1978). Etude taxonomique d'un groupe complexe d'espèces des genres *Phaseolus* et *Vigna Papilionaceae*) sur la base de données morphologiques et polliniques, traitées par l'analyse informatique [Taxonomic study of a complex group of species of the genera *Phaseolus* and *Vigna Papilionaceae*) on the basis of morphological and pollen data, processed by computer analysis]. *Boissiera.* v. 28, p. 1-273.

Ormond, J. G. P. (2006). *Glossário de termos usados em atividades agropecuárias, florestais e ciências ambientais* [*Glossary of terms used in agricultural, forestry and environmental sciences activities*]. Rio de Janeiro. BNDS, p. 316.

Queiroz, L. P. (2009). *Leguminosas da Caatinga.* Universidade Estadual de Feira de Santana [*Leguminosae of the Caatinga*]. Royal Botanic Gardens; Associação Plantas do Nordeste, p. 467.

Ribeiro, J. F. (1998). *Cerrado: Matas de galeria* [*Closed: Matas de galeria*]. Embrapa-CPAC, Planaltina, p.164.

Silva, R. R., Fortuna-Perez, A. P.; Tozzi, A. M. G. A. (2007). Novas ocorrências de Leguminosae para o Mato Grosso do Sul, Brasil [New occurrences of Leguminosae for Mato Grosso do Sul, Brazil]. *Rodriguésia*. v. 58, p. 249-254.

Simon, M. F.; Grether, R.; Queiroz, L. P.; Särkinen, T. E., Dutra, V. F. & Hughes, C. E. (2011). The evolutionary history of *Mimosa* (Leguminosae): toward a phylogeny of the sensitive plants. *American Journal of Botany* 98: 1201-1221.

Sprent, J. I., Ardley, J. K., James, E. K. (2013). From North to South: A latitudinal look at legume nodulation processes. *S. Afr. J. Bot*. 89: 31–41.

Strassburg, B. B. N.; Brooks, T.; Feltran-Barbieri, R.; Iribarrem, A.; Crouzeilles, R.; Loyola, R.; Latawiec, A. E.; Oliveira-Filho, F. J. B.; Scaramuzza, C. A. M.; Scarano, F. R.; Soares-Filho, B.; Balmford, A. (2017). Moment of truth for the Cerrado hotspot. *Nature Ecology & Evolution*. v. 1, n. 0099, p. 1-3.

Tropicos. (2018). Tropicos.org. *Jardim Botânico de Missouri [Missouri Botanical Garden*]. 10 de fevereiro de 2018 disponible in [February 10, 2018 Available in] http://www.tropicos.org.

Yahara, T.; et al. (2013). Global legume diversity assessment: Concepts, key indicators, and strategies. *Taxon*. v. 62, p. 249–266.

In: Recent Advances in Plant Research
Editors: B. Welch and M. Wilkerson

ISBN: 978-1-53614-170-2

Chapter 3

FRUIT WASTE AS A NATURAL RESOURCE OF ANTIOXIDANTS

Ante Lončarić*[*], *Lahorka Budić and Tihomir Kovač
Faculty of Food Technology Osijek,
Josip Juraj Strossmayer University of Osijek, Osijek, Croatia

ABSTRACT

The awareness of consumers regarding the functional food led to an increasing demand for functional foods especially for foods rich in antioxidants, which have a wide range of activities such as antitumoral, antiviral, cardioprotective and antimutagenic activities. Phenolic compounds are an essential part of the human diet and are of considerable interest due to their antioxidant activities. Wastes from fruits represent a major disposal problem for the industry concerned. The by-products, which are results of the fruit processing, are rich in valuable compounds and represent a novel, natural and economic sources of flavorings, colorants, polyphenols and natural antioxidants, which can be used in the food industry as a source of natural food additives. This chapter is underlying the great potential of the most important by-products of apples, grapes and berry fruits processing as a source of antioxidants. Some novel and practical aspects of extraction of natural antioxidants will be discussed in this chapter.

Keywords: fruit wastes, polyphenols, antioxidants, apples, grapes, berries

* Corresponding Author Email: ante.loncaric@ptfos.hr.

1. Introduction

Food waste is produced in all the phases of food life cycle such as agricultural production, industrial manufacturing, processing and distribution (Kumar et al., 2017). Close to 50% of all fruits and vegetables in the European Union go to waste as reported by the Food and Agriculture Organization (FAO) of the UN. These losses occur throughout the entire food chain, from "farm to fork," but the predominant losses are linked to the production of produce (EU; HLPE). The Waste Framework Directive given by the European Commision emphasized the importance of prevention of waste generation as well as the exploitation of wastes by reuse and recycling (Directive 2008/98/EC). In the same Directive the European Commission provides the definition for the term *food waste* as "food (including inedible parts) lost from the food supply chain, not including food diverted to material uses such as bio-based products, animal feed, or sent for re-distribution." In order to diminish the bad effects of this situation, some scientists saw an opportunity to develop eco-efficient, innovative and sustainable processes to reuse plant-derived by-products as sources of bioactive compounds. In this chapter, we will discuss the exploitation of apple, grape and berry wastes as sources of natural antioxidants.

Over the last few decades, plant by-products have been used as feed or as fertilizers. However, a demand for feed could depend on agricultural yields and could vary between years. The second problem is disposing of plant waste which is aggravated by legal restrictions. Thus, the attitude towards wastes changed, especially due to the growing interest in protecting the environment, but also due to the increasing awareness of the benefits deriving from their exploitation. The by-products, which resulted from the fruit processing, are rich in valuable compounds and represent a novel, natural and economic sources of flavoring, colorants, polyphenols and natural antioxidants, which can be used in the food industry as a source of natural food additives.

2. Plant Polyphenols

Polyphenols are plants' secondary metabolites, commonly found in virtually all types of fruits. The group of polyphenol compounds can be divided into several sub-groups according to their structural characteristics, while those commonly found in fruits can be categorized into three groups: non-flavonoids, flavonoids and phenolic acids (Tsao, 2010). Flavones, flavanones, flavonols, flavanols and phenolic acids are the most common phenolics in plant-based foods while anthocyanins represent the predominating group of flavonoids of red fruits such as berries (Figure 1) (Heinonen et al., 1998; Häkkinen et al., 1999). In plants, polyphenols are produced mainly from two primary synthesis pathways, shikimate and acetate/polyketide. Both acetic and shikimic acid are derived from glucose metabolism (Claudia et al., 2008). Polyphenols have a key function in plant growth and development through the attraction of animal vectors for pollination and seed dispersal, stimulation of Rhizobium bacteria for nitrogen fixation, promotion of pollen tube growth, and the resorption of mineral nutrients from senescing leaves. Furthermore, polyphenols can enhance their tolerance to a variety of abiotic stressors They are employed as agents of defense against herbivores and pathogens, and they form the basis for allelopathic interactions with other plant species.

However, polyphenols have also shown to have a great impact on enhancing human health and preventing disease (Herrmann, 1976; Steinmetz and Potter, 1991; Block et al., 1992; Steinmetz and Potter, 1996; Ness and Powels, 1997; Scalbert et al., 2005). Some of these protective effects are attributed to polyphenols, potential health-promoting agents, because of their ability to act as antioxidants and free radical scavengers (Lin and Liang 2000; Rice-Evans et al., 2005; Zhang and Tsao, 2016; Yeh et al., 2017). Dietary polyphenols are one of the most important groups of natural antioxidants and chemopreventive agents found in human diets, including fruits and foods and beverages derived from fruits. Despite the fact that consumers are aware of diet-related health problems (Gilbert, 1997), still, a large percentage of the population lack a generous intake of foods such as fruits. Thus, dietary supplements and food fortification may be an alternative to the consumption of such compounds that may have health benefits. Globally, among the polyphenol market consumers, there is a number of reasons for shifting trends toward diets that are rich in fruits. One of the reasons is the higher number of aged people in Europe and Japan. Furthermore, increased health awareness and the onset of various metabolic problems due to aging (Cherniack 2011; Visioli et al., 2011). Transparency Market Research, a global market intelligence group, predicted an increase in polyphenol market due to the increasing demands and market size. This study indicates that in 2018 polyphenol market will reach USD 873.7 million. This estimation is based on revenue (USD million) and volumes (tons) for the period between 2012 and 2018, calculated with an annual growth rate of 6.1% (Transparency Market Research 2016).

Foods enriched with natural plant antioxidants has now assumed the status of "functional foods," which means that such foods should be capable of providing the additional physiological benefit, such as preventing or delaying the onset of chronical diseases, as well as meeting basic nutritional requirements. Manufacturers and food technologists, who follow these trends and are willing to meet the requirements of the consumers for safe and antioxidant-rich foods, are introducing natural antioxidants during the production and/or processing. The scheme represents the health benefits of food fortified with natural antioxidants (Figure 2). As such, polyphenols from fruit wastes have a great commercial potential as agents of the production of food supplements and in developing novel functional food formulations.

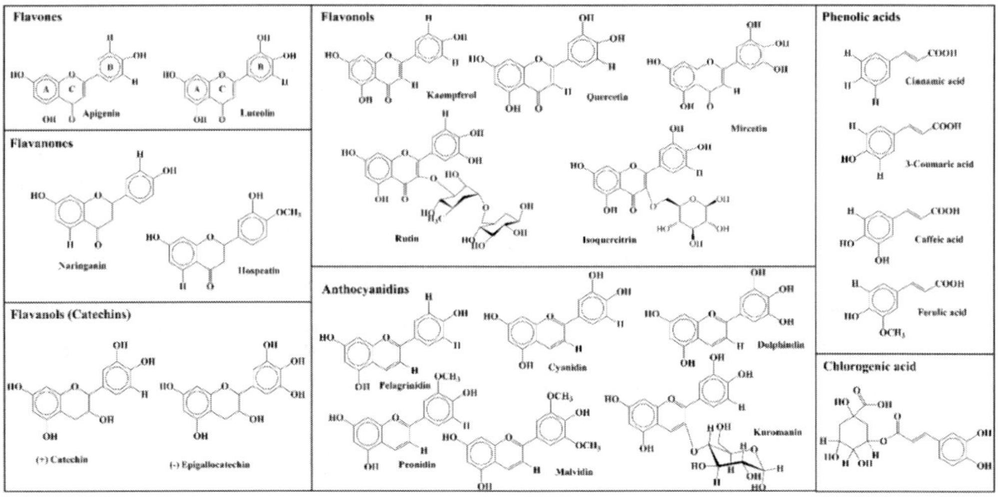

Figure 1. Most common phenolic compounds present in fruits.

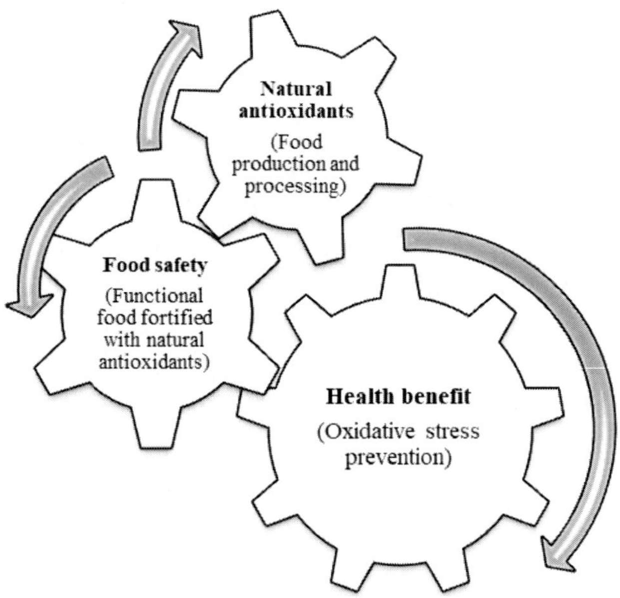

Figure 2. The health benefits of food fortified with natural antioxidants (adapted from Djilas et al., 2009).

2.1. Natural Antioxidants

The antioxidant potential of phenolic compounds depends on the number and arrangement of the hydroxyl groups in the molecules of interest. There are variations in degrees between different phenolic compounds when it comes to their effectiveness as antioxidants, and these variations have been largely attributed to the particular chemical structures and bioavailability of polyphenols. For example, phenolic acids are hydroxyl derivatives of aromatic carboxylic acids which have a single phenolic ring and can be further divided into two main types, the benzoic acids and cinnamic acids, based on the C1-C6 or C3-C6 backbone. On the other hand, flavonoids contain two phenolic rings, ring A and ring B, linked by a three-carbon bridge that is usually an oxygenated heterocycle ring C, having a common C6-C3-C6 skeleton structure (Zhang and Tsao, 2016). Diphenols in the ortho - and para -forms possess antioxidant activity, which is enhanced when ethyl or n-butyl groups substitute hydrogen atoms in the rings. The presence of one or more of the following structural characteristics is known to enhance the antioxidant properties of flavonoids:

- o -diphenolic group in the B ring
- 4-oxo function that exists in conjugation with a 2,3 double bond
- OH-groups in the 3rd and 5th positions.

Furthermore, plant polyphenols could exist as glycosides, acylglycosides and other conjugated forms rather than aglycones. Aglycones are generally more active in term of neutralizing free radicals and other reactive oxygen species (ROS) (Zhang and Tsao, 2016). Phenolic glycosides in human foods are less absorbed than their respective aglycones in the digestive tract. Thus, the form of dietary polyphenols may affect the outcome of their health benefits. Gallic acid and isoflavones are the most well-absorbed polyphenols, followed by catechins, flavanones, and quercetin glucosides, but with different kinetics. Proanthocyanidins, galloylated tea catechins, and anthocyanins are the least well-absorbed polyphenols (Manach et al., 2005).

Synthetic antioxidants are sufficiently effective in retarding oxidationand maintaining freshness. Besides, they are relatively cost-effective. They are usually readily available in large amounts and at high purity. However, an introduction of synthetic antioxidants requires extensive safety testings, which is not the case with natural antioxidants. A major advantage of using natural antioxidants is that they are generally limited only by the degree of flavor, color or cost that can be tolerated to give the maximal performance (Nahas, 2012). In particular, fruit wastes could be an alternative source for obtaining natural antioxidants which are considered completely safe in comparison with synthetic antioxidants. The idea of substitution of synthetic antioxidants with natural ones is supported by the consumers concern for safety of synthetic antioxidant (Djilas et al., 2009). Thus, new aspects concerning the use of these by-products for further exploitation of the production of food additives or supplements with high nutritional values have gained increasing interest because these high-value products and their recovery may be economically attractive. The research focus is disease prevention and health benefits of natural antioxidants. Fruits are usually promoted for their content in such antioxidants.

However, none of these antioxidants have made it yet on a commercial, large-scale as food preservatives. Nutraceutical and functional food markets are still more financially attractive for producers.

In order to carry out a meaningful comparison of the antioxidant activity of foods or commercial products and for the provision of quality standards for regulatory issues and health claim (Shahidi & Ho, 2007), different antioxidant assays were developed. These methods can be classified according to two types of mechanisms: SET - single electron transfer and HAT - hydrogen atom transfer methods (Huang et al., 2005; Prior et al., 2005; Lopez-Alarcon and Denicola, 2013). Single electron transfer based methods measure the ability of an antioxidant to transfer one electron in order to reduce target charged compounds, such as radicals and metal ions, while hydrogen atom transfer based method detects the ability of an antioxidant to quench free radicals by hydrogen donation, which is more relevant to the radical chain-breaking antioxidant capacity (Dong-Ping et al., 2017). According to the developed assays, fruits, as well as fruit wastes, showed high antioxidant capacity (Fu et al., 2010; Lu et al., 2011; Li et al., 2016; Deng et al., 2012; Rajbhar et al., 2015).

3. Antioxidants from Fruit Wastes

Bioactive compounds represent an excellent pool of molecules for the production of functional foods and food additives (Joana Gil-Chávez et al. 2013). Fruits are rich in several bioactive components and as such represent the simplest form of functional foods. In Europe, the fruit production is over 74 million tonnes. The production includes 16 crops, tomatoes, grapes, apples, pears, peaches (including nectarines) and apricots, cherries, strawberries, raspberries, blackcurrants and other berries and currants, lemons (including limes), grapefruit (including pomelos), oranges, plums and other (easy peeler) citrus such as mandarins, tangerines and clementines. (European Fruit Juice Association Liquid Fruit 2014).

The production of fresh fruits and fruit products involves numerous phases: pre-cooling, washing and disinfection, peeling, trimming, deseeding, cutting to specific sizes, sorting defects, dipping, drying, storage, packaging, labeling and distribution (James and Nagramsak, 2011). All these actions result in high amounts of waste materials. According to European Fruit Juice Association, fruit juice and nectar consumption in the EU was 9.6 billion liters in 2015 while globally it was 38.5 billion liters (European Fruit Juice Association Liquid Fruit 2016). Fruit juice processingleaves between 25 and 35% mass of the raw material called fruit pomace. Fruit pomace is one of the most common by-products in the fruit industry. Fruit pomace is the solid residue that remains after the extraction of juice from fruits and is composed of fruit skin, seeds and, in some cases, stems. Apple, grape and berry pomace are known to possess high antioxidant activities. Therefore, these value-adding components isolated from such wastes should be exploited as natural antioxidants for the formulation of functional foods or can serve as additives in food products to extend their shelf-life (Kalogeropoulos et al., 2012, Ibrahim et al., 2017).

3.1. Antioxidants from Apple Pomace

The average annual apple harvest of the European Union is about 10 million tons, with Poland, France, Italy, and Germany as the most productive countries (European Commission, 2014). Apples are fruits with a well-known positive impact on human health. Some of these positive effects are the result of the high polyphenol content. Apples contain five major groups of polyphenol compounds: hydroxycinnamic acids, flavan-3-ols, anthocyanidins, flavonols, and dihydrochalcones (Vrhovsek et al., 2004; Tsao et al., 2005.). However, single polyphenol compounds could vary up to 30% from year to year in the same cultivar (Guyot et al., 2003). Apple flesh contains catechins, procyanidins, phloridzin, phloretin glycosides, caffeic acid and chlorogenic acid. The peel possesses all of these compounds and has additional flavonoids, such as quercetin glycosides, that are not found in the flesh, (Wolfe et al., 2003). Generally, the polyphenol content, as well as antioxidant activity, is higher in apple peel than in the flesh (Burda et al., 1996; Wolfe et al., 2003). Apple peel also contains some specific antioxidant compounds such as quercetin glycosides and cyanidin-3-O-galactoside (Huber and Rupasinghe, 2009). Major polyphenol compounds in apples are chlorogenic acid, p-coumaroylquinic acid, cyanidin- 3-galactoside, catechin, epicatechin, procyanidins B1 and B2; quercetin galactoside, glucoside, rhamnoside, arabinoside, and xyloside phloridzin and phloretin-2′-xyloglucoside (Tsao et al., 2005).

Apple pomace is the by-product of processing for the recovery of apple juice and amounts to up to 25–35% of the processed fruit (Schieber et al., 2003). In 2014, more than three million tons of apples were used for processing (Association EFJ, 2018), of which about 800000 tons of apple pomace might have accrued. Therefore, profitable ways of disposal or a value-added use of apple pomace is of economic and environmental interest. Apple pomace as a by-product could be used for various purposes such as; fuel purposes, pectin extraction, biotransformation, source of fiber, etc. (Shalini and Gupta, 2010). However, it was also shown that apple pomace had a very high concentration of polyphenols which as natural antioxidants could be used for the production of novel functional foods. Apple pomace consists of apple peels, leftover flesh, core with seeds, and stems. It is a valuable source of polyphenols relevant for its antioxidant activity, as flavanols and flavonols and the antioxidant activity of apple pomace can be predicted by the contents of phloridzin, procyanidin B2, rutin + isoquercitrin, protocatechuic acid and hyperin (García et al., 2009). The moisture content of apple pomace after pressing is about 70–85% (Kara and Doymaz, 2015; Brushlyanova et al., 2013; Almeida-Trasvina et al., 2014; Junget al., 2015). In order to avoid microbial spoilage and to sustain adequate pomace quality, the water content should be immediately decreased to less than 10% (Lavelli and Corti, 2011). Fresh apple pomace showed the highest moisture content but the shortest drying period to reach equilibrium moisture content. Furthermore, Shalini (2010) reported that the reduction of pomace layer thickness and increasing the temperature could linearly increase drying rates. Increasing the temperature also leads to a reduction of equilibrium moisture content at the endpoint of the drying, and influences the composition and number of chemical constituents in apple pomace such as polyphenols (Lavelli and Corti, 2011). Except on microbial spoilage, lower water activity has an influence on polyphenol stability by increasing it through inactivation of the degrading enzymes, polyphenol oxidases, and peroxidases. Some drying methods, such as freeze-drying, additionally inhibit the formation of thermal degradation products such as 5-hydroxymethylfurfural from sugars (Schieber et al., 2001). However, the method used for drying pomace has less impact on the content of polyphenols than the originating apple cultivar and seasonal or environmental conditions during apple growth, which are the major factors that mediate the presence and content of polyphenols in apple pomace (Sato et al., 2011; Guyot et al., 2003). Total phenolic content of apples, determined by the Folin-Ciocalteu method in fresh fruits, could range between 66.2 and 211.9 mg/100 g, which demonstrates the wide range of polyphenolic content of different apple cultivars. Polyphenol range in apple pomace was reported to be 262 to 856 mg of total phenols/100 g and the content is higher compared to fresh fruit, due to the water loss during the pressing procedure (Sato et al., 2011). Higher content of polyphenols in apple pomace is the result of their predominant location in the apple peel (Krawitzky et al., 2014). Sanchez-Rabaneda et al. (2004) and Ramirez-Ambrosi et al. (2013) conducted comprehensive research on the identification of apple pomace polyphenols (Table 1), and representatives of some structural classes of apple pomace polyphenols are shown in Figure 3. The concentration of polyphenols in apple pomace is also influenced by juice production. It was shown that the pomace obtained from clear juice production had a higher concentration of polyphenols than pomace obtained from cloudy juice production. Furthermore, the use of pectinases in clear juice production cleaves polyphenols from pectin and/or other cell wall structures resulting in higher number of polyphenols available for extraction. However, polyphenol recovery was higher in the apple pomace from cloudy juice production as it accrues higher pomace mass (Waldbauer et al., 2017).

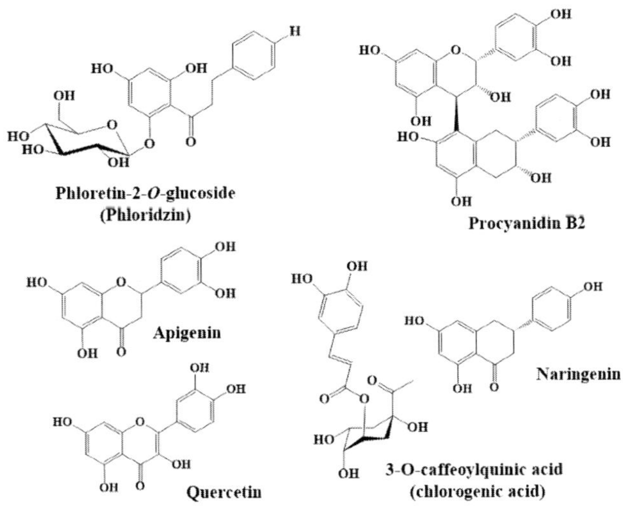

Figure 3. Polyphenols of apple pomace.

Table 1. Polyphenol compounds found in apple pomace (adopted form Waldbauer et al., 2017)

Polyphenol name	**Polyphenol group**
cyanidin-3-O-galactoside, cyanidin-3-O-hexoside	Anthocyanins
phloretin, phloretin-2-O-glucoside, phloretin-2-O-xylosyl-glucoside, phloretin-pentosylhexoside, phloretin-hexosyl-hexoside, 3-hydroxyphloretin-2′-O-xylosyl-glucoside, 3-hydroxyphloretin-2′-O-glucoside	Dihydrochalcons
(+)-catechin, (−)-epicatechin, procyanidin B1, procyanidin B2, procyanidin B3, procyanidin B5, procyanidin C1, ((epi)cat) trimer	Flavanols
((epi)cat) tetramer	
apigenin, chrysoeriol, luteolin, luteolin-7-O-galactoside, luteolin-7-O-glucoside	Flavons
eriodictyol, eriodictyol-hexoside, hesperidin-O-pentoside, naringenin, naringenin-7-O-glucoside, naringenin-7-O-neohesperidoside, naringenin-7-O-rutinoside, naringenin-O-glucuronide, naringenin-O-hexoside	Flavanons
isorhamnetin-3-O-galactoside, isorhamnetin-3-O-glucoside, isorhamnetin-3-O-rutinoside, isorhamnetin-3-O-arabinopyranoside, isorhamnetin-3-O-arabinofuranoside, isorhamnetin-3-O-rhamnoside, kaempferol-O-glucoside, quercetin, quercetin 3-O-diglucoside	Flavonols
quercetin-3-O-arabinofuranoside, quercetin-3-O-rabinopyranoside, quercetin-3-O-galactoside, quercetin-3-O-glucoside, quercetin-3-O-rhamnoside, quercetin-3-O-rutinoside quercetin-3-O-xylanoside, quercetin-O-hexoside, quercetin-O-pentosyl-hexoside, quercetin-O-pentoside, quercetin-O-xylosyl-pentoside, rhamnetin, rhamnetin-3-O-glucoside	
caffeic acid-O-hexoside	Hydroxycinnamic acids
3-O-caffeoylquinic acid, 4-O-caffeoylquinic acid, 5-O-caffeoylquinic acid, x-O-caffeoylquinic acid, dicaffeoylquinic acid	
ferulic acid-O-hexoside	
p-coumaric acid-O-hexoside, 4-O-p-coumaroylquinic acid, 5-O-p-coumaroylquinic acid, p-coumaroylquinic acid, sinapic acid-O-glucoside	
protocatechuic acid, salicylic acid	Other

3.2. Antioxidants from Grape Pomace

Grapes are one of the most extensively cultivated crops with almost 63 million tons produced worldwide, and the vast majority of the total grape production (75%) is used to produce wine (FAOSTAT 2013). The result of that production are millions of tons of grape pomace, which represents a management issue from both ecological and economical point of view (Fontana et al., 2013). Grape pomace is the residue of pressed grapes, small pieces of stalks, and yeast cells from the wine fermentation process. Grape pomace was used for the production of "wine alcohol" or other distilled spirits, liquors, and liqueurs, fertilizers or as animal feed (Silva et al., 2000; Arvanitoyannis et al., 2006; Gonzalez-SanJose 2014). However, grape pomace represents a polyphenol-rich dietary matrix and as such it is exploited in Europe as the enocyanins, food colorant E-163.

This exploitaion, is also supported by the fact that grape skin is especially high in polyphenols. Grape polyphenols could be classified into two groups of flavonoids and non-flavonoids. The flavonoids include flavan-3-ols (catechin), flavonols (quercetin) and anthocyanins, while non-flavonoids encompass hydroxybenzoic acids (gallic acid), hydroxycinnamic acids and stilbenes (resveratrol) (Yang, Martinson, & Liu, 2009). In the grape berry, the flavonoids, such as anthocyanins and resveratrol, are mainly localized in the skins, (Palomino et al., 2000), while flavan-3-ols (catechins and proanthocyanidins) are present both in the skins and in the seeds (Yang, Martinson, et al., 2009). Anthocyanins are also mostly accumulated in the skins of red grapes, predominately the 3-O-glucosides of malvidin, cyanidin, delphinidin, peonidin, and petunidin. Malvidin, the reddest of all anthocyanins, is the major anthocyanin in dark red vinifera grapes, with higher proportions of cyaniding in red grapes. Recently, it was discoveredthat in grape by-products the most prominent phenolic compounds are present in the insoluble bound form de Camargo et al. (2014). In red pomace anthocyanins and flavanols are predominant polyphenols, while in white grapes flavanols are the most abundant phenols. Flavanols are mainly located in the seeds. Their level ranges from 56% to 65% of the total flavanols of grapes, against 14% to 21% present in grape skins. Grape seeds are a rich source of monomeric phenolic compounds such as (+)-catechins, (−)-epicatechin, (+)-gallocatechins, (−)-epigallocatechin, and their dimeric, trimeric, and tetrameric proanthocyanidins (Boashan et al., 1998). The polyphenol composition of grape pomace has been well studied, with high differences in reported results (Kammerer et al., 2004; Peralbo-Molina and Luque deCastro 2013; Teixeira et al., 2014). These differences could have been associated with factors that affect phenolic composition of grapes as well as with diverse enological practices on the extraction of phenolic compounds during the winemaking process (Andrades et al., 1995; Perez-Magarino et al., 1999; Perez-Magarino and Gonzalez-SanJose 2006; Revilla and Gonzalez-SanJose 2002; Perez-Magarino et al., 2009). In order to use such by-products, grape pomace should be stabilized in order to prevent degradation. Some of the techniques which are used are:

- removing oxygen by compacting the wine pomace,
- spraying different acids (sulfuric, tartaric, or phosphoric acids) over the pomace
- using gamma-irradiation in combination with other synthetic preservatives
- drying of pomace

Drying is the most used technique for preservation of grape pomace. Freeze drying is usually considered as a gentle drying technique due to the low thermal stability of bioactive compounds. However, in comparison to oven-dried samples, freeze-dried samples exhibited higher losses of bioactive compounds during their subsequent storage conditions. This loss is most probably caused by the porosity of freeze-dried products. That porosity increases air contact and their susceptibility to oxidation. Furthermore, heat treatment over long periods of time may induce the release of certain low molecular weight compounds that increase the level of certain phenolic compounds. Recently, infrared drying, in comparison to convective drying,was proposed as suitable technique to retain the highest bioactivity but this technology as freeze-drying is more expensive than conventional drying methods (Javier Garcıa-Lomillo and Marıa Luisa Gonzalez-SanJose, 2017). Some examples of enrichment of food products with grape pomace polyphenols are shown in Table 2.

3.3. Antioxidants from Berry Pomace

Berries are recognized as an important component of healthy diets due to their bioactive compounds. The processing of berries into juice generally results in approximately 70–80% of the target product and 20–30% of by-product which contains mainly seedsand skin. During the juice production, berries are squeezed, treated with pectinases and heated prior to juice extraction. Berry juice processing residue is pomace that contains high amounts of bioactive compounds. Usually, berry pomace is used as animal feed, as compost, for biogas production or it is simply discarded. However, since many phenolic compounds, that have been located in the skins and seeds of the berries, are retained in pomace, it represents a good source of natural antioxidants and colorant. The anthocyanins are responsible for the characteristic color of various berry fruits. Among different berry varieties, the highest amount of anthocyanins have bilberries, with a ratio of 30:36:13 for cyanidin-, delphinidin-, and malvidin-3-O-glycosides (Du et al., 2004; Wang et al., 1997). Bilberry anthocyanosides represent all possible combinations of 5 anthocyanidins (cyanidin, delphinidin, peonidin, petunidin, malvidin) with 3 sugar moieties (3-O-arabinosides, 3-Oglucosides, 3-O-galactosides) (Madhavi et al., 1998; Martinelli et al., 1986). Except for anthocyanins, bilberry also contains flavonols and

Table 2. Examples of enrichment of food products with grape pomace polyphenols (adopted from Javier Garcıa-Lomillo and Marıa Luisa Gonzalez-San Jose, 2017)

Category	Foodstuff	Grape pomace product	Content [g/100 g]	Fortification with	References
Cereal products	Bread	Seed flour	2.5 to 10	Phenols	Hoye and Ross (2011)
			5 to 25	Phenols/fiber	Aghamirzaei et al. (2015)
			2.5 to 7.5	Phenols	Meral and Doğan (2013)
		Whole grape pomace powder	5 to 10	Fiber/phenols	Smith and Yu (2015)
		Skin grape pomace powder	4 to 10	Phenols/fiber	Mildner-Szkudlarz et al. (2011)
	Biscuits	Seed grape pomace powder	5	Phenols Mineral	Aksoylu et al. (2015)
		Whole grape pomace powder	10-30	Phenols/fiber	Mildner-Szkudlarz et al. (2013)
	Bread an muffins	Whole grape pomace powder	5-20	Phenols /fiber	Walker (2013)
	Muffins	Skin grape pomace powder	20	Phenols	Mildner-Szkudlarz et al. (2015)
	Breakfast cereals	Grape pomace flours	5-20	Phenols/fiber	Oliveira et al. (2013)
	Bars, Pancakes, and Noodles	Seed grape pomace flour	5-30	Phenols	Rosales Soto et al. (2012)
	Pasta	Grape pomace extract	-	Phenols	Marinelli et al. (2015)
Dairy products	Fermented milk	Grape pomace extract	0.1	Phenols	Aliakbarian et. al (2015)
		Grape pomace flour	0.01 to 0.05	Phenols	Frumento et al. (2013)
	Yogurt	Seed grape pomace extract	0.05 to 0.01	Phenols	Chouchouli et al. (2013)
		Whole grape pomace powder and extract	1 to 3	Phenols/fiber	Tseng and Zhao (2013)
		Whole grape pomace extract	1	Phenols	Karaaslan et al. (2011)
		Skin grape pomace flour	6	Phenols	Marchiani et al. (2016)
	Cheese	Whole grape pomace powder	0.8 - 1.6	Phenols	Marchiani et al. (2015)
		Whole grape pomace extract	0.0005	Phenols	Han et al. (2011)
		Skin grape pomace, whole grape pomace and seed grape pomace extracts	0.1 - 0.3	Phenols	Felix da Silva et al. (2015)
	Ice cream	Grape seed extract	0.4	Phenols	Sagdic et al. (2012a)
Others	Seafood	Skin grape pomace powder	3	Fiber/phenols	Ribeiro et al. (2013)
	Puree	Skin grape pomace powder	3.2	Phenols	Lavelli et al. (2014)
	Infusions	Skin grape pomace extract	50 to 100	Phenols	Bekhit et al. (2011)

Table 3. Polyphenol compounds found in different berries (adopted from Häkkinen et al., 1999)

		Berries type										Reference
		Blueberry	Cranberry	Lingonberry	Bilberry	Gooseberry (green)	Black currant	Red currant	White currant	Strawberry	Red raspberry	
Flavonoids and phenolic acids from berries (mg fresh weight/100 g)	Kaempferol	0±0.6	0±0.3	<0.1	0	<0.01	<0.01±1.0	≤ 2.0 0.01±0.2	0.2	1.2	<0.01	Bilyk and Sapers (1986)
	Quercetin	10.5±16.0 2.4±2.9	7.3±25 17.2	3.4	3.2	<0.01	3.3±6.8	2.0±4.0; 2.7 0.2±1.0	0.28 0.3	0.86	2.9	Hertog et al. (1992) Daniel et al.
	Myricetin	0.9±6.9	1.1±2.4; 7.7	0	0	N.D.	5.5	<0.01 0.01±0.02	<0.1	N.D.	0	(1989) Wildanger and Herrmann
	p-Coumaric acid	0±0.7	N.D.	N.D.	D	0.6±2.5	5.6±12.0	0.7±1.4	2.0±3.8	0.6±17	0.7±2.0	(1973) Starke and
	Caffeic acid	8.3±42.2	N.D.	N.D.	D	2.4±3.2	1.4±93	0.8±1.6	1.3±2.1	<0.05±1.4	0.6±1.0	Herrmann (1976)
	Ferulic acid	0.2±0.8	N.D.	N.D.	D	<0.05±0.20	0.7±1.9	0.3±0.8	0.3±1.0	0	0.3±1.7	Stöhr and
	p-Hydroxy benzoic acid	0	N.D.	N.D.	D	0	0±0.6	1.0±2.3	0.5±1.0	1.0±3.6	1.5±2.7	Herrmann (1975a) Herrmann (1976)
	Gallic acid	0	N.D.	N.D.	D	0	3.0±6.2	<0.1	0.3±1.0	0.5±4.4	1.9±3.8	Stöhr and Herrmann (1975b)
	Ellagic acid	<10.0*	12.0*	N.D.	N.D.	N.D.	N.D.	N.D.	N.D.	63.0*	150.0*	Mosel and Herrmann (1974)

N.D. – no data available; * – mg dry weight/100 g; D. – determined, no concentration observed.

phenolic acids (Jamison, 2003; Brenneisen and Steingger, 1981). Blackberries are a rich source of anthocyanins and other polyphenol antioxidants. Anthocyanins which were identified in blackberries were cyanidin 3-glucoside, cyanidin 3-rutinoside, malonic acid acylated cyaniding 3-glucoside and xylose-cyanidin derivative (Siriwoharn et al., 2004). Gallic, caffeic, ferulic, p-coumaric and ellagic acids were identified from phenolic acids in blackberries. Another berry with the high content of polyphenols is blueberry. Several studies have confirmed that blueberries contain high antioxidant activity compared with other fruits (Cao et al., 1996; Prior et al., 1998; Vinson et al., 2001; Wang et al., 1996). However, polyphenols and antioxidant activity of blueberry are mostly in the skins. For example, fifteen anthocyanins were detected in blueberries and they exist almost exclusively in the skins. Most antioxidant activities were attributed to delphinidin-3-O-galactoside, cyanidin-3-O-galactoside, delphinidin-3-O-arabinoside, petunidin-3-O-galactoside, malvidin-3-O-galactoside, malvidin-3-O-arabinoside, 5-Oferuloylquinic acid and traces of a quercetin-O-diglucoside (Borges et al., 2009). The blueberry skin contains also cinnamic acids and flavonol glycosides, with a small amount of gallic and syringic acid. The flesh, on the other hand, consists of only cinnamic acid with chlorogenic acid as the main phenolic compound. Some authors recognize that blueberry processing waste is high in antioxidant activity and total phenols and that it has potential to be a good source of natural colorants and nutraceuticals (Lee & Wrolstad, 2004).

Cranberries were the world's first government-approved fruit used for the treatment of urinary tract infections (Vinson et al., 2008). Cranberry contains simple phenolic acids and flavonoids such as anthocyanins, proanthocyanidins, as well as flavonols (Gregoire et al., 2007). Fifteen phenolic acids were identified in cranberry fruit in free and bound forms (Zuo et al., 2002). Isolated cranberry flavonols included myricetin-3-β-galactoside, myricetin-3-α-arabinofuranoside, quercetin-3-β-galactoside, quercetin-3-β-glucoside, quercetin-3-rhamnospyranoside, and quercetin-3-O-(6″-p-benzoyl)-β-galactoside (Singh et al., 2009; Yan et al., 2002). Considering anthocyanins, cranberry consists largely of galactoside and arabinoside conjugates of cyaniding and peonidin (Singh et al., 2009). Furthermore, in strawberries, the principal red pigment is pelargonidin-3-glucoside (Lopez-Serrano & Barcelo, 1996). The cinnamic acids (p-coumaric, cafeic, and ferulic acids) represent a major proportion of the phenolic content of strawberries and black currant (Häkkinen et al., 1996). Significant amounts of flavanols and flavonols, myricetin, quercetin and kaempferol, are also present. The flavonol glycoside profile does not differ greatly among the various Rosaceae fruits, for example, strawberries, raspberries and blackberries (Boyles and Wrolstad, 1993; Rommel and Wrolstad, 1993; Withy et al., 1993) and is dominated by quercetin, kaempferol and their glycosides. Table 3 shows some of the flavonoid and phenolic acids from different berries, which have been reported in other researches. These flavonoid and phenolic acids have been proposed to have beneficial effects on health as antioxidants.

After pressing, skins and seeds remain in the pomace and the fractions of antioxidants, that are associated with cell wall compounds, stay together with them Some authors reported that proanthocyanidin content of chokeberry pomace is higher than that of the berries. The study on blueberries concluded that 18% of the anthocyanins content of the fresh berries remains in the pomace while cranberry pomace contains 13–17% of the total anthocyanins more than juice itself(Mayer-Miebach et al., 2012; Skrede et al., 2000; White et al., 2011). Berry pomace has a moisture content of approximately 50% and it contains various digestible nutrients. Therefore it is highly prone to microbial spoilage. Consequently, an initial processing of pomace is necessary to ensure a satisfactory shelf life. Drying is the most used technique for the stabilization of pomace following milling and conditioning. To obtain pomace powders with a high amount of bioactive compounds it is important to adjust processing conditions such as drying method, time and temperature especially because the anthocyanins are easily degraded (Skrede et al., 2000). After drying, berry pomace should contain 5–6% (w/w) moisture (Sojka & Krol, 2008; Sojka et al., 2013a; Reque et al., 2014).

4. Extraction of Natural Antioxidants

Bioactive components, such as polyphenols, present in agro-industrial waste can be recovered using various solvents and methods of extraction. Frequently, traditional methods of sample preparation, separation, detection, and identification and are being replaced by advanced techniques. Generally, efficiency of extraction method depends on chemical nature of the substance, sample particle size and by the presence of interfering substances. On the other hand, extraction yield depends on extraction time, temperature, solvent-to-feed ratio, the number of repeated extractions of the sample, as well as on the choice of extraction solvents. Polyphenols can be extracted from fresh, frozen or dried plant material. Usually, before extraction, the material is pre-treated by milling, grinding, drying and homogenization. The solvents, that are usually used for extraction of polyphenols, are ethanol, methanol, water or mixtures of these. Modern green extraction techniques for polyphenol extraction include microwave extraction, ultrasonic extraction and supercritical fluid extraction.

4.1. Extraction Solvents

Solvent extraction is frequently used to extract polyphenols compounds from plant sources since they are efficient and easy to use. For extraction, the solvent is chosen according to the type of polyphenol compound that is required. Chemical extraction depends on the type of solvent, solvent polarity, extraction time and temperature, as well as on the chemical composition and physical characteristics of the samples (Wissam et al., 2012). Furthermore, the extraction of polyphenols is challenged by their chemical structure and their interaction with other food components. Generally, it was shown that low viscosity solvents accelerate mass transfer (Ajila et al., 2011). Some of the properties of most used solvents for polyphenol extraction are given in Table 4. The hydrophilic polyphenols including aglycones, glycosides, and oligomers, are extracted using water, polar organic solvents such as methanol, ethanol, acetonitrile and acetone, or their mixture with water. Less polar flavonoids (e.g., isoflavones, flavanones, methylated flavones, and flavonols) are extracted with chloroform, dichloromethane, diethyl ether, or ethyl acetate. Glycosides have increased water solubility of polyphenols and aqueous-alcoholic solutions are suitable for their extraction. Polyphenols are more stable at low pH as the acidic condition helps polyphenols to stay neutral. Thus, they are more easily extracted into organic solvents. Flavan-3-ols such as catechins, proanthocyanidins, and condensed tannins can be extracted directly with water, methanol, ethanol, acetone, or ethyl acetate. On the other hand, certain flavanone and chalcone glycosides are difficult to dissolve in methanol, ethanol, or alcohol-water mixtures. Flavanone solubility is highly depended on pH of water-containing solutions. The extraction of anthocyanins could be obtained by using an acidified organic solvent such as methanol or ethanol.

Nowadays, however,, one of the most important trends in the extraction of bioactive compounds is the search for new, environmentally friendly food grade solvents and their application. One of the mostly explored environmentally friendly solvent is certainly ethyl lactate. Ethyl lactate meets eight of the 12 so-called principles of green chemistry (Pereira et al., 2011). The U.S. FDA and the EU have approved its use in food products. Ethyl lactate has a relatively high flash point and is colorless, environmentally benign, and completely biodegradable (releasing CO_2 and water) (Lores et al., 2015). Research conducted by Lores et al. in 2015 showed the strong extraction capacity of ethyl lactate, which is comparable to or even better than that of other commonly used extraction solvents, with the advantage of being compatible with food, cosmetics and pharmaceuticals.

4.2. Extraction Methods

Traditional methods for the extraction of polyphenols include solid-liquid extraction simply by means of solvent application and leaching. Conventional solvent extraction encompasses Soxhlet extraction, percolation, and maceration extraction. Conventional solvent extraction exposes suitably sized raw material to different organic solvents, which takes up soluble components of interest such as polyphenols. Samples are then usually centrifuged and filtered to remove solid residue, and the extract could be used as additive, food supplement or for the preparation of functional foods. However, conventional solvent extraction has some disadvantages such as excessive consumption of time, energy, and pollution of solvents (Cravotto et al., 2011). In order to overcome this issue, researches explore more cost-effective and greener techniques for the extraction of polyphenols from a wide range of plant matrices and their by-products (Figure 4).

Table 4. The properties of the most commonly used solvents for polyphenol extraction

	Acetone	Acetonitrile	Chloroform	Ethanol	Ethyl acetate	Methanol	Water
Boiling point (°C)	56.2	81.6	61.2	78.5	77.0	64.6	100.0
Density (g/mL)	0.786	0.786	1.498	0.789	0.894	0.791	0.998
Polarity Index	5.10	6.20	4.40	5.20	4.30	6.60	10.20

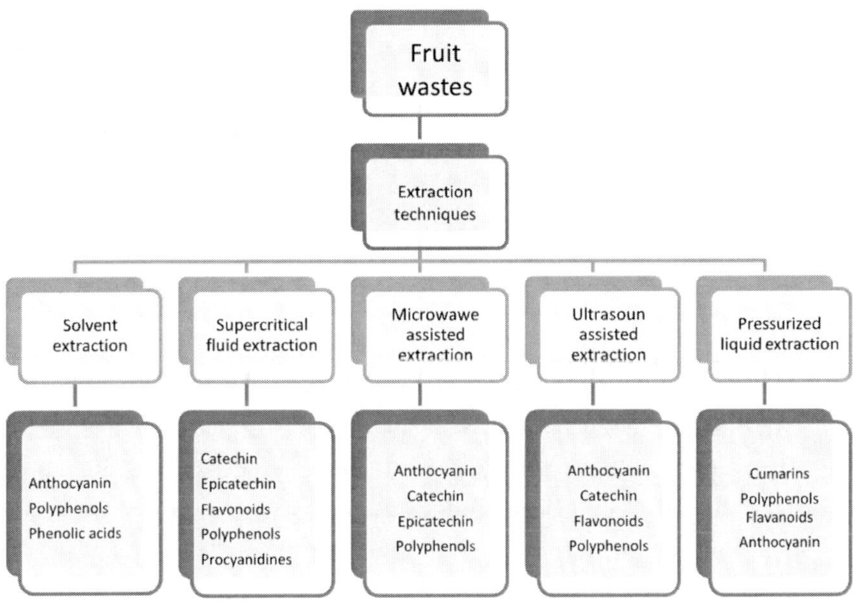

Figure 4. Green techniques for the extraction of polyphenols from fruit wastes.

4.2.1. Microwave-Assisted Extraction

Microwave-assisted extraction was developed around 1980's and since then it has become one of the more popular and cost-effective extraction methods (Rajbhar et al., 2014). It is an advantageous method due to shorter extraction time, higher extraction rate, less requirement of solvent and lower costs over the traditional method of extraction of compounds. Such advantages are related to a selective/localized heat transfer due to the high microwave energy absorption by water found inside plant cellular tissues. Different factors, such as solvent nature, solvent to solid ratio, microwave power level, irradiation time and a number of irradiation cycles, influence microwave-assisted extraction. Bai et al. (2010) studied microwave-assisted extraction technology of polyphenols from apple pomace. The 62.1% ethanol was used as the solvent, the microwave power level of 650.4

W for 53.7 s, the solid/liquid ratio (g/mL) was 1:22.9 and the result was total phenol content of only 62.7 mg of gallic acid equivalent per 100 g. However, in the study on microwave-assisted extraction of phenolic antioxidants from red delicious and Jonathan apple pomace conducted by Chandrasekar et al., (2015) the total phenol content was 1580.7 mg of gallic acid equivalent per 100 g and the used extraction parameters were: 60% of ethanol as solvent, the microwave power level of 735 W for 149 s and the solid/liquid ratio (g/mL) was 1:22.9. Simić et al. (2016) optimized the microwave-assisted extraction of total polyphenolic compounds from chokeberries. They found out that conditions which provide the highest possible total phenol yield and which could be used for the scale-up procedure were the most economic ones. The 300 W, 53.6% ethanol concentration and 5 min procedure resulted in a total phenol content yield of 420.1 mg GAE/100 g fresh plant material. Furthermore, Routray and Orsat (2014) evaluated extraction efficiencies of flavonoids extracted by microwave-assisted extraction, ultrasonic assisted extraction, and solvent extraction methods from the leaves of the Bluetta variety of blueberry. They used different microwave power levels (10% to 20%) with extraction times spanning from 4 to 16 min and 80 mL solvent used in combination with a 97:3 (v/v) ratio of 30% ethanol and 1.5 M citric acid. They also found out the correlation between microwave power and extraction time with a higher yield of flavonoids (Routray and Orsat, 2014).

4.2.2. Ultrasonic-Assisted Extraction

Ultrasound-assisted extraction has emerged as a promising technique that fulfills the required criteria as an inexpensive green extraction technique which depends on destructive effects of ultrasonic waves. In order to conduct sonication for extraction purposes, the operational frequencies of ultrasound waves generally range from 20 kHz or above. Such high frequencies are helpful when intensifying mass transfer, cell disruption, penetration and capillary effects. In combination with high-temperature ultrasound, extraction is even more efficient since high temperature increases the solubility, diffusivity and pressure, which help the waves to penetrate the tissue and transport contents in a variety of solvents, both organic and inorganic. Drosou et al. (2015) examined the extraction yield of red grape pomace by three different extraction methods using water, water: ethanol (1:1) and ethanol as solvents. They concluded that ultrasonic-assisted extraction water: ethanol extracts were richer in phenolic compounds (up to 438,984 mg/mL GAE in dry extract) compared to conventional Soxhlet extraction and microwave-assisted extraction. In studies conducted on ultrasound extraction of polyphenols from apple pomace the use of 50% of ethanol as solvent for 45 min at 40.1°C with ultrasound power of 0.142 W/g and solid/liquid ratio (g/mL): 1:6.7 showed the highest yield (971.3 mg of gallic acid equivalent per 100 g) of total phenol content (Virot et al., 2010). Ghafoor et al., (2009) reported optimal extraction conditions for anthocyanins from grape seeds. The parameters were as follows: ethanol concentration of

53% at 53°C for 29 min, while keeping the ultrasonic power and frequency constant, 250 W and 40 kHz, respectively. In addition, degradation of phenolic acids was reported to take place at high temperatures for longer periods of time under ultrasound assisted extraction. Therefore, the ultrasound-assisted extraction technique should be used carefully by optimizing all parameters for extraction in order to avoid thermal degradation of phenolic compounds. Ma et al., (2008) reported maximum yields of phenolic acids using the following optimized operational parameters: 20 min of extraction time at 303.15 K temperature and 8 W ultrasonic power.

4.2.3. Supercritical Fluid Extraction

The supercritical fluid can be defined as a fluid existing in such a phase in which it possesses features of both liquids and gases above its characteristic critical temperature and pressure. Supercritical fluids combined the properties of gaseous and liquid simultaneously and thus improved mass transfer rate of solutes during extraction. The properties of the supercritical fluid such as viscosity, density, and solvation can be modified by varying temperature/pressure (Pereira and Meireles 2009). Supercritical fluid extraction has been extensively investigated for isolation of polyphenols from plant material, particularly the use of carbon dioxide as a solvent has the increasing interest for the recovery of such compounds suited for application in the food industry (De Melo et al., 2014). European Food Safety Authority and the United States Food and Drug Administration have assigned carbon dioxide a *generally recognized as safe* (GRAS) status. Carbon dioxide represents an ideal solvent for the extraction of polyphenols because it is non-toxic, non-explosive, readily available and easy to remove from extracted products (Herrero et al., 2010). Furthermore, carbon dioxide is an environmentally friendly solvent with low oxidative and thermal impact. Some factors that influenced supercritical fluid extraction are the type of fluid and particle size. Fluid selection is one of the crucial factors governing SFE efficiency for polyphenol extraction. Since polyphenols are thermolabile, supercritical water is not a good choice in spite of high extraction yields in the case of polar compounds. However, in spite of the fact that polyphenols possess a low degree of solubility in supercritical carbon dioxide, the addition of small amounts of organic modifiers could solve this problem (King 2014; Azmir et al., 2013). Addition of 5–30% ethanol increase the solubility of quercetin, catechin, epicatechin, resveratrol, o-,m-,p- isomers of coumeric acid, caffeic acid, ferulic acid in supercritical carbon dioxide (Murga et al., 2003; Choi et al., 1998; Berna et al., 2001; Chafer et al., 2004; Berna et al., 2001a, Chafer et al., 2002). Particle size is another factor which can influence the extraction efficiency of polyphenols. Particle size influences mass transfer of solutes during the extraction process, thereby affecting polyphenol recovery rates. After the research conducted on supercritical fluid extraction of polyphenols from apple pomace Adil et al. (2007) concluded that extracts from apple pomace, obtained by subcritical carbon dioxide + ethanol, was higher than the extracts obtained by ethanol. The process of extraction indicated that lower amounts of polyphenols were extracted. However, these polyphenols extracted by subcritical carbon dioxide + ethanol, were more active when compared to polyphenols obtained by ethanol extraction. They also reported that optimum ethanol concentration was 20% ethanol and extraction conditions were as follows: pressure of 54.6–57MPa and temperature 55.7–58.4°C. Supercritical or subcritical carbon dioxide extraction with ethanol and/or methanol as a co-solvent has been applied to extract polyphenols from grape seeds (Palma and Taylor, 1999; Murga et al., 2000), from wine industry by-products (Louli et al., 2004), grape peel (Ghafoora et al., 2010) etc. Liza et al., (2010) studied extraction of antioxidants from pomaces of blueberry, cranberry and raspberry pressing wastes using conventional and supercritical solvents. In their study the maximal extraction yield (5.8%) from raspberry was observed at 150–200 bar whereas from blueberry it was only 1.4%. They concluded that extraction was more selective from blueberries, since the extracts collected in the second separator contained 9 g of gallic acid equivalents per 100 g extracts. Barjaktarović et al. (2005) reported a maximum on the extraction yield from juniper berries about 4% at 100 bar and 40°C.

Conclusion

Nowadays, when fruit production is larger than ever and constantly growing, we are facing the growing problem of generating a large amount of wastes. Food wastes are recognized as a good source of high- value compounds which can be exploited for the development of new functional ingredients. The scientific research in the field of valorization of food industry by- products highlighted the fruit by-products, especially pomaces after fruit pressing as a good source of natural antioxidants and food colorants. Apple, grape and berry pomaces proved to be an excellent source of polyphenols. The practical examples of extraction of polyphenols from fruit wastes, in innovative and unconventional ways such as microwaveultrasound-assisted extractions and supercritical fluid extraction, showed to be more cost-effective and greener for the recovery of polyphenols.

References

Adil Hasbay, I., Cetin, H.I., Yener, M.E., Bayındırlı, A. (2007). Subcritical (carbon dioxide + ethanol) extraction of polyphenols from apple and peach pomaces, and determination of the antioxidant activities of the extracts. *J. Supercritical Fluids*. 43, 55–63.

Aghamirzaei, M., Peighambardoust, S.H., Azadmard-Damirchi, S., Majzoobi, M. (2016). Effects of grape seed powder as a functional ingredient on flour physicochemical characteristics and dough rheological properties. *J. Agric. Sci. Technol.* 17, 365–73.

AIJN Market report (2016). European Fruit Juice Association. *Liquid Fruit.* http://www.aijn.org/files/default/aijn2016-market-report. pdf; Accessed 2018-01-22.

Ajila, C.M., Brar, S.K., Verma, M. (2011). Extraction and analysis of polyphenols: Recent trends. *Crit Rev Biotechnol.* 3, 227-49.

Aliakbarian, B., Casale, M., Paini, M., Casazza, A.A., Lanteri, S., Perego, P. (2015). Production of a novel fermented milk fortified with natural antioxidants and its analysis by NIR spectroscopy. *LWT – Food Sci. Technol.* 62, 376–83.

Almeida-Trasvina, F., Medina-Gonzalez, S., Ortega-Rivas, E., Salmeron-Ochoa, I., Perez-Vega, S. (2014). Vacuum drying optimization and simulation as a preservation method of antioxidants in apple pomace. *J. Food Process Eng.* 37, 575-587.

Andrades Rodrıguez, M.S., Gonzalez-SanJose, M.L. (1995). Influencia climaticaen la maduracion de la uva: estudio de cultivares de la Rioja y dc Madrid [Climatic influence on the ripening of the grape: study of cultivars from Rioja and Madrid]. *Zubia Monografico.* 7, 79-102.

Arvanitoyannis, I.S., Ladas, D., Mavromatis, A. (2006). Potential uses and applications of treated wine waste: a review. *Intl. J. Food Sci. Technol.* 41:475–87.

Association EFJ (2014). *Market Report 2014.* [Available at http:// aijnorg/files/ default/aijn2014-fullpdf.; Accessed: 2018-01-25]

Bai, X., Yue, T., Zhang, H., Gao, C. (2010). Optimization of microwave-assisted extraction technology of polyphenols from apple pomace by response surface method. *Zhongguo Shipin Xuebao* 10, 169–177.

Baoshan Sun, M., Spranger, I. (2005). Review: Quantitative extraction and analysis of grape and wine proanthocyanidins and stilbenes. *Ciência Téc Vitiv.* 20, 59–89.

Barjaktarović, B., Sovilj, M., Knez, Ž. (2005). Chemical composition of Junuperus communis L. fruits supercritical CO_2 extracts: dependence on pressure and extraction time. *J. Agric. Food Chem.* 53, 2630.

Bekhit, A.E.D., Cheng, V.J., McConnell, M., Zhao, J.H., Sedcole, R., Harrison, R. (2011). Antioxidant activities, sensory and anti-influenza activity of grape skin tea infusion. *Food Chem.* 129, 837–45.

Berna, A., Chafer, A., Monton, J.B., Subirats, S. (2001). High-pressure solubility data of system ethanol (1) + catechin (2) + CO2 (3). *J. Supercrit. Fluids.* 20, 57–162.

Berna, A., Chafer, A., Monton, J.B. (2001a). High-pressure solubility data of system reveratrol (1) + ethanol (2) + CO2(1). *J. Supercrit. Fluids.* 19133–139.

Bilyk, A., Sapers, G.M. (1986). Varietal differences in the quercetin, kaempferol, and myricetin contents of highbush blueberry, cranberry, and thornless blackberry fruits. *J. Agric. Food Chem.* 34, 585-588.

Block, G., Patterson, B., Subar, A. (1992). Fruit, vegetables, and cancer prevention: a review of the epidemiological evidence. *Nutr. Cancer.* 18, 1-29.

Borges, G., Degeneve, A., Mullen, W., Crozier, A. (2009). Identification of flavonoid and phenolic antioxidants in black currants, blueberries, raspberries, red currants, and cranberries. *J. Agric. Food Chem.* 58,3901– 3909.

Boyles, M.J., Wrolstad, R.E. (1993). Anthocyanin composition of red raspberry juice: influences of cultivar, processing and environmental factors. *J. Food Sci.* 58, 1135-1141.

Brenneisen, V.R., Steingger, E. (1981). Analysis of polyphenols from *Vaccinium myrtillus* L. (Ericaceae) fruit. *Pharma. Acta Helvetiae.* 56, 180–185.

Brushlyanova, B., Petrova, T., Penov, N., Karabadzhov, O., Katsharova, S. (2013). Drying kinetics of different fruit pomaces in a heat pump dryer. *Bulg. J. Agric. Sci.* 19, 780-782.

Burda, S., Oleszek, W., Lee, C.Y. (1990). Phenolic compounds and their changes in apples during maturation and cold storage. *J. Agri. Food Chem.* 38, 945–948.

Cao, G., Sofic, E., Prior, R.L. (1996). Antioxidant capacity of tea and common vegetables. *J. Agric. Food Chem.* 44, 3426–3431.

Chafer, A., Berna, A., Monton, J.B., Munoz, R. (2002). High-pressure solubility data of system ethanol (1) + epicatechin (2) + CO2 (3), *J. Supercrit. Fluids.* 24, 103–109.

Chafer, A., Fornari, T., Berna, A., Stateva, R.P. (2004). Solubility of quercetin in supercritical CO2+ ethanol as a modifier: measurements and thermodynamic modeling. *J Supercrit. Fluids.* 32, 89–96.

Chandrasekar, V., Martín-González, M.F.S., Hirst, P., Ballard, T.S. (2015). Optimizing microwave-assisted extraction of phenolic antioxidants from red delicious and Jonathan apple pomace. *J. Food Proc. Eng.* 38, 571–582.

Choi, E.S., Noh, M.J., Yoo, K.P. (1998). Solubilities of o-, m- and p-coumeric acidisomers in carbon dioxide at 308.15–323.15 K and 8.5–25 MPa. *J. Chem. Eng. Data.* 43, 6–8

Chouchouli, V., Kalogeropoulos, N., Konteles, S.J., Karvela, E., Makris, D.P., Karathanos, V.T. (2013). Fortification of yoghurts with grape (*Vitis vinifera*) seed extracts. *LWT – Food Sci. Technol.* 53, 522–9.

Claudia, A., Graciela, E., Rosana, F. (2008). Total Polyphenol Content and Antioxidant Capacity of Commercially Available Tea (*Camellia sinensis*) in Argentina. *J. Agric. Food Chem.* 56, 9225-9229.

Daniel, E.M., Krupnick, A.S., Heur, Y-H., Blinzler, J.A., Nims, R.W., Stoner, G.D. (1989). Extraction, stability, and quantitation of ellagic acid in various fruits and nuts. *J Food Compos. Anal.* 2, 338-349.

de Camargo, A.C., Regitano-d'Arce, M.A.B., Biasoto, A.C.T., Shahidi, F. (2014). Low molecular weight phenolics of grape juice and winemaking by-products: Antioxidant activities and inhibition of oxidation of human low-density lipoprotein cholesterol and DNA strand breakage. *J. Agri. Food Chem.* 62, 12159–12171.

Deng, G.F., Shen, C., Xu, X.R., Kuang, R.D., Guo, Y.J., Zeng, L.S., Gao, L.L., Xi, Lin., Jie-Feng, X., Xia, E.Q., Li, S., Wu, S., Chen, F., Ling, W.H., Li, H.B. (2012). Potential of Fruit Wastes as Natural Resources of Bioactive Compounds. *Int. J. Mol. Sci.* 13, 8308-8323

Directive 2008/98/EC of the European Parliament and of the Council of 19 November 2008 on waste and repealing certain Directives (Text with EEA relevance) (2008). [http://eur-lex.europa.eu/legal-content/HR/TXT/PDF/?uri=CELEX:32008L0098&from=EN; Accessed: 2018-01-20]

Djilas, S., Čanadanović-Brunet, J., Ćetković, G. (2009). By-products of fruits processing as a source of phytochemicals. *Chem. Ind. & Chem. Engineer.* 15(4), 191−202.

Drosou, C., Kyriakopoulou, K., Bimpilas, A., Tsimogiannis, D., Krokida, M. (2015). A comparative study on different extraction techniques to recover red grape pomace polyphenols from vinification byproducts. *Ind. Crops Prod.* 75, 141–149.

Du, Q., Jerz, G., Winterhalter, P. (2004). Isolation of two anthocyanin sambubiosides from bilberry (Vaccinium myrtillus) by high-speed counter-current chromatography. *J. Chromatography. A.* 1045:59–63.

EU actions against food waste (2018). [Available from: http://ec.europa.eu/food/safety/food_waste/eu_actions/index_en.htm; Accessed: 2018- 01- 20]

European Commission (2014) Director-General for Communication. *The European Union Explained: Agriculture.* Luxembourg: Publications Office of the European Union.

Felix da Silva, D., Matumoto-Pintro, P.T., Bazinet, L., Couillard, C., Britten, M. (2015). Effect of commercial grape extracts on the cheese-making properties of milk. *J. Dairy Sci.* 98, 1552–62.

Frumento, D., Santo, A.P.E., Aliakbarian, B., Casazza, A.A., Gallo, M., Converti, A., Perego, P. (2013). Development of milk fermented with *Lactobacillus acidophilus* fortified with *Vitis vinifera* marc flour. *Food Technol. Biotechnol.* 51, 370–5.

Fu, L., Xu, B.T., Xu, X.R., Qin, X.S., Gan, R.Y., Li, H.B. (2010). Antioxidant Capacities and Total Phenolic Contents of 56 Wild Fruits from South China. *Molecules.* 15, 8602-8617.

Fu, L., Xu, B.T., Xu, X.R., Qin, X.S., Gan, R.Y., Li, H.B. (2011). Antioxidant capacities and total phenolic contents of 62 fruits. *Food Chem.* 2129(2), 345-350.

Gao, L., Mazza, G. (1994). Quantitation and distribution of simple and acylated anthocyanins and other phenolics in blueberries. *J. Food Sci.* 59, 1057-1059.

García, Y.D., Valles, B.S., Lobo, A.P. (2009). Phenolic and antioxidant composition of by-products from the cider industry: Apple pomace. *Food Chem.* 117, 731–738.

Ghafoor, K., Choi, Y.H., Jeon, J.Y., Jo, I.H. (2009). Optimization of ultrasound-assisted extraction of phenolic compounds, antioxidants, and anthocyanins from grape (*Vitis vinifera*) seeds. *J. Agric. Food Chem.* 57(11), 4988–94.

Ghafoora, K., Park, J., Choi, Y.H. (2010). Optimization of supercritical fluid extraction of bioactive compounds from grape (*Vitis labrusca* B.) peel by using response surface methodology. *Innov. Food Sci. Emerg.* 11, 485–490.

Gonzalez-San Jose, M.L., Garcıa-Lomillo, J., Del Pino-Garcıa, R., Rivero-Perez, D., Ortega-Heras, M., Muniz, P. (2014). *Seasoning products from wine pomace with interesting preservative and healthful properties.* Amsterdam, Holland: World Bulk Wine Exhibition SLU.

Gregoire, S., Singh, A.P., Vorsa, N., Koo, H. (2007). Influence of cranberry phenolics on glucan synthesis by glucosyltransferases and Streptococcus mutans acidogenicity. *J. Applied Microbiol.* 103, 1960–1968.

Guyot, S., Marnet, N., Sanoner, P., Drilleau, J.F. (2003). Variability of the polyphenolic composition of cider apple (*Malus domestica*) fruits and juices. *J. Agric. Food Chem.* 51, 6240-6247.

Hakkinen, S., Mykkanen, H., Karenlampi, S., Heinonen, M., Torronen, R. (1996) HPLC method for screening of flavonoids and phenolic acids in berries: phenolic profiles of strawberry and black currant. In Kumpulainen J. T. & Salonen J. T. (Eds.). Natural Antioxidants and Food Quality in Atherosclerosis and Cancer Prevention (pp. 298-300). Cambridge: Royal Society of Chemistry.Häkkinen, S.H., Heinonen, M., Kärenlampi, S., Mykkänen, H., Ruuskanen, J., Törrönen, R., 1999. Screening of selected flavonoids and phenolic acids in 19 berries. *Food Resear. Int.* 32, 345–353.

Han, J., Britten, M., St-Gelais, D., Champagne, C.P., Fustier, P., Salmieri, S., Lacroix, M. (2011). Polyphenolic compounds as functional ingredients in cheese. *Food Chem.* 124, 1589–94.

Heinonen, I.M., Meyer, A.S., Frankel, E.N. (1998). Antioxidant activity of berry phenolics on human low-density lipoprotein and liposome oxidation. *J. Agric. Food Chem.* 46, 4107–4112.

Herrmann, K. (1976). Flavonols and flavones in food plants: a review, *J. Food Technol.* 11, 433-448.

Hertog, M.G.L., Hollman, P.C.H., Katan, M.B. (1992). Content of potentially anticarcinogenic favonoids of 28 vegetables and 9 fruits commonly consumed in The Netherlands. *J. Agri. Food Chem.* 40, 2379-2383.

HLPE (2014). *Food losses and waste in the context of sustainable food systems.* A report by the High Level Panel of Experts on Food Security and Nutrition of the Committee onWorld Food Security. Rome, Italy. [Available from: http://www.fao.org/3/a-i3901e.pdf; Accessed: 2018- 01- 20]

Hoye, C., Ross, C.F. (2011). Total phenolic content, consumer acceptance, and instrumental analysis of bread made with grape seed flour. *J Food Sci.* 76, S428–S36.

Huang, D.J., Ou, B.X., Prior, R.L. (2015). The chemistry behind antioxidant capacity assays. *J. Agric. Food Chem.* 53, 1841–1856.

Huber, G.M., Rupasinghe, H.P.V. (2009). Phenolic profiles and antioxidant properties of apple skin extracts. *J. Food Sci.* C693–C700.

Ibrahim, U.K., Kamarrudin, N., Suzihaque, M.U.H., Abd Hashib, S. (2017). Local Fruit Wastes as a Potential Source of Natural Antioxidant: An Overview. In *Proceedings IOP Conference Series: Materials Science and Engineering* 206 (p. 1-9).

James, J.B., Nagramsak, T. (2010). *Processing of fresh-cut tropical fruits and vegetables: a Technical Guide.* Bangkok: Food and Agriculture Organization of the United Nations Regional Office for Asia and the Pacific.

Jamison, J.R. (2003). Bilberry fruit (*Vaccinium myrtillus*). In Jamison JR (Ed.). *Clinical guide to nutrition and dietary supplements in disease management.* (p. 449–451) London, UK: Churchill Livingstone.

Joana Gil-Chávez, G., Villa, J.A., Fernando Ayala-Zavala, J., Basilio Heredia, J., Sepulveda, D., Yahia, E.M., González-Aguilar, G.A. (2013). Technologies for extraction and production of bioactive compounds to be used as nutraceuticals and food ingredients: an overview. *Compr. Rev. Food Sci. Food Saf.* 12, 5–23.

Jung, J., Cavender, G., Zhao, Y. (2015). Impingement drying for preparing dried apple pomace flour and its fortification in bakery and meat products. *J. Food Sci. Technol.* 52, 5568-5578.

Kammerer, D., Claus, A., Carle, R., Schieber, A. (2004). Polyphenol screening of pomace from red and white grape varieties (*Vitis vinifera* L.) by HPLC-DAD-MS/MS. *J. Agric. Food Chem.* 52, 4360–7.

Kara, C., Doymaz, I. (2015). Effective moisture diffusivity determination and mathematical modelling of drying curves of apple pomace. *Heat Mass Transf.* 51, 983-989.

Karaaslan, M., Ozden, M., Vardin, H., Turkoglu, H. (2011). Phenolic fortification of yogurt using grape and callus extracts. *LWT – Food Sci Technol.* 44, 1065–72.

Kiokias, S., Varzakas, T., Oreopoulou, V. (2008). In vitro activity of vitamins, flavanoids, and natural phenolic antioxidants against the oxidative deterioration of oil-based systems. *Crit. Rev. Food Sci. Nutr.* 48, 78–93.

Krawitzky, M., Arias, E., Peiro, J.M., Negueruela, A.I., Val, J., Oria, R. (2014). Determination of color, antioxidant activity, and phenolic profile of different fruit tissue of Spanish 'Verde Doncella' apple cultivar. *Int. J. Food Prop.* 17, 2298–2311.

Krishan, K., Yadav, A.N., Vinod, K., Pritesh, V., Harcharan, S.D. (2017). Food waste: a potential bioresource for extraction of nutraceuticals and bioactive compounds. *Bioresour. Bioprocess.* 4, 18.

Lavelli, V., Corti, S. (2011). Phloridzin and other phytochemicals in apple pomace: stability evaluation upon dehydration and storage of dried product. *Food Chem.* 129: 1578-1583.

Lavelli, V., Sri Harsha, P.S.C., Torri, L., Zeppa, G. (2014). Use of grapemaking by-products as an ingredient for tomato puree: the effect of particle size on product quality. *Food Chem.* 152, 162–8.

Lee, J., Wrolstad, R.E. (2004). Extraction of anthocyanins and polyphenolics from blueberry processing waste. *J. Food Sci.* 69, 564–573.

Li, Y., Zhang, J.J., Xu, D.P., Zhou, T., Zhou, Y., Li, S., Li, H.B. (2016). Bioactivities and Health Benefits of Wild Fruits. *Int. J. Mol. Sci.* 17(8):1258

Lin, J., Liang, Y. (2000). Cancer chemoprevention by tea polyphenols. *Proc. Natl. Sci. Counc. ROC(B).* 24(1), 1–13.

Liza, E., Díaz-Reinoso, L.B., Moure, A., Elvira, M., Domínguez, Z.H. (2010). Extraction of antioxidants from several berries pressing wastes using conventional and supercritical solvents. *Eur. Food Res. Technol.* 231, 669–677.

Lopez-Alarcon, C., Denicola, A. (2013). Evaluating the antioxidant capacity of natural products: A review on chemical and cellular-based assays. *Anal. Chim. Acta.* 763, 1–10.

Lopez-Serrano, M., Barcelo, A.R. (1996). Purification and characterization of a basic peroxidase isoenzyme from strawberries. *Food Chem.* 55, 133-137.

Louli, V., Ragoussis, N., Magoulas, K. (2004). Recovery of phenolic antioxidants form wine industry by-products. *Bioresour. Technol.* 92, 201–208.

Ma, Y.Q., Ye, X.Q., Fang, Z.X., Chen, J.C., Xu, G.H., Liu, D.H. (2008). Phenolic compounds and antioxidant activity of extracts from ultrasonic treatment of satsuma mandarin (*Citrus unshiu Marc.*) peels. *J Agric Food Chem.* 56(14), 5682–90.

Madhavi, D.L., Bomser, J., Smith, M.A.L., Singletary, K. (1998). Isolation of bioactive constituents from *Vaccinium myrtillus* (bilberry) fruits and cell cultures. *Plant Sci.* 131, 95–103.

Manach, C., Williamson, G., Morand, C., Scalbert, A., Rémésy, C. (2005). Bioavailability and bio-efficacy of polyphenols in humans. I. Review of 97 bioavailability studies. *Am. J. Clin. Nutr.* 3(1), 230-242.

Marchiani, R., Bertolino, M., Belviso, S., Giordano, M., Ghirardello, D., Torri, L., Piochi, M., Zeppa, G. (2016). Yogurt enrichment with grape pomace: effect of grape cultivar on physicochemical, microbiological and sensory properties. *J Food Qual.* 39, 77–89

Marchiani, R., Bertolino, M., Ghirardello, D., McSweeney, P.L.H., Zeppa, G. (2015). Physicochemical and nutritional qualities of grape pomace powder-fortified semi-hard cheeses. *J Food Sci Technol.* 53, 1–12.

Marinelli, V., Padalino, L., Nardiello, D., Del Nobile, M.A., Conte, A. (2015). New approach to enrich pasta with polyphenols from grape marc. *J Chem.* Article ID 734578, 1–8

Martinelli, E.M., Baj, A., Bombardelli, E. (1986). Computer-aided evaluation of liquid-chromatographic profiles for anthocyanins in Vaccinium myrtillus fruits. *Anal. Chim. Acta.* 191, 275–281.

Mayer-Miebach, E., Adamiuk, M., Behsnilian, D. (2012). Stability of Chokeberry bioactive polyphenols during juice processing and stabilization of a polyphenol-rich material from the by-product. *Agriculture.* 2, 244–258.

Meral, R., Dŏgan, I.S. (2013). Grape seed as a functional food ingredient in bread-making. Intl *J Food Sci Nutr.* 64, 372–9.

Mildner-Szkudlarz, S., Bajerska, J. (2013). Protective effect of grape by-product-fortified breads against cholesterol/cholic acid diet-induced hypercholesterolaemia in rats. *J Sci Food Agric.* 93,3271–8.

Mildner-Szkudlarz, S., Siger, A., Szwengiel, A., Bajerska, J. (2015). Natural compounds from grape by-products enhance nutritive value and reduce formation of CML in model muffins. *Food Chem.* 172, 78–85.

Mildner-Szkudlarz, S., Zawirska-Wojtasiak, R., Szwengiel, A., Pacy´nski, M. (2011). Use of grape by-product as a source of dietary fibre and phenolic compounds in sourdough mixed rye bread. *Intl J Food Sci Technol.* 46, 1485–93.

Mosel. H-D, Herrmann, K. (1974). The phenolics of fruits. IV. The phenolics of blackberries and raspberries and their changes during development and ripeness of the fruits. *Z Lebensm Unters Forsch* 154, 324-327.

Murga, R., Ruiz, R., Beltŕan, S., Cabezas, J.L. (2000). Extraction of natural complex phenols and tannins from grape seeds by using supercritical mixtures of carbon dioxide and alcohol. *J. Agric. Food Chem.* 48,3408–3412.

Murga, R., Sanz, M.T., Beltran, S., Cabezas, J.L. (2003). Solubility of three hydroxycinnamic acids in supercritical carbon dioxide. *J Supercrit. Fluids.* 27, 239–245

Nahas (2012) Natural antioxidants as food and beverage ingredients. Chapter 5. In: Baines D. & Seal R. (Eds). *Natural food additives, ingredients and flavourings.* (pp. 100–126) Cambridge, UK: Woodhead Publishing.

Ness, A.R., Powles, J.W. (1997). Fruit and vegetables, and cardiovascular disease: a review. *Int. J. Epidemiol.* 26, 1-13.

Palma, M., Taylor, L.T. (1999). Extraction of polyphenolic compounds from grape seeds with near critical carbon dioxide. *J. Chromotogr. A.* 849, 117–124.

Palomino, O., Gomez-Serranillos, M.P., Slowing, K., Carretero, E., Villar, A. (2008). Study of polyphenols in grape berries by reversed-phase high-performance liquid chromatography. *J. Chromatography A.* 870, 449–451.

Peralbo-Molina, T., Luque deCastro, M.D. (2013). Potential of residues from the Mediterranean agriculture and agrifood industry. *Trends Food Sci. Technol.* 32, 16–24.

Pereira, C.S.M., Silva, V.M.T.M., Rodrigues, A.E. (2011). Ethyl lactate as a solvent: Properties, applications and production processes – a review. *Green Chem.* 13, 2658–2671.

Perez-Magarino, S., Gonzalez-SanJosé, M.L. (2006). Polyphenols and colour variability of red wines made from grapes harvested at different ripeness grade. *Food Chem.* 96, 197–208.

Perez-Magarino, S., Ortega-Heras, M., Cano-Mozo, E,, Gonzalez-SanJose, M.L. (2009). The influence of oak wood chips, micro-oxygenation treatment, and grape variety on colour, and anthocyanin and phenolic composition of red wines. *J. Food Compo. Anal.* 22, 204–211.

Perez-Magarino, S., Revilla, I., Izcara, E., Gonzalez-SanJose, M.L. (1999). Discrimination of DO wine vintages as a function of chemical composition. *Quim Anal.* 18, 133–164.

Prior, R.L., Cao, G., Martin, A., Sofic, E., McEwen, J., O'Brien, C., Lischner, N., Ehlenfeldt, M., Kalt, W., Krewer, G., Mainland, C.M. (1998). Antioxidant capacity as influenced by total phenolic and anthocyanin content, maturity, and variety of Vaccinium species. *J. Agric.Food Chem.* 46, 2686–2693.

Prior, R.L., Wu, X.L., Schaich, K. (2005). Standardized methods for the determination of antioxidant capacity and phenolics in foods and dietary supplements. *J. Agric. Food Chem.* 53, 4290–4302.

Rajbhar, K., Dawda, H., Mukundan, U. (2015). Polyphenols: methods of extraction. *Sci. Revs. Chem. Commun.* 5(1), 1-6.

Ramirez-Ambrosi, M., Abad-Garcia, B., Viloria-Bernal, M., Garmon-Lobato, S., Berrueta, L.A., Gallo, B. (2013). A new ultrahigh performance liquid chromatography with diode array detection coupled to electrospray ionization and quadrupole time-of-flight mass spectrometry analytical strategy for fast analysis and improved characterization of phenolic compounds in apple products. *J. Chromatogr A.* 1316, 78-91.

Reque, P.M., Steffens, R.S., Silva, A.M.D., Jablonski, A., Hickman Flôres, S., de Oliveira Rios, A., Voght de Jong, E. (2014). Characterization of blueberry fruits (*Vaccinium* spp.) and derived products. *Food Scechnol.* 34, 773–779.

Revilla, I., Gonzalez-SanJose, M.L. (2002). Multivariate evaluation of changes induced in red wine characteristics by the use of extracting agents. *J. Agric. Food Chem.* 50, 4525–30.

Ribeiro, B., Cardoso, C., Silva, H.A., Serrano, C., Ramos, C., Santos, P.C., Mendes, R. (2013). Effect of grape dietary fibre on the storage stability of innovative functional seafood products made from farmed meagre (*Argyrosomus regius*). *Intl J Food Sci Technol.* 48, 10–21.

Rice-Evans, C.A., Miller, N.J., Bolwell, P.G., Bramley, P.M., Pridham, J.B. (1995). The relative antioxidant activities of plant-derived polyphenolic flavonoids. *Free Radic. Res.* 22, 375–383.

Rommel, A., Wrolstad, R.E. (1993). Composition of flavonols in red raspberry juice as influenced by cultivar, processing and environmental factors. *J. Agri. Food Chem.* 41, 1941-1950.

Rosales Soto, M.U., Brown, K., Ross, C.F. (2012). Antioxidant activity and consumer acceptance of grape seed flour-containing food products. *Intl J Food Sci Technol.* 47, 592–602.

Sagdic, O., Ozturk, I., Cankurt, H., Tornuk, F. (2012). Interaction between some phenolic compounds and probiotic bacterium in functional ice cream production. *Food Bioprocess Tech.* 5, 2964–71.

Sanchez-Rabaneda, F., Jauregui, O., Lamuela-Raventos, R.M., Viladomat, F., Bastida, J., Codina, C. (2004). Qualitative analysis of phenolic compounds in apple pomace using liquid chromatography coupled to mass spectrometry in tandem mode. *Rapid Commun. Mass Spectrom.* 18, 553-563.

Sato, M.F., Rigoni, D.C., Canteri, M.H.G., Petkowicz, C.Ld.O., Nogueira, A., Wosiacki, G. (2011). Chemical and instrumental characterization of pectin from dried pomace of eleven apple cultivars. *Acta Sci. Agron.* 33, 383-389.

Schieber, A., Hilt, P., Streker, P., Endress, H.U. Rentschler, C., Carle, R. (2003). A new process for the combined recovery of pectin and phenolic compounds from apple pomace. *Innovat. Food Sci. Emerg. Tech.* 4, 99-107.

Schieber, A., Keller, P., Carle, R. (2001). Determination of phenolic acids and flavonoids of apple and pear by high-performance liquid chromatography. *J. Chromatogr. A.* 910, 265–273.

Shahidi, F., Ho, C.T. (2007). Antioxidant measurement and applications: An overview. In Shahidi F, Ho CT, (Eds.). *Antioxidant measurement and applications. ACS Symposium series* 956. (pp 2-7). Washington, DC: American Chemical Society.

Shalini, R., Gupta, D.K. (2010). Utilization of pomace from apple processing industries: a review. *J. Food Sci. Technol.* 47(4), 365-371.

Shalini, R. (2010). *Apple pomace drying: drying characteristics in apple pomace. By-Products Utilization*. Saarbrücken: LAP Lambert Academic Publishing

Silva, M.L., Macedo, A.C., Malcata, F.X. (2000). Review: Steam distilled spirits from fermented grape pomace. *Food Sci. Technol. Int.* 6, 285–300.

Simić, V.M., Rajković, K.M., Stojičević, S.S., Veličković, D.T., Nikolić, N.Č., Lazić, M.L., Karabegović, I.T. (2016). Optimization of microwave-assisted extraction of total polyphenolic compounds from chokeberries by response surface methodology and artificial neural network. *Sep Purif Technol.* 160,89-97.

Singh, A.P.,Wilson, T., Kalk, A.J., Cheong, J., Vorsa, N. (2009). Isolation of specific cranberry flavonoids for biological activity assessment. *Food Chem.* 11: 963–968.

Siriwoharn, T., Wrolstad, R.E., Finn, C.E., Pereira, C.B. (2004). Influence of cultivar, maturity, and sampling on blackberry (*Rubus* L. hybrids) anthocyanins, polyphenolics, and antioxidant properties. *J. Agric. Food Chem.* 52, 8021–8030.

Skrede, G., Wrolstad, R.E., Durst, R.W. (2000). Changes in anthocyanins and polyphenolics during juice processing of highbush blueberries (*Vaccinium corymbosum* L.). *J. Food Sci.* 65, 357–364.

Sojka, M., Kołodziejczyk, K., Milala, J. (2013). Polyphenolic and basic chemical composition of black chokeberry industrial by-products. *Ind. Crops Prod.B.* 51, 77–86

Sojka, M., Krol, B., 2008. Composition of industrial seedless black currant pomace. *Eur. Food Researc. Techn.* 228, 597–605.

Starke, H., Herrmann, K. (1976). The phenolics of fruits. VIII. Changes in favonol concentrations during fruit development. *Z Lebensm Unters Forsch.* 161,131-135

Steinmetz, K.A., Potter, J.D. (1996). Vegetables, fruit, and cancer prevention: A review. *J. Am. Diet. Assoc.* 96(10), 1027-39.

Stöhr, H., Herrmann, K. (1975b). The phenolics of fruits. V. The phenolics of strawberries and their changes during development and ripeness of the fruits. *Z Lebensm Unters Forsch.* 159, 341-348.

Stöhr, H., Herrmann, K. (1975a). The phenolics of fruits. VI. The phenolics of currants, gooseberries and blueberries. Changes inphenolic acids and catechins during development of black currants. *Z Lebensm Unters Forsch.* 159, 31-37.

Teixeira, A., Baenas, N., Dominguez-Perles, R., Barros, A., Rosa, E., Moreno, D.A., Garcia-Viguera, C. (2014). Natural bioactive compounds from winery by-products as health promoters: a review. *Intl. J. Mol. Sci.* 15, 15638–78.

Transparency Market Research (2016) *Polyphenols market by product (grape seed, green tea, apple and others), by application (functional beverages, functional food, dietary supplements and others)–global industry analysis, size, share, growth, trends and forecast (2012-18).* [Available from: http://www.transparencymarketresearch.com/ poly phenol-market.html. Accessed 2018 January 09]

Tsao, R., Yang, R., Sockovie, E., Khanizadeh, S. (2005). Which polyphenolic compounds contribute to the total antioxidant activities of apple? *J. Agri. Food Chem.* 53, 4989–4995

Tsao, R. (2010). Chemistry and biochemistry of dietary polyphenols. *Nutrients.* 2:1231-1246.

Verghese, K., Lewis, H., Lockrey, S., Williams, H. (2013). *The role of packaging in minimising food waste in the supply chain of the future.* Final report. Melbourne, Australia: CHEP Australia.

Vinson, J.A., Bose, P., Proch, J., Al Kharrat, H., Samman, N. (2008). Cranberries and cranberry products: Powerful in vitro, ex vivo, and in vivo sources of antioxidants. *J. Agric. Food Chem.* 56, 5884–5891.

Vinson, J.A., Su, X., Zubik, L., Bose, P. (2001). Phenol antioxidant quantity and quality in foods: Fruits. *J. Agric. Food Chem.* 49, 5315–5321.

Virot, M., Tomao, V., Le Bourvellec, C., Renard, C.M.C.G., Chemat, F. (2010).Towards the industrial production of antioxidants from food processing byproducts with ultrasound-assisted extraction. *Ultrason Sonochem.* 17, 1066–1074.

Vrhovsek, U., Rigo, A., Tonon, D., Mattivi, F. (2004). Quantitation of polyphenols in different apple varieties. *J. Agric. Food Chem.* 52, 6532-6538.

Waldbauer, K., McKinnon, R., Kopp, B. (2017). Apple Pomace as Potential Source of Natural Active Compounds. *Planta Med.* 83(12/13), 994-1010.

Wang, H., Cao, G., Prior, R.L. (1997). Oxygen radical absorbing capacity of anthocyanins. *J. Agri. Food Chem.* 45, 304–309.

Wang, H., Cao, G., Prior, R.L. (1996). Total antioxidant capacity of fruits. *J. Agric. Food Chem.* 44, 701–705.

Wan-Ju, Y., Shih-Min, H., Wei-Hwa, L.,Chi-Hao, W. (2017). Polyphenols with antiglycation activity and mechanisms of action: A review of recent findings. *J. Food Drug Analy.* 25(1), 84-92.

White, B.L., Howard, L.R., Prior, R.L. (2011). Impact of different stages of juice processing on the anthocyanin, flavonol, and procyanidin contents of cranberries. *J. Agric. Food Chem.* 59, 4692–4698.

Wildanger, W., Herrmann, K. (1973). The phenolics of fruits. II. The favonols of fruits. *Z. Lebensm. Unters. Forsch.* 151:103-108.

Wissam, Z., Ghada, B., Wassim, A., Warid, A. (2012). "Effective Extraction of Polyphenols and Proanthocyanidins from Pomegranate's Peel. *Int. J. Pharm. Pharmaceut. Sci.* 4(3), 675-682.

Withy, L.M., Nguyen, T.T., Wrolstad, R.E., Heatherbell, D.A. (1993). Storage changes in anthocyanin content of red raspberry juice concentrate. *J. Food Sci.* 58, 190-192

Wolfe, K., Wu, X., Liu, R.H. (2003). Antioxidant activity of apple peels. *J. Agri. Food Chem.* 51, 609–614.

Xu, D.P., Li, Y., Meng, X., Zhou, T., Zhou, Y., Zheng, J., Zhang, J.J., Li, H.B. (2017). Natural Antioxidants in Foods and Medicinal Plants: Extraction, Assessment and Resources. *Int. J. Mol. Sci.* 18(1), 96.

Yan, X., Murphy, B.T., Hammond, G.B., Vinson, J.A., Neto, C.C. (2002). Antioxidant activities and antitumor screening of extracts from cranberry fruit (*Vaccinium macrocarpon*). *J. Agric. Food Chem.* 50, 5844–5849.

Yang, J., Martinson, T.E., Liu, R.H. (2009). Phytochemical profiles and antioxidant activities of wine grapes. *Food Chem.* 116, 332–339.

Zhang, H., Tsao, R. (2016). Dietary polyphenols, oxidative stress and antioxidant and anti-inflammatory effects. *Curr. Opin. Food Sci.* 8, 33-42.

Zuo, Y., Wang, C., Zhan, J. (2002). Separation, characterization, and quantitation of benzoic and phenolic antioxidants in American cranberry fruit by GC−MS. *J. Agric. Food Chem.* 50, 3789–3794

In: Recent Advances in Plant Research
Editors: B. Welch and M. Wilkerson
ISBN: 978-1-53614-170-2

Chapter 4

OCCURRENCE OF GRAY MOLD ROT DISEASE ON VEGETABLE CROPS CAUSED BY *BOTRYTIS CINEREA* IN MALAYSIA

***A. Tijjani*[1,3,*] *and A. Khairulmazmi*[1,2]**
[1]Department of Plant Protection, Faculty of Agriculture, Universiti Putra Malaysia, UPM Serdang, Selangor Darul Ehsan, Malaysia
[2]Institute of Plantation Studies, Universiti Putra Malaysia, UPM Serdang, Selangor Darul Ehsan, Malaysia
[3]Department of Crop Production, Faculty of Agriculture and Agricultural Technology, Abubakar Tafawa Balewa University Bauchi, Bauchi State, Nigeria

ABSTRACT

The gray mold rot is caused by *Botrytis cinerea* Persoon: Fries [teleomorph *Botryotinia fuckeliana* (de Bary) Whetzel]. It is a serious disease causing severe damages on fruits and vegetables at both pre and post-harvest periods. *Botrytis cinerea* is an ascomycetous fungus as well as a necrotrophic plant pathogen that was ranked second in the top 10 most devastating fungal plant pathogens worldwide. Due to its broad host range, the fungus infects more than 200 plant species of agricultural importance such as cucumbers, strawberries, grapes, tomatoes, cut flowers and ornamental plants. The fungus occurs mostly where its host plants are cultivated, under climatic conditions ranging from cool temperate zones to subtropical regions. The fungus is highly concentrated in the temperate areas between latitudes 25 and 30. In Malaysia, the occurrence of this disease has been widespread in the greenhouses of the commercial farms in the Cameron Highland area of Pahang, Malaysia. This hindered the growing industries, hence detection, and characterization of the pathogen using molecular approaches provide an essential tool for establishing disease management programs.

* Corresponding Author Email: tijjaniahmadu72@gmail.com.

Since the pathogen is highly destructive, in this book chapter we will try to highlight the disease distribution and hosts range, pathogenesis of the pathogen (*B. cinerea*), symptoms of the disease, disease cycle and epidemiology of the pathogen (*B. cinerea*) as well as the current status of the disease (gray mold rot) in Malaysia.

Keywords: occurrence, *Botrytis cinerea*, gray mold disease, molecular identification

1. Introduction

The gray mold disease caused by *Botrytis cinerea* Persoon: Fries [teleomorph *Botryotinia fuckeliana* (de Bary) Whetzel], is a serious disease of tomato (*Solanum lycopersicum*) in Malaysia and in most areas of the world. The fungus causes severe damages on fruits and vegetables at both pre- and post-harvest periods (Aktaruzzaman et al., 2014; González-Fernandez et al., 2014; Vercesi et al., 2014; Kim et al., 2016). *Botrytis cinerea* is an ascomycetous fungus as well as a necrotrophic plant pathogen (Staats et al., 2005), ranked second after *Magnaporthe oryzae* in the top 10 devastating fungal plant pathogens worldwide due to its broad host range (Dean et al., 2012; Elad et al., 2015; Shaw et al., 2016). The fungus is capable of infecting more than 200 plant species, including important agricultural crops, and harvested commodities such as grapes,
cucumbers, tomatoes, strawberries, cut flowers, bulb flowers, and ornamental plants (Kars et al., 2005; Parvu et al., 2010). The wide host range of the fungus leads to great losses in quantity and quality in the field, at harvest, after harvest, on transit and at storage (Elad, 2003). The fungus kills susceptible host cell via the secretion of lytic enzyme or toxic compound. It also produces an array of virulence factors that can destroy or conquer the defence mechanisms of the host (Makovitzki, 2007; Parvu et al., 2010; Williamson et al., 2007). The disease results in reduced plant growth, premature leaves defoliation and fruit rot, and consequently reduces yield. The disease can be destructive enough to cause extensive yield losses with subsequent significant economic risks to the strawberry and tomato growers if not properly and effectively managed.

2. Background of *Botrytis Cinerea*

The genus *Botrytis* was first coined by Micheli in 1929. It is a group of fungi causing important diseases to crop plants (Pande et al., 2001). The genus consists of one generalist, an opportunistic pathogen known as *B. cinerea*, usually infecting senescence, wounded and weakened plants (Elad, 1989; Elad, 1997; Shaw et al., 2016; Elad et al., 2016). All other species infect bulb plants and are taken into consideration as host range with narrow specialists except *B. fabae*, which infects *Vicia*, *Lens*, *Pisum and Phaseolus* species all belonging to the family *Fabaceae* (Jarvis 1977; Staats et al., 2005). Among the species in the genus, *B. cinerea* is the most destructive (Dean et al., 2012; Herzog et al., 2015; Shaw et al., 2016).

Botrytis cinerea is a specie name copied from the Latin word for "grape like ashes"; "grapes" meaning "the bunch-like structure of the conidia on conidiophores", and "ashes" meaning "grayish appearance of the conidia masses" (Liddel et al., 1940). *Botrytis cinerea* is characterized by fast growing colonies, hyaline at first and later changed to gray with abundant aerial mycelial growth. The conidia were abundant hyaline, smooth one-celled, ellipsoidal, globose, ovate, flat at one end, and borne on branching tree-like conidiophores (Aktaruzzaman et al., 2014; Al-Mahmooli et al., 2015; Javed et al., 2017). The fungus also produces highly resistant survival structure called sclerotia which serve as a survival mechanism usually found in old cultures. The sclerotia were large, black, spherical or irregular in shapes. Both asexual and sexual forms were observed in *B. cinerea* with the anamorph referring to the asexual form. The sexual form comprises the spermatization of sclerotia with subsequent generation of apothecia. The apothecia of *B. fulkeliana* are the teleomorp phase of *B. cinerea* (Figure 1) (Williamson et al., 2007; Beever and Weeds, 2004).

3. World Disease Distribution and Host Range

Botrytis cinerea is a necrotrophic pathogen which triggers the host to induce plant cell death by host tissue colonisation (Liana et al., 2013; Williamson et al., 2007; Elad et al., 2016). The fungus occurred worldwide mostly where susceptible host plants are cultivated, under climatic conditions ranging from cool temperate zones to subtropical regions (Javis, 1977; Herzog et al., 2015). It is highly prevalent in the temperate areas between latitudes 25 and 30; infecting many plants species especially fruits and vegetables with substantial loses in quality and yield (Vercesi et al., 2014).

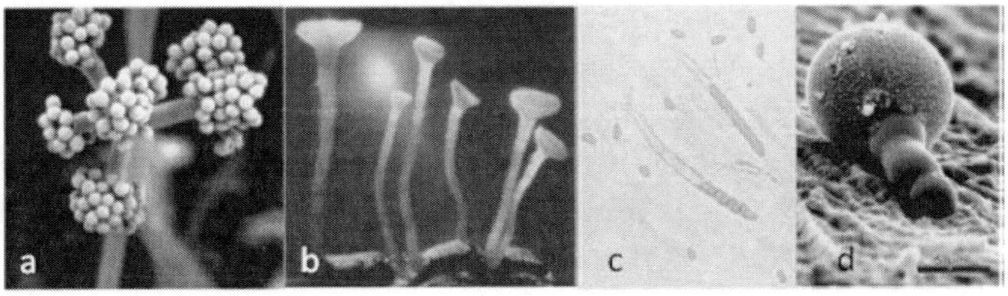

Figure 1. (a) *B. cinerea* conidia borne on conidiophore (b) Apothecia *Botrytinia fuckeliana* (c) Showing asci with each containing eight ascospores (d) Germinating conidium on the surface of rose petal. Source: Williamson et al. (2007).

Diseases caused by this group of fungi have been reported to have occurred in the field and greenhouses, in many countries globally, namely Europe (Samuel et al., 2012); Southern Europe (Albertini et al., 2002; Milicević et al., 2006; Rajaguru and Shaw, 2010); Eastern Europe (Vaczy et al., 2008; Samuel et al., 2012); Britain (Atwell et al., 2015); Germany (Herzog et al., 2015; Kretschmer and Hahn, 2008); France (Giraud et al., 1999; Fournier et al., 2005; Walker et al., 2011); Southern France (Bardin et al., 2014); Australia (Isenegger et al., 2008); Netherlands (Staats et al., 2005); Denmark (Corwin et al., 2016); Italy (Vercesi et al., 2014); southern Italy (De Miccolis- Angelini et al., 2014); New Zealand (Leroux et al., 2004; Johnstoon et al., 2014); northern Portugal (Fernández-González et al., 2012); Serbia (Tanovic et al., 2009); Canada (van Kan et al., 2010); Greece (Samuel et al., 2012); Colombia and Romania (Davidson et al., 2007); Spain (González et al., 2009; Gindro et al., 2014); Hungry (Fekete et al., 2012); America (Esterio et al., 2011); United States in the San Joaquin Valley of California (Smilanick et al., 2010); Chile (Esterio et al., 2011); Venezuela, Vietnam (Davidson et al., 2007); Argentina (Muñoz et al., 2010); southern Brazil (de Bem et al., 2015); China (Zhou et al., 2014; Xue et al., 2017); Central China (Li et al., 2015); korea (Aktaruzzaman et al., 2014; Aktaruzzaman et al., 2016); Japan (Saito and Takayanagi, 2009); India and Nepal (Kumari et al., 2014); Pakistan (Javed et al., 2017); Bangledash (Bakr et al., 1998) Tunusia (Karchani-Balma et al., 2008); Western Cape of South Africa (Wessels et al., 2013); Egypyt (Wahab, 2015); northern Algeria (Adjebli et al., 2015); Zimbabwe, Morocco, Swaziland, Malawi and Tanzania (Porter et al., 1984; Allen and Lenne, 1998; Davidson et al., 2007).

The phytopathogenic fungus, *B. cinerea* with necrotrophic lifestyle (Fekete et al., 2012), has been ranked as the second most destructive fungal pathogen (Dean et al., 2012). The fungus caused gray mold disease in more than 200 host plants species mostly dicots, worldwide (Williamson *et al.* 2007; Fekete et al., 2012; Kumari et al., 2014). Being a necrotrophic ascomycete, it displays the capacity to kill host cells by producing toxins, reactive oxygen species (ROS) and the induction of a plant produced oxidative burst (van Kan, 2003; van Kan, 2006; van Kan et al., 2010). Due to its wide host range and broad adaptability, there are several reports showing that the fungus is a serious pathogen of agricultural products such as fruits, vegetables, ornamental crops, orchard and field crops (Javis, 1977; Aguayo et al., 2011). For example strawberry and raspberry (Carisse et al., 2014); grape vine (Herzog et al., 2015); tomato (Decognet et al., 2009; Kuzmanovska et al., 2012); grape vine and bramble (Fournier and Giraud, 2008); Japanese plum (Ferrada et al., 2015); plum fruits (Aminifard and Mohammadi, 2013); vineyard grapes (Elmer and Reglinski, 2006; Jacometti et al., 2010); apple and kiwi fruit (Samuel et al., 2012); sweet basil (Yermiyahu et al., 2006); egg plants (Wu et al., 2008); cucumber, egg plants and pepper (Elad et al., 2016); table grapes (Romanazzi et al., 2012). Vegetables such as Lettuce (Leyronas et al., 2015), Legumes such as lentils (Davidson et al., 2007); chickpeas (Pande et al., 2002; Pande et al., 2010); pigeon pea, faba beans, common beans, peanuts (Allen and Lenne, 1998; Davidson et al., 2007). Important crops like *Chrysanthemum morifolium* (Hobbs and Waters, 1964), *Arabidopsis thaliana* (Windram et al., 2012; Windram et al., 2016) were also infected by *B. cinerea.* Other small fruits and vegetables host attacked by gray mold pathogen (*B. cinerea*) after harvest include asparagus, apple, artichoke, beet, broccoli, Brussels, sprouts, blackberry, black-eyed pea, cauliflower, celery, currant, carrot, chives, cucumber, endive, gooseberry, kohlrabi, kale, lemons, lettuce, onion, okra, pears, parsnip, pea, peaches, pumpkin, plums, quince, red chicory, rhubarb, rutabaga, sunflower, squash, turnip, shallot, sweet cherry and sweet potato (Droby and Lichter, 2007).

4. PATHOGENESIS OF *B. CINEREA*

Pathogenesis is a process that a pathogen undergoes to cause infection in the host (Mehrotra and Agarwal, 2003). The ability of the pathogen to cause infection (disease) referred to as pathogenicity. The success of this ability depends on the host-pathogen interaction, environmental variations, pathogen genetic diversity and virulence (Atwell et al., 2015; Corwin et al., 2016). Virulence is the ability of the pathogen to reproduces inside the host and the extent of infection it causes. *Botrytis cinerea* acquires virulence factors which help it to complete the successive cyclic events (stages) of its necrotrophic life style (Kumari et al., 2014; Li et al., 2012). Pathogenesis in the fungus was enhanced by the roles played by the virulence factors shown in Table 1. They include extracellular enzymes, as well as phytotoxic metabolites and proteins (Nakajima and Akutsu, 2014). The virulence factors can degrade plant material or overcome the host defence mechanism (Staats et al., 2007). In addition to the virulence factors, Elad (1989) reported that *B. cinerea* developed three forms of structures from dispersed conidia, namely: germ tube apices, infection cushions and appressoria, which were used in the host penetration.

The processes of infection exhibited by *B. cinerea* take place in distinctive and consecutive stages. They include pre-penetration (attachment to the host), penetration, formation of lesion and its enlargement, host tissue killing and sporulation (van Kan, 2006; Choquer et al., 2007). *Botrytis cinerea* as an opportunistic fungus enters the host through stomata, natural openings or attacks weakened, wounded and senescence tissues through wound (Elad and Evensen, 1995). As necrotrophic fungus, it has complex approaches to overcome plant defence mechanisms (van Kan, 2006), that are arbitrated by toxin, lytic enzymes, necrosis-secreted proteins, stress-persuaded reactive oxygen species (ROS), low-molecular-weight compounds like oxalic acid and a huge group of secondary metabolites (Choquer et al., 2007; Williamson et al., 2007; Leroch et al., 2013). Penetration through stomata was achieved via germ tube apices whereas direct

Table 1. *Botrytis cinerea* Virulence factors

Factors	Pathogenesis activity	References
Lipase	Host cell surface penetration	Commenil et al. 1999 Commenil et al. 1998
Cutinase		Van der Vlugt-Bergmans et al. 1997
Cerato-platanin family protein	Killing of host cells	Frias et al. 2011
Botrydial and related metabolites		Dalmais et al. 2011 Colmenares et al. 2002)
Xylanase		Noda et al. 2010
Nep1-like proteins		Schouten et al. 2008
NADPH oxidases		Segmuller et al. 2008
Botrydial derivatives		Reino et al. 2004
Cu–Zn-superoxide dismutase		Rolke et al. 2004
Botcinins E and F		Tani et al. 2006
Sesquiterpenoid metabolites		Dura´n-Patro´n et al. 2000
Endopolygalacturonase1		Have et al. 1998
Botcinolide		Cutler et al. 1993
BcPG2	Conversion of host tissue into fungal biomass	Kars et al. 2005
Pectin methylesterase		Valette-Collet et al. 2003
BcPGl		Have et al. 1998
Aspartic protease		Movahedi and Heale, 1990
BcBirl	Overcoming host defence responses	Shlezinger et al. 2011a, b
ATP-binding cassette transporters		Stefanato et al. 2009

Source: Nakajima and Akutsu (2014) with modifications.

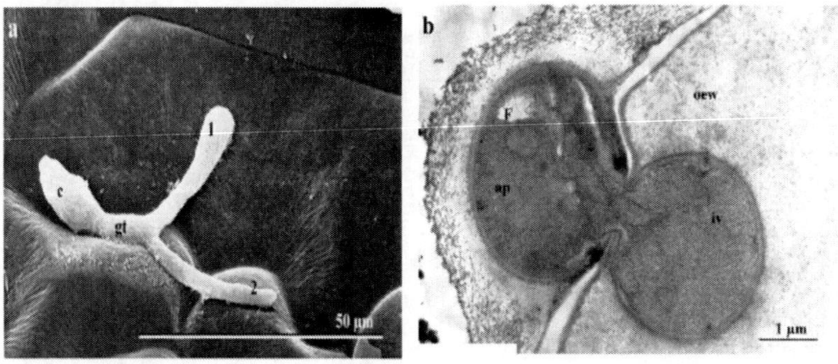

Source: Choquer et al. (2007).

Figure 2. (a) Penetration of an infected host via appressorium (b) Hypha depicted with or without swollen tips.

penetration is accomplished via appressoria (Elad et al., 2007). Infection may be latent, destructive, limited or widely developed (Shaw et al., 2016).

Infection starts with conidial attachment to the surface of the susceptible host (Leroch et al., 2013), initiated by the germlings secretion of an extracellular matrix that mediated the attachment by hydrophobicity interaction between the host and conidial surfaces (van Kan, 2003; Choquer et al., 2007). A crucial step next to the conidial attachment to the surface of the host is the germination of the conidia. This was followed by adhesion of the conidia to the plant surface, prior to the release of different extracellular enzymes. The enzymes mediated easier penetration by rupturing the structural barriers of the plant (Choquer et al., 2007). Next step is the penetration but *B. cinerea* does not uses mechanical pressure, instead, produces enzymes that can degrade cells and caused oxidative burst (Govrin and Levine, 2000). The fungus grows via the polarised tip elongation, leading to the development of appresorium-like structure (Figure 2a), depicted with hyphae with or without swollen tips (Figure 2b). It grows on the host surface and expands into small bulbous cell aggregates, a claw-like structure that penetrates and colonises the plant tissue (Aguayo et al., 2011).

5. Symptoms of Diseases Caused by *B. Cinerea*

Among the diseases caused by *B. cinerea* on important agricultural crops are the leaf blight, shoot blight, leaf spots, bulb rots and fruit rots in the field and at storage. The fruit rot disease is the most important, because it causes severe pre and postharvest losses (Wang et al., 2010; Liana et al., 2013; Zhou et al., 2014). Due to a wide host range of *B. cinerea*, the symptoms of the diseases it causes cannot be generalised across plant tissues and organs (Williamson et al., 2007). Under favourable conditions, the fungus infects and colonises wounded stems, leaves, flowers and fruits both at green and ripe stages. On young plants, gray mold rot appears as gray-brown velvety mold covering stem or leaves, and this cause the above stem to wilt. Apart from the devastating effect on the stem, the fungus infects leaves at maturity or senescence causing a wedge–shaped chesnut brown lesion with yellow margin. This spreads to the node on vegetative stem giving rise to a pale brown rapid growing lesion. The lesions are large gray-brown, slightly water-soaked usually covered with grayish mycelium near the stem end of the tomato (Williamson et al., 2007; Andrea, 2013). Depending on the environmental conditions, symptoms observed on the flowers are tiny "pock" marks and full scale soft rots which spread to the adjacent developing fruits (Williamson et al., 2007).

Figure 3. (a) Tomato fruit showing large gray mold lesions covered by mycelium (edis.ifas.ulf.edu). (b) Tomato fruit with gray mold from carlyx (dwpictures.com.au). (c) Gray mold (*B. cinerea*) on tomato (gardener.wikia.com).

In tomato, when small green fruit was infected directly by airborne spores instead of other sources of infection, the fruit does not show decay symptoms but rather develops a white circular rings called "ghost spots". On ripe fruit, a watery lesion with a tan centre appears and eventually leads to a soft rot with occasional abortion of the infected fruit (Pernezny, 1997). Additionally, a grayish mold was formed at the breaks on the skin surface or sparsely over the surface of the infected fruit (Andrea, 2013). Infected fruits kept in storage at high humidity, often decay and show typical gray coating of spores and sometimes white mycelium appear on the surface. Also, tomatoes held in cold storage for a prolonged period are at high risk of gray mold infection, especially damaged fruits (Andrea, 2013). Figure 3 (a-c) showed the symptoms of gray mold rots on tomato fruits.

6. Disease Cycle and Epidemiology

The fungus *B. cinerea* remains a serious obstacle in the production of numerous important crops. Although efforts have been directed towards understanding the etiology of the gray mold disease, still there are quite a number of epidemiological facets that remain elusive or not clear (Bardin et al., 2014). One important thing is the source and geographic distribution of conidia in different growing environments. Little information has been presented following investigation of this fungus by population genetics. Part of their investigations has showed tremendous level of genetic variations and hence draws conclusions on populations of *B. cinerea* to have a narrow clonal spread (Nicot et al., 1996; Vaczy et al., 2008; Fournier and Giraud, 2008; Bardin et al., 2014). The conidia are easily dispelled by windy conditions and turbulent air currents (Holz et al., 2004; Bardin et al., 2014; Shaw et al., 2016) and this enabled *B. cinerea* to exhibits polycyclic life style i.e., to complete numerous cycles of development within a short period of time or single growing season.

The fungus produces highly resistant survival structure called sclerotia that serve as a survival mechanism usually found in old cultures. The fungus overwinters in infected host tissues as sclerotia or intact mycelia, both of which germinate under favourable conditions, to produce conidiophores that bearing conidia. The conidia are dispersed by wind and rain water to cause new infections elsewhere (Williamson et al., 2007). Climatic factors such as temperatures and relative humidity affect the availability of the conidia and its dispersal. One or two short germ tubes and obvious terminal appresorium were produced by dry conidia (inoculated) before achieving penetration by the fungus into undamaged host cuticle on leaves, flowers and fruits (Williamson et al., 2007). The fungus is capable of developing appresoria (Tenberge, 2004), and disease development can occur at 18-24°C and continues even at 0°C (Andrea, 2013). The disease was also favoured by sufficient humidity in the canopy, and most severe in acidic to neutral sandy soil with high soil moisture (Stall, 1991). Fresh infection usually begins with wind-disseminated conidia that germinate and penetrate the surface of plant tissue through either wounds or by penetration directly to the cuticles and cell walls of epidermal cells (Holz et al., 2004; Leroch et al., 2013). The tissue of susceptible host was killed rapidly and depraved by different mechanisms such as release of toxins, secretion of cell wall-degrading enzymes and other necrosis-inducing factors (van Kan, 2006; Leroch et al., 2013). The life cycle continuous again (Figure 4).

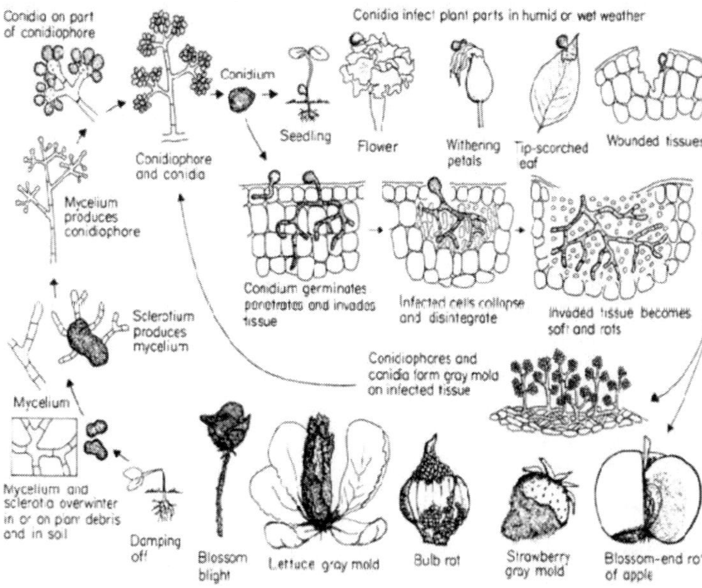

Source: (Agrios, 2005).

Figure 4. Show the cyclic events in the life cycle of *Botrytis cinerea* in different host.

7. Current Status of the Gray Mold Rots Disease in Malaysia

In Malaysia, the gray mold disease is mostly found in the cool areas, particularly in the Cameron Highlands, Pahang and Kundasang Highland, Sabah. The disease was first reported in Malaysia in the early 2000 on the strawberry. Although, at that time, the disease was highly prevalent in the field, detailed study on the impact and epidemiology were not investigated. During that period, the study on *B. cineria* was focused on the biological control. Later, in 2015 serious outbreak of gray mold disease was reported in the Cameroon Highland on tomato fruits (Figures 5a-b), as well as on other crops such as eggplant. The disease infected tomato fruits severely under storage condition, in the cold room (size = 3 x 2m, T° = 7 and RH = 90%) of the Bayern Farms, Kampung Raja (GPS coordinates of the location: N 4.5589203014089; E 101.400050609587) (Figure 6).

Figure 5. Shows gray mold rot symptoms on tomato fruits caused by *B. cinerea* detected in a cold room from greenhouse located in Bayern Farm Kampung Raja in Malaysia.

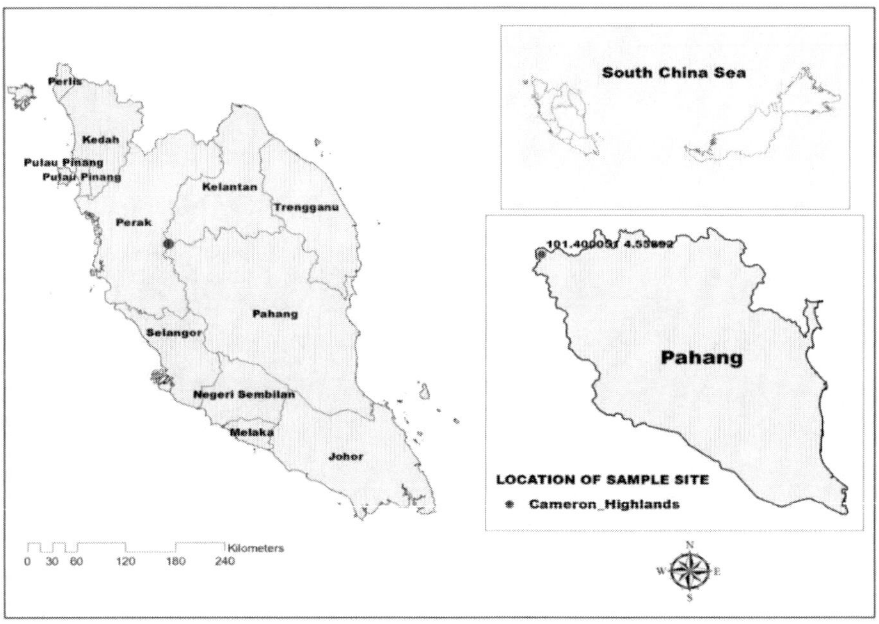

Figure 6. Map of Malaysia showing the location of Bayern Farm, Kampung Raja (red dot) in Cameron Highlands where the samples were collected.

Detailed studies on the morphological, pathological and molecular characteristics of the *B. cinerea* were conducted, to further confirm that it is the causative agent of the gray mold rot disease on the tomato fruits. Based on the morphological characteristics, the study revealed that the isolates were *B. cineria*. The cultural colour of the isolates was white or hyaline at early stage, turning to light gray and later dark gray with abundant aerial mycelial growth. The mycelium were branched, septate, white to gray, warty, fluffy, cottony or flat, and with conidiophore arising from it. Hyphal swelling and rupture were sometimes observed. Colony edge was well defined, branched and raised (Figure 7).

The conidiophores were septate, relatively straight and branched towards the apex. They bear the ampulla, where conidia are supported by fine denticles. The ampullae observed were spherical, subspherical, clavate and somehow lobate in shapes (Figure 8a-d). Immediately after abscission of the conidia from the ampulla, a rounded flat scar was left on the ampulla. As the conidiophores aged, the walls become deep brown and some isolates have thicker and darker conidiophores. The conidia were smooth one-celled, usually produced or developed over the surface of the medium. They differ in size, and shape with some ovate, globose, subglobose, pyriform, ellipsoidal, narrowly ellipsoidal and flat in one part. They were pale brown or hyaline with a slight protuberant hilum. Conidia measured 7.5-17.1 × 5.2-10.5 μm and were supported by branching tree-like conidiophores.

On the basis of pathogenicity assay, all the isolates developed the gray mold disease symptoms but they showed variations in their levels of severity as observed from 2-5 days after inoculation (Figure 9a-e). For the ripe tomato fruits, disease severity ranged from 36 to 97% (Figure 10) with maximum per cent observed in BCH07 (97%) and minimum in BCH04 (36%) isolates. Similarly, in the unripe tomato fruits, disease severity ranged from 20 to 50% (Figure 11) with maximum per cent observed in BCH07 (50%) and minimum in BCH04 (20%).

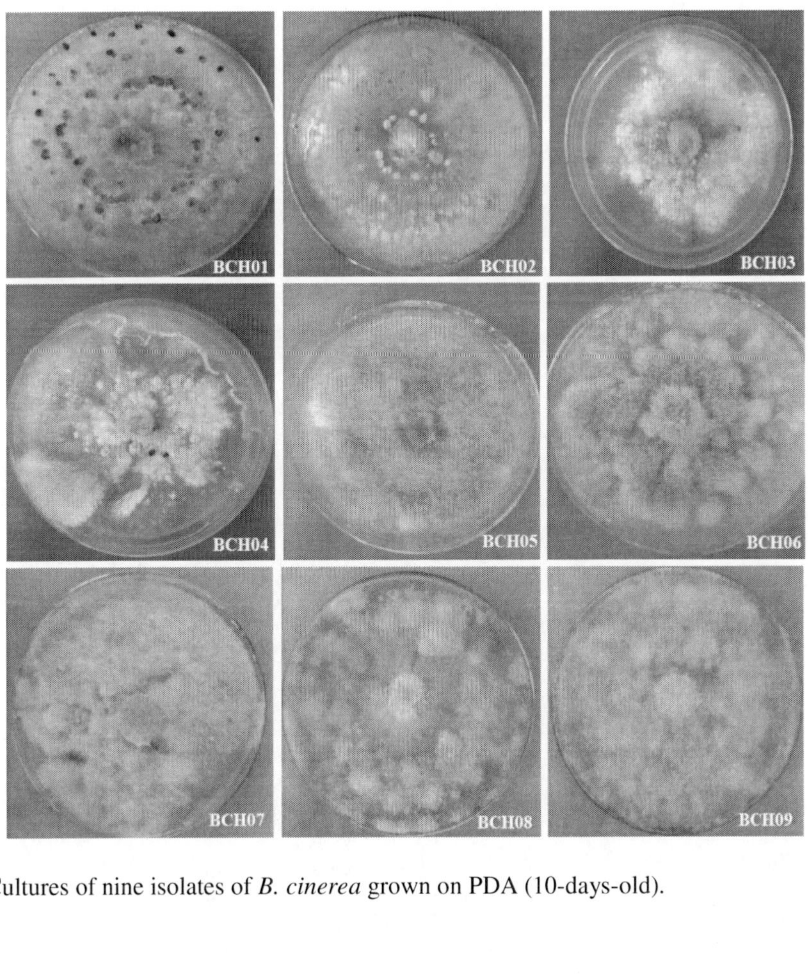

Figure 7. Cultures of nine isolates of *B. cinerea* grown on PDA (10-days-old).

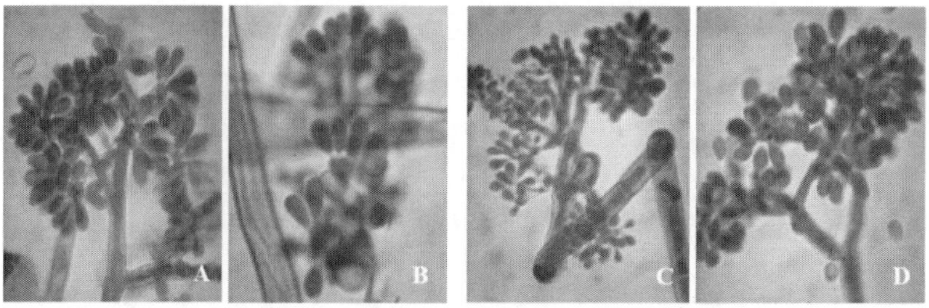

Figure 8. (A-D) Patterns of conidiophores and ampullae in *B. cinerea* isolates (A) spherical (B) whole shape (C) subspherical and (D) Clavate.

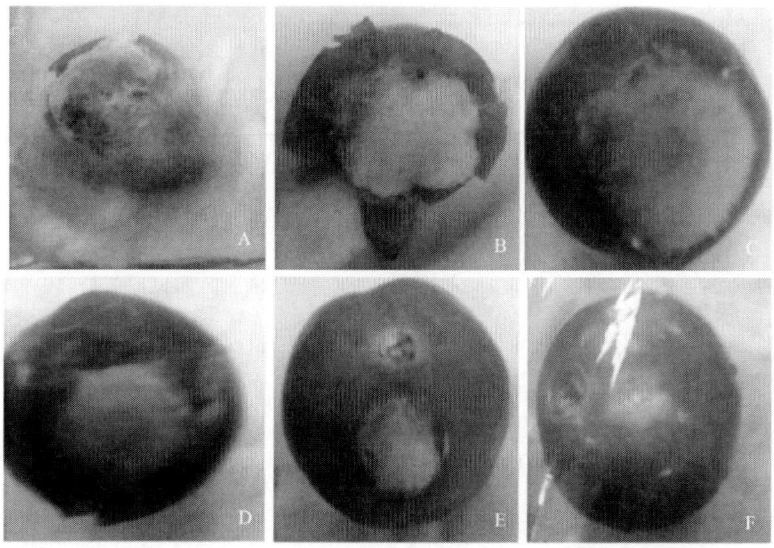

Figure 9. (a) Highly pathogenic (b & c), moderately pathogenic (d &e), Less pathogenic and (f) Control.

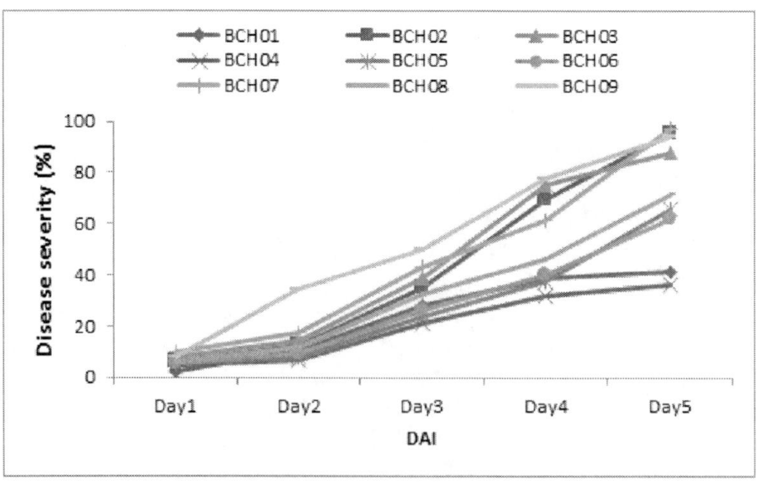

*BCH= *Botrytis cinerea* Cameron Highlands; DAI = Day after Inoculation.

Figure 10. Progress curve of disease severity of the gray mold disease on ripe tomato fruits inoculated with various isolates of *Botrytis cinerea*, 5 days after inoculation.

On the basis of disease severity levels on the inoculated tomato fruits, the isolates were grouped into three main groups in comparison to the control (Figure 9f), namely; highly pathogenic (isolates 2,3,7, and 9) with disease severity levels ranging from 88% to 97% (Figure 9a), moderately pathogenic (isolates 5, 6 and 8) with disease severity levels ranging from 61% to 70% (Figure 9b & c) and less pathogenic (isolates 1 and 4) with disease severity of 36% and 41% (Figure 9d & e) respectively. Similar outcomes were found based on the area under disease progress curve (AUDPC) (Table 2) under ripe tomato fruits, where BCH09 isolate was found to have the highest AUDPC value (221.39 unit^2), and BCH04 isolate with the lowest AUDPC value (82.83 unit^2). This shows that isolate BCH09 was highly pathogenic compared to isolate BCH04 which was less pathogenic. Under the unripe fruits, isolates BCH02 and BCH07 were observed to have almost similar AUDPC values (106.10 unit^2) and (106.9 unit^2), compared to the low AUDPC value (46.85 unit^2) observed in BCH04 also under inoculated unripe fruits.

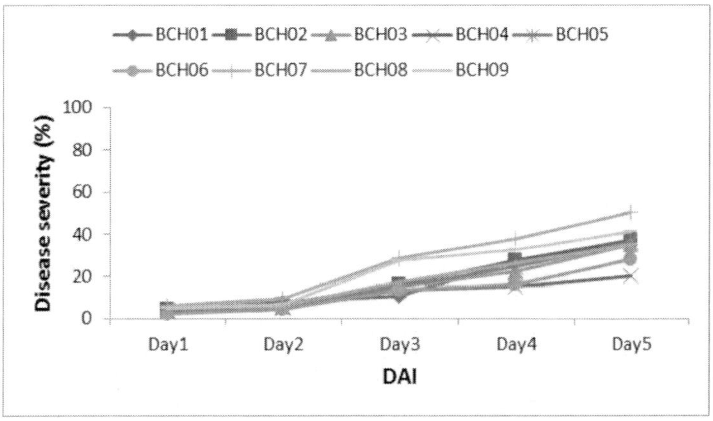

*BCH= *Botrytis cinerea* Cameron Highland; DAI = Days after inoculation.

Figure 11. Progress curve of disease severity of the gray mold disease on unripe tomato fruits inoculated with various isolates of *Botrytiscinerea*, 5 days after inoculation.

Table 2. Area under disease progress curve (AUDPC) in tomato fruits inoculated with isolates of *B. cinerea* 5 days after inoculation (DAI)

Isolate	**AUDPC (unit^2)**	
	Ripe tomato	**Unripe tomato**
BCH01	100.55	68.10
BCH02	170.72	106.10
BCH03	179.94	69.40
BCH04	82.83	46.85
BCH05	107.96	67.53
BCH06	112.13	50.71
BCH07	180.50	106.9
BCH08	132.50	72.55
BCH09	221.39	92.47
Control	0	0

*BCH = *Botrytis cinerea* Cameron Highland.

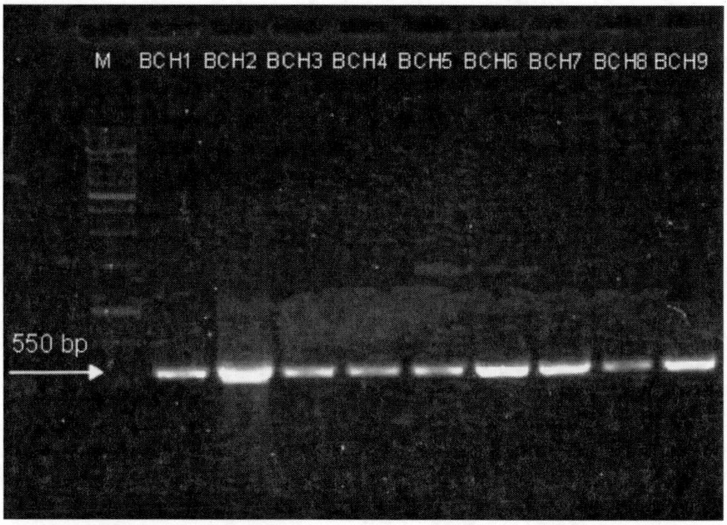

*BCH = *Botrytis cinerea* Cameron Highland.

Figure 12. Gel electrophoresis showing representative bands of PCR product from ITS region. Size of DNA ladder (M) used is 1kb. The amplification fragments were approximately 550 bp.

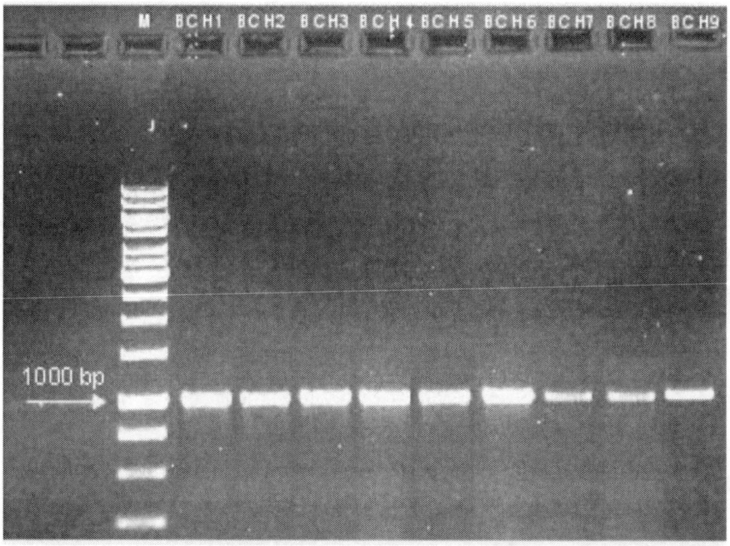

*BCH = *Botrytis cinerea* Cameron Highland.

Figure 13. Gel electrophoresis showing representative bands of PCR product from G3PDH region. Size of DNA ladder (M) used is 1kb. The amplification fragments were approximately 980bp.

Based on the molecular analysis, the partial sequence of ITS region was approximately 550 bp band size (Figure 12). BLASTn search in NCBI nucleotide database with the ITS gene sequence region, showed that the isolates belong to the genus of *Botryotinia* and *Sclerotiniaceae* family and were 99-100% identical to *B. cinerea.* The GenBank accession numbers of the isolates and their consensus sequences were given in

Table 3. BLAST nucleotide search results of nine isolates of *B. cinerea* for ITS and G3PDH gene

Isolate code	ITS REGION			G3PDH REGION		
	Accession no. isolate	Accession no. of closest species	Maximum identity	Accession no. isolate	Accession no. of closest species	Maximum identity
BCH01	KU992693	KT723007	100	KY201456	KP120868	100
BCH02	KU992694	KT582110	100	KY201457	KM016535	100
BCH03	KU992695	KT723007	99	KY201458	AB971644	100
BCH04	KU992696	KT723007	99	KY201462	KT003231	99
BCH05	KU992692	KT723007	100	KY201463	KM016535	99
BCH06	KU992697	KP151609	100	KY201459	CP009819	99
BCH07	KU992698	KP151609	100	KY201460	CP009819	100
BCH08	KU992699	KT723007	100	KY201461	KP120868	100
BCH09	KU992700	KP050616	100	KY201464	CP009819	100

*Accession equivalent obtained from GenBank database: http://www.ncbi.nml.nih.gov/Blast *BCH=*Botrytis cinerea* Cameron Highlands.

(Table 3). Further confirmation was done with the analysis of Glyceraldehyde-3-phosphate dehydrogenase gene (G3PDH) region. It was observed from the gel electrophoresis that the bands were approximately 980 bp in size (Figure 13), and were the targeted band size for G3PDH gene region. BLASTn search in NCBI nucleotide database showed that the isolates have 90-100% homology with the reference isolate of *B. cinerea.* The GenBank accession numbers of the isolates and their consensus sequences were also given in (Table 3).

Phylogenetic analysis using Molecular Evolutionary Genetics Analysis (MEGA) version 6.0 (Tamura et al., 2013) revealed that all the nine isolates considered in this study clustered together in one clade (clade 1) based on ITS region. The isolates were found to have high similarity with the reference isolates of *B. cinerea* (GenBank accession no. KT582111 and KT723007), that clustered also in clade 1 but in different sub-clade. Reference isolates *Coletotrichum gloeosporioides* (GU174548) and *C. gloeosporioides* (AY266378) were used as out group for ITS region and clearly separated from *B. cinerea* (Figure 14). Similarly, the phylogram (Figure 15) shows the phylogenetic relationship of the nine isolates based on G3PDH region. Findings, revealed that all isolates were closely related to *B. cinerea* (KU570455) by clustering together in the same clade. However, reference isolate *Sclerotinia sclerotiorum* (AJ705044) was used as out group for G3PDH region and clearly separated from *B. cinerea.*

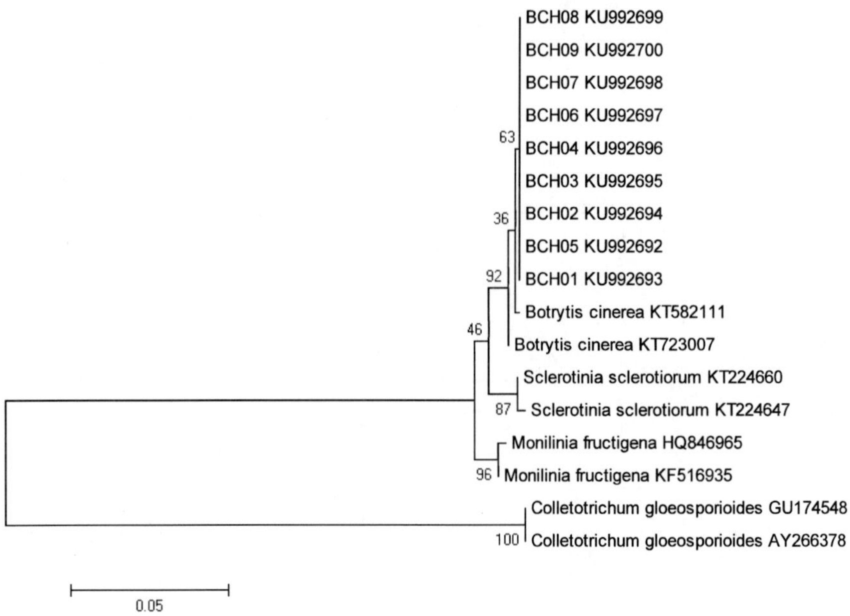

Figure 14. Phylogenetic tree of *B. cinerea* isolates based on ITS gene sequences using neighbour-joining likehood method. Bootstrap values were indicated by numbers at the nodes from a test of 1000 replications. Scale bar indicates the number of nucleotide substitutions.

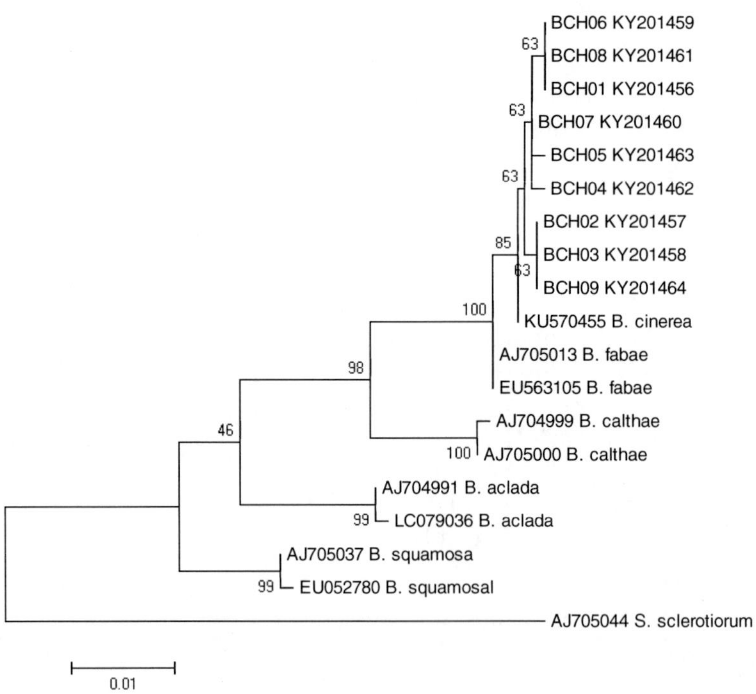

Figure 15. Phylogenetic tree of *B. cinerea* isolates based on G3PDH gene sequences using maximum likehood method. Bootstrap values were indicated by numbers at the nodes from a test of 1000 replications. Scale bar indicates the number of nucleotide substitutions.

CONCLUSION

It is essential to increase the production of tomato worldwide and Malaysia in particular due to increasing demands by the consumers in the food chain. However, occurrence of gray mold rot disease on tomato caused by *B. cinerea*, hinders the sustainable production of the crop. The disease can be destructive enough to cause extensive yield losses with subsequent significant economic backwards to the growers, if not properly and effectively managed. In order to manage the pathogen appropriately, the first step is to identify the pathogen and monitor the areas where there are high epidemics of the disease. This will help in the selection and implementation of plans on how to manage the pathogen for sustainable production of tomato.

This chapter has discussed many studies which mostly reported the disease distribution and hosts range, pathogenesis of *B. cinerea*, symptoms of the disease, disease cycle and epidemiology of *B. cinerea* as well as the current status of gray mold disease in Malaysia. Thc morphological and molecular procedures are important tools for an accurate identification of the pathogen in tomato fruits. Molecular identification provides information that improves the understanding of *B. cinerea* epidemiology, such as localization of the infection and latency within the plant. It also provides information; not only for establishing management strategies, but also for choosing the less contaminated lots of tomato fruits to be stored for a long period of time.

REFERENCES

Adjebli, A., Leyronas, C., Aissat, K. & Nicot, P. C. (2015). Comparison of *Botrytis cinerea* populations collected from tomato greenhouses in Northern Algeria. *J Phytopathol.*, *163*(2), 124-132.

Agrios, G. N. (2005). *Plant Pathology* (5th Edit). New york, Elsevier Academic Press.

Aguayo, C., Riquelme, J., Valenzuela, P. D. T., Hahn, M. & Moreno, E. S. (2011). Bchex virulence gene of *Botrytis cinerea*: characterization and functional analysis. *J. Gen. Plant. Pathol.*, *77*(4), 230-238.

Aktaruzzaman, M., Afroz, T., Kim, B. S. & Shin, H. D. (2016). First Report of Gray Mold Disease of Sponge Gourd (*Luffa cylindrica*) Caused by *Botrytis cinerea* in Korea. *Res. Plant Dis.*, *22*(2), 107-110.

Aktaruzzaman, M., Kim, J. Y., Xu, S. J. & Kim, B. S. (2014). First Report of Postharvest Gray Mold Rot on Carrot Caused by *Botrytis cinerea* in Korea. *Res. Plant Dis.*, *20*(2), 129-131.

Albertini, C., Thébaud, G. Fournier, E. & Leroux, P. (2002). Eburicol 14α-demethylase gene (cyp51) polymorphism and speciation in *Botrytis cinerea. Mycol. Res.*, *106*, 1171–1178.

Allen, D. J. & Lenné, J. M. (1998). *The Pathology of Food and Pasture Legumes*. CAB International, Wallingford, UK.

Al-Mahmooli, I. H., Al-Fahdi, A. R., Al-Sadi, A. M. & Deadman, M. L. (2015). First Report of Root Rot and Crown Necrosis Caused by *Pythium aphanidermatum* on *Phaseolus vulgaris* in Oman. *Plant Dis.*, *99*(3), 419-419.

Aminifard, M. H. & Mohammadi, S. (2013). Essential oils to control *Botrytis cinerea in vitro* and *in vivo* on plum fruits. *J. Sci. Food Agricul.*, *93*(2), 348-353.

Andrea, A. (2013). *Biosecurity guidelines for postharvest greenhouse tomatoes: Prevention of postharvest and storage rot.*

Atwell, S., Corwin, J. A., Soltis, N. E., Subedy, A., Denby, K. J. & Kliebenstein, D. J. (2015). Whole genome resequencing of *Botrytis cinerea* isolates identifies high levels of standing diversity. *Front. Microbiol.*, *6*, 996. http://doi.org/10.3389/fmicb.2015.00996.

Bakr, M. A., Rahman, M. L., Hossain, M. S. & Ahmed, A. U. (1998). Steps towards management of *Botrytis* gray mold of chickpea in Bangladesh. In *4. Working Group Meeting to Discuss Collaborative Research on Botrytis Gray Mold of Chickpea, Joydebpur, Gazipur (Bangladesh)*, 23-26, Feb 1998. ICRISAT.

Bardin, M., Decognet, V. & Nicot, P. C. (2014). Remarkable predominance of a small number of genotypes in greenhouse populations of *Botrytis cinerea. Phytopathology.*, *104*(8), 859-864.

Beever, R. E. & Weeds, P. L. (2004). Taxonomy and genetic variation of *Botrytis* and *Botryotinia.* In: Elad, Y., Williamson, B., Tudzynski, P. and Delen, N. (eds). *Botrytis, Biology, Pathology and Controls.* Kluwer Academic Publisher. Netherland. pp. 29-52.

Carisse, O., Tremblay, D. M. & Lefebvre, A. (2014). Comparison of *Botrytis cinerea* airborne inoculum progress curves from raspberry, strawberry and grape plantings. *Plant Pathol.*, *63*(5), 983-993.

Choquer, M., Fournier, E., Kunz, C., Levis, C., Pradier, J. M., Simon, A. & Viaud, M. (2007). *Botrytis cinerea* virulence factors: new insights into a necrotrophic and polyphageous pathogen. *FEMS Microbiology Letters*, *277*(1), 1-10.

Colmenares, A. J., Aleu, J., Duran-Patron, R., Collado, I. G. & Hernandez-Galan, R. (2002). The putative role of botrydial and related metabolites in the infection mechanism of *Botrytis cinerea. J. Chem. Ecol.*, *28*(5), 997-1005.

Commenil, P., Belingheri, L. & Dehorter, B. (1998). Antilipase antibodies prevent infection of tomato leaves by *Botrytis cinerea*. *Physiol. Mol. Plant Pathol.*, *52*, 1–14.

Comménil, P., Belingheri, L., Bauw, G. & Dehorter, B. (1999). Molecular characterization of a lipase induced in *Botrytis cinerea* by components of grape berry cuticle. *Physiol. Mol. Plant Pathol.* *55*(1), 37-43.

Corwin, J. A., Subedy, A., Eshbaugh, R. & Kliebenstein, D. J. (2016). Expansive Phenotypic Landscape of *Botrytis cinerea* shows Differential Contribution of Genetic Diversity and Plasticity. *Mol. Plant Microbe Interact.*, *29*(4), 287-298.

Cutler, H. G., Jacyno, J. M., Harwood, J. S., Dulik, D., Goodrich, P. D. & Roberts, R. G. (1993). Botcinolide: a biologically active natural product from *Botrytis cinerea. Biosci Biotechnol Biochem.*, *57*, 1980–1982.

Dalmais, B., Schumacher, J., Moraga, J., Le Pecheur, P., Tudzynski, B., Collado, I. G. & Viaud, M. (2011). The *Botrytis cinerea* phytotoxin botcinic acid requires two polyketide synthases for production and has a redundant role in virulence with botrydial. *Mol. Plant Pathol.*, *12*(6), 564-579.

Davidson, J. A., Pande, S., Bretag, T. W., Lindbeck, K. D. & Krishna-Kishore, G. (2007). Biology and management of *Botrytis spp.* in legume crops. In *Botrytis: biology, pathology and control*. Springer Netherlands., pp. 295-318.

de Bem, B. P., Bogo, A., Everhart, S., Casa, R. T., Gonçalves, M. J., Marcon Filho, J. L. & da Cunha, I. C. (2015). Effect of Y-trellis and vertical shoot positioning training systems on downy mildew and *Botrytis* bunch rot of grape in highlands of southern Brazil. *Sci. Hortic.*, *185*, 162-166.

De Miccolis Angelini, R. M., Rotolo, C., Masiello, M., Gerin, D., Pollastro, S. & Faretra, F. (2014). Occurrence of fungicide resistance in populations of *Botryotinia fuckeliana* (*Botrytis cinerea*) on Table grape and strawberry in southern Italy. *Pest Manag. Sci.*, *70*(12), 1785-1796.

Dean, R., Van Kan, J. A., Pretorius, Z. A., Hammond-Kosack, K. E., Di Pietro, A., Spanu, P. D. & Foster, G. D. (2012). The Top 10 fungal pathogens in molecular plant pathology. *Mol. Plant Pathol.*, *13*(4), 414-430.

Decognet, V., Bardin, M., Trottin-Caudal, Y. & Nicot, P. C. (2009). "Rapid change in the genetic diversity of *Botrytis cinerea* populations after the introduction of strains in a tomato glasshouse." *Phyto-pathology.*, *99*(2), 185-193.

Droby, S. & Lichter, A. (2007). Post-harvest Botrytis infection: etiology, development and management. In *Botrytis: Biology, pathology and control*. Springer Netherlands., pp. 349-367.

Duran-Patron, R., Hernandez-Galan, R. & Collado, I. G. (2000). Secobotrytriendiol and related sesquiterpenoids: new phytotoxic metabolites from *Botrytis cinerea. J. Nat. Prod.*, *63*, 182–184.

Elad, E., Williamson, B., Tudzynski, P. & Delen, N. (2007). *Botrytis: Biology, Pathology and Control*, Springer Netherlands., pp. 393.

Elad, Y. (1997). Response of plants to infection by *Botrytis cinerea* and novel means involved in reducing their susceptibility to infection. *Biol. Rev.*, *72*, 381-422.

Elad, Y. (1989). Effect of abiotic conditions on development of grey mould of rose and scanning electron microscopy. *Phytopathol. Mediterr.*, *28*, 122-130.

Elad, Y. (2003). Biocontrol of foliar pathogens: mechanisms and application. *Commun. Agric. Appl. Biol. Sci.*, *68*, 17-24.

Elad, Y. & Evensen, K. (1995). Physiological aspects of resistance to *Botrytis cinerea. Phytopathology.*, *85*(6), 637-643.

Elad, Y., Pertot, I., Prado, A. M. C. & Stewart, A. (2016). Plant hosts of *Botrytis* spp. In *Botrytis–The Fungus, The Pathogen and Its Management in Agricultural Systems*, Springer International Publishing., pp. 413-486.

Elmer, P. A. G. & Reglinski, T. (2006). Biosuppression of *Botrytis cinerea* in grapes. *Plant Pathol.*, *55*(2), 155-177.

Esterio, M., Muñoz, G., Ramos, C., Cofré, G., Estévez, R., Salinas, A. & Auger, J. (2011). Characterization of *Botrytis cinerea* isolates present in Thompson seedless table grapes in the Central Valley of Chile. *Plant dis.*, *95*(6), 683-690.

Fekete, É., Fekete, E., Irinyi, L., Karaffa, L., Árnyasi, M., Asadollahi, M. & Sándor, E. (2012). Genetic diversity of a *Botrytis cinerea* cryptic species complex in Hungary. *Microbiol. Res.*, *167*(5), 283-291.

Fernández-González, M., Rodríguez-Rajo, F. J., Jato, V., Aira, M. J., Ribeiro, H., Oliveira, M. & Abreu, I. (2012). Forecasting ARIMA models for atmospheric vineyard pathogens in Galicia and Northern Portugal: *Botrytis cinerea* spores. *Ann. Agr. Env. Med.*, *19*(2), 255-262.

Ferrada, E. E., Zoffoli, J. P., Diaz, G. A. & Latorre, B. A. (2015). First report of *Botrytis cinerea* causing blossom blight on Japanese plums in Chile. *Plant Dis.*, *99*, 888.

Fournier, E. & Giraud, T. (2008). Sympatric genetic differentiation of a generalist pathogenic fungus, *Botrytis cinerea*, on two different host plants, grapevine and bramble. *J. Evol. Biol.*, *21*, 122–32.

Fournier E., Giraud, T. & Brygoo, Y. (2005). Partition of the *Botrytis cinerea* complex in France using multiple gene genealogies. *Mycologia.*, *97*, 1251–1267.

Frıas, M., Gonzalez, C. & Brito, N. (2011). BcSpl1, a cerato-platanin family protein, contributes to *Botrytis cinerea* virulence and elicits the hypersensitive response in the host. *New Phytol.*, *192*, 483–495.

Gindro, K., Lecoultre, N., Molino, L., de Joffrey, J. P., Schnee, S., Voinesco, F. & Dubuis, P. H. (2014). Development of rapid direct PCR assays to identify downy and powdery mildew and grey mould in *Vitis vinifera* tissue. *J. Int. Sci. Vigne Vin.*, *48*, 261-268.

Giraud, T., Fortini, D., Levis, C., Lamarque, C., Leroux, P., Lobuglio, K. & Brygoo, Y. (1999). Two sibling species of the *Botrytis cinerea* complex, transposa and vacuma, are found in sympatry on numerous host plants. *Phytopathology.*, *89*(10), 967-973.

González, G., Moya, M., Sandoval, C. & Herrera, R. (2009). Genetic diversity in Chilean strawberry (*Fragaria chiloensis*): differential response to *Botrytis cinerea* infection. *Span. J. Agric. Res.*, *7*(40), 886–95.

González-Fernández, R., Aloria, K., Valero-Galván, J., Redondo, I., Arizmendi, J. M. & Jorrín-Novo, J. V. (2014). Proteomic analysis of mycelium and secretome of different *Botrytis cinerea* wild-type strains. *J. Proteomics.*, *97*, 195-221

Govrin, E. M. & Levine, A. (2000). The hypersensitive response facilitates plant infection by the necrotrophic pathogen *Botrytis cinerea. Curr. Biol.*, *10*(13), 751-757.

Hall, T. A. (1999). BioEdit: a user-friendly biological sequence alignment editor and analysis program for windows 95/98/NT. *Nucleic Acids Symposium Series*, *41*, 95–98.

Have, A. T., Mulder, W., Visser, J. & van Kan, J. A. (1998). The endopolygalacturonase gene Bcpg1 is required for full virulence of *Botrytis cinerea. Mol. Plant Microbe Int.*, *11*(10), 1009-1016.

Herzog, K., Wind, R. & Töpfer, R. (2015). Impedance of the grape berry cuticle as a novel phenotypic trait to estimate resistance to *Botrytis cinerea. Sensors*, *15*(6), 12498-12512.

Hobbs, E. L. & Waters, W. E. (1964). Influence of nitrogen+potassium on susceptibility of *Chrysanthemum morifolium* to *Botrytis cinerea. Phytopathology*, *54*(6), 674.

Holz, G., Coertze, S. & Williamson, B. (2004). The ecology of *Botrytis* on plant surfaces. In: *Botrytis*: *Biology, Pathology and Control.* Y. Elad, B. Williamson, P. Tudzynski, and N. Delen, eds. Kluwer Academic Publishers, Dordrecht, The Netherlands., pp. 9-27.

Isenegger, D. A., Ades, P. K., Ford, R. & Taylor, P. W. J. (2008). Status of the *Botrytis cinerea* species complex and microsatellite analysis of transposon types in South Asia and Australia. *Fungal Divers.*, *29*, 17-26.

Jacometti, M. A., Wratten, S. D. & Walter, M. (2010). Review: alternatives to synthetic fungicides for *Botrytis cinerea* management in vineyards. *Austr. J. Grape Wine R.*, *16*(1), 154-172.

Jarvis, W. R. (1977). *Botryotinia and Botrytis species: Taxonomy, Physiology and Pathogenicity, A guide to the Literature.* Monograph No. 15, Canada Department of Agriculture, Ottawa, Canada.

Javed, S., Javaid, A., Anwar, W., Majeed, R. A., Akhtar, R. & Naqvi, S. F. (2017). First Report of *Botrytis* Bunch Rot of Grapes Caused by *Botrytis cinerea* in Pakistan. *Plant Dis.*, *101*(6), 1036.

Johnstoon, P. R., Hoksbergen, K., Park, D. & Beever, R. E. (2014). Genetic diversity of *Botrytis* in New Zealand vineyards and the significance of its seasonal and regional variation. *Plant Pathol.*, *63*(4), 888-898.

Karchani-Balma, S., Gautier, A., Raies, A. & Fournier, A. (2008). Geography, plants, and growing systems shape the genetic structure of Tunisian *Botrytis cinerea* populations. *Phytopathology.*, *98*, 1271–1279.

Kars, I., McCalman, M. E. L. Y. S. I. A., Wagemakers, L. & van Kan, J. A. (2005). Functional analysis of *Botrytis cinerea* pectin methylesterase genes by PCR-based targeted mutagenesis: *Bcpme1* and *Bcpme2* are dispensable for virulence of strain B05. 10. *Mol. Plant Pathol.*, *6*(6), 641-652.

Kim, J. O., Shin, J. H., Gumilang, A., Chung, K., Choi, K. Y. & Kim, K. S. (2016). Effectiveness of Different Classes of Fungicides on *Botrytis cinerea* Causing Gray Mold on Fruit and Vegetables. *Plant Pathol. J.*, *32*, 570-574.

Koch, R. (1882). Uber Die Milzbrand empfung. Eine Entgegnung auf den von Pasteur in Genf Gehaltenen Vortrag. [About the anthrax vaccine. A reply to the lecture held by the Pasteur in Geneva] Reprint 1912. *Gesammelte Werke von Robert Koch*, *1*, 207-231.

Kretschmer, M. & Hahn, M. (2008). Diversität von *Botrytis-cinerea*-Isolaten eines deutschen Weinbergs. [Fungicide resistance and genetic diversity of *Botrytis cinerea* isolates of a German vineyard.] *J. Plant Dis. Protect*, *115*(6), 214-219.

Kumari, S., Tayal, P., Sharma, E. & Kapoor, R. (2014). Analyses of genetic and pathogenic variability among *Botrytis cinerea* isolates. *Microbiol. Res.*, *169*(11), 862-872.

Kuzmanovska, B., Rusevski, R., Jankuloski, L., Jankulovska, M., Ivic, D. & Bandzo, K. (2012). Phenotypic and genetic characterization of *Botrytis cinerea* isolates from tomato. *Genetika-Belgrade.*, *44*(3), 633-647.

Leroch, M., Kleber, A., Silva, E., Coenen, T., Koppenhöfer, D., Shmaryahu, A. & Hahn, M. (2013). Transcriptome profiling of *Botrytis cinerea* conidial germination reveals upregulation of infection-related genes during the prepenetration stage. *Eukaryot. Cell.*, *12*(4), 614-626.

Leroux, P., Elad, Y., Williamson, B., Tudzynski, P. & Delen, N. (2004). *Chemical control of Botrytis and its resistance to chemical fungicides*. New Zealand: Kluwer Academic Publishers, pp. 95–217.

Leyronas, C., Bryone, F., Duffaud, M., Troulet, C. & Nicot, P. C. (2015). Assessing host specialization of *Botrytis cinerea* on lettuce and tomato by genotypic and phenotypic characterization. *Plant Pathol.*, *64*(1), 119-127.

Li, B., Wang, W., Zong, Y., Qin, G. & Tian, S. (2012). Exploring pathogenic mechanisms of *Botrytis cinerea* secretome under different ambient pH based on comparative proteomic analysis. *J. Proteome Res.*, *11*(8), 4249-4260.

Li, N., Zhang, J., Yang, L., Wu, M. D. & Li, G. Q. (2015). First report of *Botrytis pseudocinerea* causing gray mold on tomato (*Lycopersicon esculentum*) in Central China. *Plant Dis.*, *99*(2), 283.

Liana, D., Ongouya, M., Aijie, W., Zhenzhu, Z., Kehinde, O., Chen, X. & Wang, A. (2013). Enhancement of defence responses by *Clonostachys rosea* against *Botrytis cinerea* in tomato. *Sci. Res.*, *4*(12), 1–7.

Liddell, H. G., Scott, R. & Jones, H. S. (1940). *A Greek-English lexicon*/compiled by Henry George Liddell and Robert Scott.

Makovitzki, A., Viterbo, A., Brotman, Y., Chet, I. & Shai, Y. (2007). Inhibition of fungal and bacterial plant pathogens *in vitro* and in planta with ultrashort cationic lipopeptides. *Appl. Environ. Microbiol.*, *73*, 6629-6636.

Mehrotra, R. S. & Agarwal, A. (2003). *Plant Pathology* (2nd edn). New Delhi, India. Teta McGraw-Hill Publishing Company.

Milicevic, T., Topolovcc-Pintarić, S. Cvjctković, B. Ivić, D. & Duralija, B. (2006). Sympatric subpopulations of *Botrytis cinerea* on strawberries based on the content of transposable elements and their connection with resistance to botryticides. *Acta Hortic.*, *708*, 115–118.

Movahedi, S. & Heale, J. B. (1990). The roles of aspartic proteinase and endo-pectin lyase enzymes in the primary stages of infection and pathogenesis of various host tissues by different isolates of *Botrytis cinerea* Pers ex. Pers. *Physiol. Mol. Plant Pathol.* *36*(4), 303-324.

Munoz, C., Talquenca, S. G., Oriolani, E. & Combina, M. (2010). Genetic characterization of grapevine-infecting *Botrytis cinerea* isolates from Argentina. *Rev. Iberoam. Micol.* 27(2), 66-70.

Nakajima, M. & Akutsu, K. (2014). Virulence factors of *Botrytis cinerea*. *J. Gen. Plant Pathol.*, *80*(1), 15-23.

Nicot, P. C., Mermier, M., Vaissiere, B. E. & Lagier, J. (1996). Differential spore production by *Botrytis cinerea* on agar medium and plant tissue under near-ultraviolet light-absorbing polyethylene film. *Plant Dis.*, *80*, 555-558.

Noda, J., Brito, N. & González, C. (2010). The *Botrytis cinerea* xylanase Xyn11A contributes to virulence with its necrotizing activity, not with its catalytic activity. *BMC Plant Biol.*, *10*(1), 38.

Palavouzi, S. C., Tzamos, S., Paplomatas, E. & Thomidis, T. (2015). First report of *Cytospora punicae* isolated from pomegranate plants with symptom of collar rot in Northern Greece, *J Plant Pathol.*, *97* (1), 216.

Pande, S., Sharma, M., Kishore, G. K., Shivram, L. & Mangala, U. N. (2010). Characterization of *Botrytis cinerea* isolates from chickpea: DNA polymorphisms, cultural, morphological and virulence characteristics. *Afr. J. Biotechnol.*, *9*(46), 7961-7967.

Pande, S., Singh, G., Narayana, J., Bakr, M., Chaurasia, P., Joshi, S. & Gowda, C. (2001). Integrated management of *Botrytis* gray mold of chickpea. *India: International Crops Research Institute for the Semi-Arid Tropics*.

Pande, S., Singh, G., Rao, J. N., Bakr, M. A., Chaurasia, P. C. P., Joshi, S., Johansen, C., Singh, S. D., Kumar, J., Rahman, M. M. & Gowda, C. L. L. (2002). *Integrated management of Botrytis gray mold of chickpea.* Information Bulletin no. 61. ICRISAT Patancheru, Andhra Pradesh, India.

Parvu, M., Parvu, A. E., Craciun, C., Barbu-Tudoran, L., Vlase, L., Tamas, M. & Molnar, A. M. (2010). Changes in *Botrytis cinerea* conidia caused by *Berberis vulgaris* extract. *Not. Bot. Horti Agrobo.*, *38*(3), 15.

Pernezny, K. (1997). *Tomato Production Guide for Florida: Disease Control*. Institute of Food and Agricultural Sciences, University of Florida Cooperative Extension Service, Gainesville, U.S.A, EDIS.

Porter, D. M., Smith, D. H. & Rodriguez-Kabana, R. (1984). *Compendium of Peanut Diseases*. American Phytopathological Society Press, St. Paul, Minnesota, USA.

Rajaguru, B. A. P. & Shaw, M. W. (2010). Genetic differentiation between hosts and locations in populations of latent *Botrytis cinerea* in southern England. *Plant Pathol.*, *59*, 1081–1090.

Reino, J. L., Hernández-Galán, R., Durán-Patrón, R. & Collado, I. G. (2004). Virulence–toxin production relationship in isolates of the plant pathogenic fungus *Botrytis cinerea*. *J. Phytopathol.*, *152*(10), 563-566.

Rolke, Y., Liu, S., Quidde, T., Williamson, B., Schouten, A., Weltring, K. M. & Tudzynski, P. (2004). Functional analysis of H2O2-generating systems in *Botrytis cinerea*: the major Cu-Zn-superoxide dismutase (BCSOD1) contributes to virulence on French bean, whereas a glucose oxidase (BCGOD1) is dispensable. *Mol. Plant Pathol.*, *5*(1), 17-27.

Romanazzi, G., Lichter, A., Gabler, F. M. & Smilanick, J. L. (2012). Recent advances on the use of natural and safe alternatives to conventional methods to control postharvest gray mold of table grapes. *Postharvest Biol. Tec.*, *63*(1), 141-147.

Saito, S., Suzuki, S. & Takayanagi, T. (2009). Nested PCR-RFLP is a high-speed method to detect fungicide-resistant *Botrytis cinerea* at an early growth stage of grapes. *Pest Manag. Sci.*, *65*(2), 197-204.

Samuel, S., Veloukas, T., Papavasileiou, A. & Karaoglanidis, G. S. (2012). Differences in frequency of transposable elements presence in *Botrytis cinerea* populations from several hosts in Greece. *Plant Dis.*, *96*, 1286–1290.

Schouten, A., van Baarlen, P. & van Kan, J. A. L. (2008). Phytotoxic Nep1- like proteins from the necrotrophic fungus *Botrytis cinerea* associate with membranes and the nucleus of plant cells. *New Phytol.*, 177:493–505.

Segmuller, N., Kokkelink, L., Giesbert, S., Odinius, D., van Kan, J. & Tudzynski, P. (2008). NADPH oxidases are involved in differentiation and pathogenicity in *Botrytis cinerea*. *Mol. Plant Microbe In.*, *21*(6), 808-819.

Shaw, M. W., Emmanuel, C. J., Emilda, D., Terhem, R. B., Shafia, A., Tsamaidi, D. & Van Kan, J. A. (2016). Analysis of cryptic, systemic *Botrytis* infections in symptomless hosts. *Front. Plant Sci.*, *7*, 1-14.

Shlezinger, N., Doron, A. & Sharon, A. (2011a). Apoptosis-like programmed cell death in the grey mould fungus *Botrytis cinerea*: genes and their role in pathogenicity. *Biochem. Soc. Trans.*, *39*, 1493–1498.

Shlezinger, N., Minz, A., Gur, Y., Hatam, I., Dagdas, Y. F., Talbot, N. J. & Sharon, A. (2011b). Anti-apoptotic machinery protects the necrotrophic fungus *Botrytis cinerea* from host-induced apoptotic-like cell death during plant infection. *PLoS Pathog*, *7*(8), 1002185.

Smilanick, J. L., Mansour, M. F., Mlikota Gabler, F., Margosan, D. A. & Hashim-Buckey, J. (2010). Control of postharvest gray mold of Table grapes in the San Joaquin Valley of California by fungicides applied during the growing season. *Plant Dis.*, *94*(2), 250-257.

Staats, M., van Baarlen, P. & van Kan, J. A. (2005). Molecular phylogeny of the plant pathogenic genus *Botrytis* and the evolution of host specificity. *Mol. Biol. Evol.*, *22*(2), 333-346.

Staats, M., van Baarlen, P., Schouten, A., van Kan, J. A. & Bakker, F. T. (2007). Positive selection in phytotoxic protein-encoding genes of *Botrytis* species. *Fungal Genet. Biol.*, *44*(1), 52-63.

Stall, R. (1991). Gray mold. In T. Jones, JB; Jones, JP; Sall, RE and Zitter (Ed.), *Compendum of tomato diseases*. American Phytopathological Society Press, St Paul Mn., pp. 16–17.

Stefanato, F. L., Abou-Mansour, E., Buchala, A., Kretschmer, M., Mosbach, A., Hahn, M. & Schoonbeek, H. J. (2009). The ABC transporter BcatrB from *Botrytis cinerea* exports camalexin and is a virulence factor on *Arabidopsis thaliana*. *Plant J.*, *58*(3), 499-510.

Tamura, K., Stecher, G., Peterson, D., Filipski, A. & Kumar, S. (2013). MEGA6: molecular evolutionary genetics analysis version 6.0. *Mol. Biol. Evol.*, *30*(12), 2725-2729.

Tani, H., Koshino, H., Sakuno, E., Cutler, H. G. & Nakajima, H. (2006). Botcinins E and F and Botcinolide from *Botrytis cinerea* and Structural Revision of Botcinolides. *J. Nat. Prod.*, *69*(4), 722-725.

Tanovic, B., Delibasic, G., Milivojevic, J. & Nikolic, M. (2009). Characterization of *Botrytis cinerea* isolates from small fruits and grapevine in Serbia. *Arch Biol Sci*, *61*(3), 419–29.

Vaczy, Z. K., Sandor, E., Karaffa, L., Fekete, E., Fekete, E., Arnyasi, M., Czegledi, L., Kovics, G., Druzhinina, S. I. & Kubicek, P. C. (2008). Sexual recombination in the *Botrytis cinerea* populations in Hungarian vineyards. *Phytopathology*, *98*(12), 1312-1319.

Valette-Collet, O., Cimerman, A., Reignault, P., Levis, C. & Boccara, M. (2003). Disruption of *Botrytis cinerea* pectin methylesterase gene *Bcpme1* reduces virulence on several host plants. *Mol. Plant Microbe In.*, *16*(4), 360-367.

Van der Vlugt-Bergmans, C. J. B., Wagemakers, C. A. M. & Van Kan, J. A. L. (1997). Cloning and expression of the cutinase A gene of Botrytis cinerea. *Mol. Plant Microbe In.*, *10*(1), 21-29.

Van Kan, J. A. (2006). Licensed to kill: the lifestyle of a necrotrophic plant pathogen. *Trends Plant Sci*, *11*, 247–253.

van Kan, J. A. (2003). Infection strategies of *Botrytis cinerea*. In *VIII International Symposium on Postharvest Physiology of Ornamental Plants*, *669*, (pp. 77-90).

van Kan, J., Duarte, J., Dekkers, E., Dyer, P. & Kohn, L. M. (2010). The *Botrytis cinerea* mating type loci. In *Book of Abstracts 15th International Botrytis symposium, Cadiz*, Spain, *30*, May–4 June 2010, (pp. 39-39).

Vercesi, A., Toffolatti, S. L., VEnTurini, G., CAmPiA, P. & Scagnelli, S. (2014). Characterization of *Botrytis cinerea* populations associated with treated and untreated cv. Moscato vineyards. *Phytopathol. Mediterr.*, *53*(1), 108-123.

Wahab, H. A. (2015). Characterization of Egyptian *Botrytis cinerea* Isolates from Different Host Plants. *Adv. Microbiol.*, *5*(3), 177-189.

Walker, A. S., Gautier, A., Confais, J., Martinho, D., Viaud, M., Le Pêcheur, P. & Fournier, E. (2011). *Botrytis pseudocinerea*, a new cryptic species causing gray mold in French vineyards in sympatry with *Botrytis cinerea*. *Phytopathology*, *101*(12), 1433-1445.

Wang, H. B., Wang, B. L., Wang, X. D., Wei, C. C., Nie, J. Y. & Liu, F. Z. (2010). The current situation of the grape industry, problems and development strategy. *China Fruits Issue*, *6*, 69–71.

Wessels, B. A., Lamprecht, S. C., Linde, C. C., Fourie, P. H. & Mostert, L. (2013). Characterization of the genetic variation and fungicide resistance in *Botrytis cinerea* populations on rooibos seedlings in the Western Cape of South Africa. *Eur. J. Plant Pathol.*, *136*(2), 407-417.

White, T. J., Bruns, T., Lee, S. J. W. T. & Taylor, J. W. (1990). Amplification and direct sequencing of fungal ribosomal RNA genes for phylogenetics. In: *PCR protocols: a guide to methods and applications*, Innis, M. A., Gelfand, D. H., Snisky, J. J and White, T. J. (Eds). Academic press, New York, USA., pp. 315-322.

Williamson, B., Tudzynski, B., Tudzynski, P., Jan, A. L. & van, kan. (2007). *Botrytis cinerea*:the cause of grey mold disease. *Mol. Plant Pathol.*, *8*(5), 561–580.

Windram, O., Stoker, C. & Denby, K. (2016). Overview of Plant Defence Systems: Lessons from Arabidopsis-*Botrytis cinerea* Systems Biology. In: *Botrytis–the Fungus, the Pathogen and its Management in Agricultural Systems*. Springer International Publishing. pp. 335-360.

Wu, D., Feng, L., Zhang, C. & He, Y. (2008). Early detection of *Botrytis cinerea* on eggplant leaves based on visible and near-infrared spectroscopy. *Transactions of the ASABE*, *51*(3), 1133-1139.

Xue, L. H., Liu, Y., Liu, H. Y. & Li, C. J. (2017). First Report of *Botrytis cinerea* Causing Fruit Rot on Kiwifruit (*Actinidia chinensis*) in China. *Plant Dis.*, *101*(6), 1054-1054.

Yermiyahu, U., Shamai, I., Peleg, R., Dudai, N. & Shtienberg, D. (2006). Reduction of *Botrytis cinerea* sporulation in sweet basil by altering the concentrations of nitrogen and calcium in the irrigation solution. *Plant Pathol.*, *55*(4), 544-552.

Zhou, Y. J., Zhang, J., Wang, X. D., Yang, L., Jiang, D. H., Li, G. Q. & Zhuang, W. Y. (2014). Morphological and phylogenetic identification of *Botrytis sinoviticola*, a novel cryptic species causing gray mold disease of table grapes (*Vitis vinifera*) in China. *Mycologia*, *106*(1), 43-56.

BIOGRAPHICAL SKETCH

Tijjani Ahmadu

Affiliation:

- Department of Plant Protection, Faculty of Agriculture, Universiti Putra Malaysia, UPM Serdang, Selangor Darul Ehsan, Malaysia.
- Department of Crop Production, Faculty of Agriculture and Agricultural Technology, Abubakar Tafawa Balewa University Bauchi, Bauchi State, Nigeria.

Education:

- Nigeria Certificate in Education (NCE, Diploma) in 1993
- B. Agric Tech. (First Class Hons.) in Crop Production (2004/2005)
- MSc Plant Pathology in 2010
- PhD Candidate in Plant Pathology, Universiti Putra Malaysia (2014-todate).

Professional Appointments:

- Lecturer at Abubakar Tafawa Balewa University P M B 0248, Bauchi, Bauchi State, Nigeria.
- Appointed as the department assistant examination officer.
- Department registration officer.
- Faculty registration officer.
- Faculty exams and records officer

Honors:

- Best graduating student in school of science in 1993
- Best graduating student in school of vocational education in 1993

- Best graduating student in educational foundation in 1993
- Best graduating student Crop Production in 2004/2005 academic session
- Best Poster Prize by the American Society for Microbiology in the International Conference of the Malaysian Society for Micro-biology (ICMSM) 2017
- Poster Presenter Winner in the International Conference of the Malaysian Society for Microbiology (ICMSM) 2017

In: Recent Advances in Plant Research
Editors: B. Welch and M. Wilkerson
ISBN: 978-1-53614-170-2

Chapter 5

THE COUNTER DEFENCE MECHANISMS OF INSECT HERBIVORES AGAINST CHEMICAL PLANT DEFENCE

Mandeep Kaur*[1,*], *PhD
[1]Crop Protection Division, Department of Agriculture,
Khalsa College Amritsar, Punjab, India

ABSTRACT

Insect herbivores use different feeding strategies to obtain nutrients and energy from plants for their growth and reproduction. Plants produce an array of toxins and defensive proteins through various metabolic pathways for their defence against insects. To cope with these defences, herbivores have developed counter defences. These responses of insect herbivores are important in insect plant interactions. Among these interactions, host plant selection and feeding behaviour are important factors and play a significant role in habitat selection and discrimination mechanisms between host species. Furthermore, insects have employed a diverse array of strategies that enable them to bypass defensive barriers, or to metabolise these chemicals after ingestion.

1. INTRODUCTION

Terrestrial plants and herbivore insects have co-existed for more than 400 million years. The interactions between them dominate the terrestrial ecology of Earth and are important determinants of plant productivity in managed and natural vegetation. The plants have developed a range of defences such as morphological, biochemical, and molecular mechanisms (War et al., 2012) to reduce the threat of injury and loss of productivity (Mitchell et al., 2016). These defences are a complex network of different

* Corresponding Author Email: mandeepkaur8@gmail.com.

pathways composed of direct and indirect defences. Direct defences include production of allelochemicals such as phenols, alkaloids, terpenoids, glucosinolates or protease inhibitors, while indirect defences include the emission of volatile organic compounds after herbivore attack. Plants are the only source of food for the survival of approximately one million or more phytophagous insect species. On average, plant production consumed by insects account for more than 10% of the annual worldwide production (Beddington, 2010; Popp et al., 2013; Sobhy et al., 2014). These insect herbivore species have different feeding strategies which lead to different types of mechanical damage on plant tissues. Among all the known herbivores, two-thirds are leaf-eating beetles (Coleoptera) or caterpillars (Lepidoptera) that cause damage to plants with chewing, snipping or tearing types of mouthparts (Schoonhoven et al., 1998). Other insect herbivores are leaf miners that feed on the soft tissue between the epidermal cell layers, while thrips with the piercing-sucking type of mouthparts, suck the liquid content from lacerated cells. Phloem-suckers such as aphids, whiteflies and other hemiptera have special stylets that are inserted between the cells and the phloem. Although, root feeding insect herbivores have not received much attention as above ground herbivores, but the majority are root-chewers and a few root borers/piercers have also been reported (Houlihan, 1969). All the above mentioned herbivorous insects need to cope with a number of barriers such as chemical defences imposed by plants to avoid herbivory that provoke deleterious effects in insects, affecting their reproductive performance (Bernays and Chapman, 1994; Bennett and Walsgrove, 1994; Schoonhoven et al., 2005). In response, insects developed strategies to overcome plant barriers such as avoidance mechanisms (Zangerl, 1990), sequestration of poison (Nishida, 2002), detoxification of toxic compounds (Scott and Wen, 2001), and alteration of gene expression pattern (Silva et al., 2001).

In the above discussed insect-plant interactions, both partners send and receive chemical cues that determine the outcome of the interaction and help in driving the coevolution of plants and insects. This much research has focused so far on describing the diversity of plant chemicals and their effects on herbivores. Less is known about the multiple mechanisms evolved by insects to overcome these chemical defences. In the present chapter, different approaches to understand food selection, biology of feeding behaviours of insect herbivores and the relative importance of different types of traits involved in defence against plants are discussed.

2. Food Selection

Host plant selection behaviour involves habitat selection and discrimination mechanisms between host species and is known to play a significant role in insect-plant interactions. Insects are endowed with an array of sensory perceptions such as, visual, chemosensory, mechano-sensory etc. By visual and chemical stimuli, they locate a suitable habitat and are attracted to the host plant. The olfactory and gustatory sensilla enable detection of these stimuli resulting in the acceptance or rejection of the host plant (Ananthakrishnan et al., 1985; Ananthakrishnan, 1986).

During host-plant selection, herbivorous insect move over plant surface and intermittent evaluation is performed by repetitive contact of legs, antennae, mouthparts or ovipositor with the plant surface (Dethier, 1982). During contact phase a series of behavioural elements and chemo receptors serve to evaluate physical and chemical plant traits (Martin et al., 2011) that could not be perceived from a distance. In particular, plant selection during flight is very quick process and made in a time scale of tens to hundreds of milliseconds (Cardé and Willis, 2008; Baker, 2009; Bruce and Pickett, 2011). This is because odour plumes are patchy in structure and insects encounter this only for fractions of a second (Riffellet al., 2009; Webster et al., 2010). For such a quick action, herbivorous insects have evolved with different chemoreceptive systems, e.g., lepidopterous larvae have two chemoreceptive systems: a gustatory and an olfactory. The gustatory system consists of two sensilla styloconica on the galea of each maxilla. Each sensillum has four chemoreceptors of considerable specificity. In each sensillum of *Pieris brassicae* there is a receptor sensitive to low concentrations of mustard oil glycosides. The olfactory organs are the antennae and maxillary palpi, and consist of 16 and 19-24 olfactory cells on each antenna and palpus respectively. The receptors are not narrowly specific to particular odour. Each one responds to many odours by differentially increasing or decreasing its rate of spontaneous firing (Schoonhoven and Dethier, 1966).

As a next step in the evaluation sequence, the insect may damage the plant and thereby release a complex mixture of primary and secondary metabolites from plants. The insect mouthparts inflict injury to host plant and this is designated as test biting, or probing in the case of piercing-sucking insects. A test bite is often smaller than a regular bite and the plant material may be kept longer in the pre-oral cavity than the regular food intake. The sensory information collected during contact is evaluated by the central nervous system. After evaluation, many species decide either to proceed or to reject the plant. The final decision of acceptance in the host-plant selection process is confirmed and further feeding or oviposition is started (Eigenbrode and Espelie, 1995).

When the available host plants to an insect herbivore vary in both their nutritional and plant secondary metabolite (PSM) composition, selection decisions should be based on trade-off between the benefits gained from obtaining an optimal balance of nutrients and the costs of ingesting PSMs (Behmer et al., 2002). Two hypotheses are discussed in the literature to explain patterns of food selection by herbivores. The first states that herbivores are physiologically constrained in their capacity to detoxify PSMs and their feeding decisions based on a capacity to dilute the intake of PSMs (Freeland and Janzen, 1974). The second hypothesis states that because plants are nutritionally variable, both spatially and temporally, the overriding criterion driving food selection is the need to acquire an appropriate balance of nutrients (Pulliam, 1975; Westoby, 1978; Rapport, 1980). It seems unlikely, however, that these two hypotheses would be mutually exclusive. Rather, herbivores probably select a mix of foods that represent a trade-off between minimizing the consumption of toxic PSMs and balancing nutrient intake.

3. Biology of Feeding Behaviour

The feeding behaviour of herbivorous insects is an important element of plant–insect interactions. The selected food items must be processed in ways that increase their nutrient quality and reduce the negative effects of toxins or other compounds that make food items less suitable for insect growth and reproduction.

The caterpillars may feed two or three times of their own weight daily. The feeding behaviour of caterpillars to suit nutrient needs is governed by osmotic pressure of blood and by the concentration of circulating nutrients. If these factors are high, then the insect do not feed and go for rest (Simpson and Simpson, 1989). Many insects eat more when essential nutrients are at low concentrations and probably this is typical of most insects, e.g., Due to low levels of protein content in plants, herbivorous insects have more difficulty in obtaining this nutrient than insects in other major insect guilds, such as predators or parasites. Proteins (i.e., amino acids) are the main macronutrients containing nitrogen and strongly affect the growth rates of insect herbivores (Mattson, 1980; Barbehenn et al., 2013). Because insect herbivores have of small size and rapid growth, so they need a high-quality protein. To obtain sufficient protein, herbivorous insects must have a large gut capacity and a fast throughput of ingested food. The insect herbivores, such as grasshoppers and a few species of caterpillars can switch among different food plants and can improve their nutrient balance. Such dietary mixing strategy has relevance for obtaining adequate levels of carbohydrates and proteins as well as minor but important nutrients, such as particular sterols, or aromatic amino acids that are used in cuticular protein.

Furthermore, herbivore's host-plant range is closely associated to its digestive physiology (Pearse et al., 2013). The midgut and salivary glands are important organs involved in establishment of interaction between an herbivore insect and its host plant (Shukle et al., 2010). For example, the ability of insects to tolerate ingested tannins, one of the most abundant secondary plant metabolite, is determined by biochemical and physiological features of their midgut. These include the release of surfactants and antioxidants, the maintenance of a high pH and the formation of a protective peritrophic membrane envelope lining the midgut which acts as a barrier (Barbehenn and Constabel, 2011). The flexibility and diversity of digestion and detoxification pathways of an insect herbivore eventually lead to wider host plant range (Dermauw et al., 2013). This also leads to the notion of specialist and generalist insect herbivores. The specialists have physiological adaptations to cope with the plant defences, which allow greater tolerance than most generalists (Cornell and Hawkins, 2003). The generalists have 'general' mechanisms to tolerate an array of plant defences and also have mechanisms to manipulate plants via highly conserved plant pathways (Krieger et al., 1971; Whittaker and Feeny, 1971).

4. Insect Defences

To cope up with the plant defences insects have evolved behavioural, physiological and biochemical mechanisms which allow them to feed on chemically defended plants (Berenbaum, 2001; Karban and Agrawal, 2002). Recent studies have improved our understanding of both behavioural mechanisms and metabolic detoxification for avoiding toxins. Two general mechanisms allow herbivores to cope with the xenobiotics from their environment: mechanisms that decrease exposure (pharmacokinetic responses) and mechanisms that decrease sensitivity (pharmacodynamic responses) (Kant et al., 2015). Together, these mechanisms determine the level of tolerance of herbivores to xenobiotics.

4.1. Pharmacokinetic Response

Pharmacokinetic responses comprise a variety of adaptations that reduce uptake, increase metabolism and allow sequestration.

4.1.1. Behaviour

Insects can avoid eating toxic plants after visual, olfactory or via contact evaluation (Chapman, 2003). Avoidance mechanisms can be either inherited or learned through plant toxin encounter. One of the inherited mechanisms is oviposition behaviour that allows females to lay eggs on suitable plants (Fox et al., 2004). However, larval performance is not always correlated with oviposition preference and larvae might have to move from site of hatching to select a suitable host plant (Cotter and Edwards, 2006). Insects can also escape to a toxin-free area by feeding selectively on plant organs that do not produce the toxin (i.e., spatial shift), or exploit that stage of plant in which the toxin is absent or present at low level (i.e., phonological shift) (Nealis and Nault, 2005). Several recent studies provide mechanistic perception into the processes involved in food selection by herbivorous insects (Chapman, 2003; Glendinning et al., 2002).

Bitter foods are also avoided by insects. Although the molecular bitterness-signaling pathways are highly diversified, but they are not always specific to each ingested compound, e.g., The larvae of tobacco hornworm *Manduca sexta* cannot discriminate between two different bitter-tasting compounds, salicin (non-toxic phenolic compound) and caffeine (toxic alkaloid), because the same bitterness-signalling pathway is activated by both compounds (Glendinning et al., 2002). Similarly, the concentrations of bitter-tasting cyanogenic glucosides below the toxicity threshold level deter the grasshoppers and weevils (Zagrobelny et al., 2004). Therefore, bitterness is not always a signal of toxicity. The similarity of pathways activated by toxic and non-toxic allelochemicals might restrict the range of host plants used by an insect, increasing the evolutionary cost of avoidance. The ecological context can also modify insect feeding behaviour. Solitarious and gregarious-phase *Schistocerca gregaria* locusts respond differently to hyoscyamine, a bitter-tasting plant alkaloid that protects against predators. Solitarious-phase locusts are deterred by the taste of this compound, whereas gregarious-phase locusts accept and even prefer food that contains this compound.

4.1.2. Manipulation of Plant Chemical Defences

Insects can also manipulate the chemical defences of the host plant before feeding. Insects that feed on plants having secretory canals often cut trenches across leaves to depressurize the canals and eliminate toxic exudation at their feeding site (Becerra, 2003; Helmus and Dussourd, 2005). This time consuming behaviour was observed in the lepidopteran polyphagous plusiine noctuid moth *Thysanoplusia orichalcea*, which was induced by allelochemicals such as lactucin from lettuce latex, myristicin from parsley oil and lobeline from cardinal flowers (Dussourd, 2003, Helmus and Dussourd, 2005).

Herbivorous insects also avoid contact with the toxin by suppressing plant defences, e.g., Hemipteran phloem feeders such as the mealybug *Phenacoccus solenopsis* (Zhang et al., 2011) and the whitefly *Bemisia tabaci* were found to suppress jasmonate (JA) defences of plants (Zhang et al., 2009), possibly by inducing antagonist salicylic acid (SA) defences (Zarate et al., 2007; Walling, 2009). The aphid *Megoura viciae* inhibits defensive phloem clogging (Will et al., 2007), whereas other aphid species were reported to suppress the oxidative burst (Bos et al., 2010). Many lepidopteran larvae with chewing mouthparts use salivary elicitors and have been reported to suppress or reduce host-plant defences (Bede et al., 2006). For example, the glucose oxidase enzyme is the main component of the tobacco earworm *Helicoverpa zea* saliva that decreases the level of nicotine induced in leaves of tobacco *Nicotiana tabacum* (Musser et al., 2002). The same enzyme in the saliva of the beet armyworm *Spodoptera exigua* decreases the transcript level of key regulatory genes involved in the early steps of the defence pathways of the *Medicago truncatula* plant (Bede et al., 2006). Another resistance mechanism, i.e., mustard oil bomb was reported in diamondback moth *Plutella xylostella* against the defence system of its crucifer host plants. This system involves the co-secretion of glucosinolates and a specific enzyme (myrosinase), each stored in separate cell compartments. Following tissue damage, glucosinolates are hydrolysed to highly toxic product, i.e., isothiocyanate by myrosinase. The diamondback moth has developed an adaptive strategy which modifies the ingested plant glucosinolates by a sulfatase digestive enzyme, preventing their hydrolysis by plant myrosinase, thus disarming the 'mustard oil bomb' (Ratzka et al., 2002). The cabbage white butterfly *Pieris rapae* has developed a different adaptive mechanism by redirecting toxic isothio-cyanate formation toward non-toxic nitrile formation with a specific gut protein (Wittstock et al., 2004). The gall-inducing insects such as sawflies decrease the level of toxic phenolic compounds inside the gall where larvae develop. The convergence in the chemical properties of galls induced on different host plant species, favouring frequent host shifts in this group of insects (Nyman and Julkunen-Tiitto, 2000).

4.1.3. Excretion, Sequestration and Further Use of Plant Toxins

The insects also avoid toxic effects of host plants by excretion, sequestration and further use of plant toxins. In many insect species, a large proportion of the accumulated toxic plant compounds can be excreted, or lost with exuvia during the moulting (Zagrobelny et al., 2004). Many secondary plant compounds under normal physiological conditions are either charged molecules or polar molecules or hydrophilic compounds (such as glycosides). The bio-membranes are often almost impermeable to these compounds. Such properties along with the rapid intestinal passage in herbivorous insects mean that many toxicants get little chance to enter the body cavity (Wink and Schneider 1990). Tobacco hornworm (*M. sexta*) larvae have evolved a number of resistance mechanisms that defend against plant allelochemicals. The larvae rapidly excrete nicotine and other ingested alkaloids before a toxic dose can accumulate inside the body. The small amounts of nicotine that do get into the hemolymph of tobacco hornworm cannot cross the ion impermeable neural sheath and are eliminated via the malpighian tubules. Fiedler et al. (1993) studied rapid excretion mechanism in polyphagous larvae of the green hairstreak (*Callophrys rubi*), which feed on plants from 10 different families and excrete all alkaloids from their host plant *Genista tinctoria* in unchanged form. They even do so when raised on a non-host, such as *Lupinus polyphyllus*. Glucosinolates are found in the faeces of *S. gregaria* after feeding on *Schouwia purpurea* (Elsayed, 1994).

Some herbivores make use of plant allelochemicals for their own defence against predators by accumulating ingested chemicals in specialized tissues or in the integument (Willinger and Dobler, 2001; Nishida 2002; Ode, 2006). This type of defence is energetically advantageous to insects (Termonia et al., 2001; Kuhn et al., 2004). Compounds can be sequestered either directly or after biotransformation, such as by oxidation and conjugation. Sequestration of compounds requires selective transport and storage capabilities that prevent the toxin interfering with the physiological processes of the insect. More than 250 insect species have been reported to sequester plant metabolites belonging to wide range of chemical classes, such as alkaloids, cyanogenic glucosides, glucosinolates, isoprenoids and cucurbitacins (Nishida, 2002; Opitz and Muller, 2009). Kuhn et al. (2004) have demonstrated the molecular basis of the sequestration of plant compounds in leaf beetles. The same enzymes with minor modifications are involved in the transport and sequestration of many plant compounds in recently evolved leaf beetle species that produce chemical defences de novo. Few insect species have evolved several ways to efficiently utilize sequestered compounds for their own defence. For example, some aphid and flea beetle species have evolved a specific enzyme myrosinase that allows them to convert glucosinolates into their toxic breakdown products in a similar way as plants do and to release these as toxic and repellent volatiles into the air for their protection (Beran et al., 2014; Rahfeld et al., 2014). The tobacco hornworm *M. sexta* is actually nicotine-tolerant and it indeed accumulates nicotine in order to deter parasitoids because parasitoids of *M. sexta* are negatively affected by the nicotine in their hosts.

4.1.4. Metabolic Resistance to Plant Toxins

The herbivorous insects have developed metabolic resistance against plant chemicals (Glendinning, 2002). Metabolic resistance often results from the overproduction of 'detoxification enzymes' that can metabolize xenobiotics. This mechanism is often associated with phenotypic plasticity, such as induction of detoxification enzymes by the presence of plant xenobiotics in the diet of the insect. However, resistance to phytotoxins may also result from specific mutations in genes encoding enzymes (Wen et al., 2005), that increase the production of specific detoxification enzymes as well as their catalytic or transport properties (Brattsten, 1992; Despres et al., 2007; Kennedy and Tierney, 2013).

The detoxification of xenobiotics usually occurs in three phases. In phase I, the xenobiotic is modified by enzymes such as cytochrome P450 monooxygenase and carboxyl/ choline esterases. These enzymes incorporate nucleophilic functional group such as hydroxyl, carboxyl or amine group) and results in a more polar/hydrophilic substance. In phase II, the metabolite resulting from phase I is conjugated to endogenous molecules such as glutathione or a sugar molecule by enzymes such as glutathione-S-transferases (GSTs) and UDP-glycosyltransferases (UGTs), which further cause increase in polarity/hydrophilicity of the compound. In phase III, the phase II conjugated xenobiotic is transported out of the cell by cellular transporters such as by ATP-binding cassette (ABC) and solute carrier (SLC) family proteins.

During phase I, the P450s play a key role and have been intensively studied (Feyereisen, 2005). A wide range of allelochemicals, including furanocoumarins, terpenoids, glucosinolates, flavonoids and alkaloids can be metabolized by this enzyme of herbivorous insects (Despres et al., 2007; Feyereisen, 2012). The involvement of P450s in furanocoumarin resistance within the lepidopteran genus *Papilio* is one of the first report relating to plant-arthropod interactions (Berenbaum, 1982). Extensive research on the detoxification of plant compounds by insects has been carried out on the furanocoumarins of the plant families Apiaceae and Rutaceae. The larvae of the lepidopteran genera *Papilio* and *Helicoverpa* can metabolize furanocumarins by cytochrome-P450-dependent monooxygenases (cytochrome P450s) of the CYP6B-subfamily (Cohen et al., 1992; Ma et al., 1994). The furanocoumarin had induced more than 30 identified members of this subfamily in insects (Li et al., 2000; Petersen et al., 2001). However, the basal and inducible expression levels of the individual cytochrome P450s differ depending on the degree of specialization of the insect (Li et al., 2001). Many other studies have isolated P450s that are related to the detoxification of plant chemicals in various herbivorous species, such as the parsnip webworm *Depressaria pastinacella* (Cianfrogna et al., 2002), *M. sexta* (Stevens et al., 2000) and several *Helicoverpa* species (Li et al., 2002a). The larvae of *H. zea*, has evolved adavanced strategy to detoxify plant compounds. The larvae can use volatile plant signal molecules (i.e., jasmonate and salicylate) that activate herbivory-induced plant defence pathways and pre-emptively overproduce P450s that can metabolize host-plant toxins (Li et al., 2002b).

Another enzyme used in phase I is carboxylesterases that can hydrolyse ester bonds from various substrates with a carboxylic ester. Although esterases have been linked to pesticide resistance by sequestration, i.e., by binding to the target substance without modifying it, but their role in plant allelochemical defence remains elusive (Despres et al., 2007; Li et al., 2007). In some species of aphids, genes were duplicated which gave rise to elevated expression of esterases, such that 3% of their total protein content (Devonshire and Sawicki, 1979; Devonshire and Moores, 1982).

Phase II detoxification constitute the conjugation of xenobiotics to GSTs. The GST enzymes usually catalyse the conjugation of glutathione to electrophilic toxic molecules, increasing their solubility and facilitating their elimination by the insect (Enayati et al., 2005). This enzyme has been linked to allochemical tolerance in several insects, but most of evidences have been obtained from in vitro studies in numerous herbivore insects (Yu, 1996). The role for GSTs in tolerance to glucosinolates was also reported in the whitefly *B. tabaci* and in the mustard-feeding specialist *Scaptomyza nigrita* (Elbaz et al., 2012; Gloss et al., 2014). The polyphagous insect *Spodoptera frugiperda* are able to metabolize a variety of glucosinolate-derived thiocyanates conjugates by GSTs (Li et al., 2007). In the aphid *Myzus persicae*, the overproduction of GSTs is probably responsible for the adaptation of the insect to glucosinolates and isothiocyanates contained in its Brassicaceae host plants (Francis et al., 2005). As in P450s, the greater diversification of GST enzymes in generalist than in specialist herbivoresis thought to reflect the ability of such herbivores to adapt to a broader range of plant chemicals (Francis et al., 2001; 2005).

Another important class of conjugation enzymes is UDP-glycosyl-transferases (UGTs), which may be wide spread enzymes that allow herbivorous arthropods to adapt to xenobiotics (Ahn et al., 2014). This class of enzyme converts lipophilic aglycones into more hydrophilic glycosides by conjugation with UDP-glucose. UGTs have been reported to be involved in resistance of lepidopterans to the alkaloid capsaicin and the detoxification of benzoxazinoids (Ahn et al., 2011; Wouters et al., 2014). Only few reports documented the role of UGTs in the detoxification of xenobiotics by insects. Ahmad and Hopkins, (1993) documented that metabolism of plant compounds in *M. sexta* by UGTs. Recently, Luque et al. (2002) documented that gene encoding another UGT (BmUGT1), involved in the degradation of flavonoids and coumarins, has been isolated in the silkworm *Bombyx mori*.

In phase III, the phase II conjugated xenobiotic is transported out of the cell by cellular transporters. In several cases, these transporters can also act as a first line of defence, preventing allelochemicals entering the cell by rapid efflux without the need for modifications (sometimes this is referred to as phase 0). Gaertner et al. (1998) reported that excretion of nicotine and other alkaloids by the malpighian tubules in *M. sexta* is mediated by an ABC transporter. This finding was based on the observation that an inhibitor verapamil of the ABC transporter, blocked nicotine transport in the Malpighian tubules of *M. sexta*. In addition, Petschenka et al. (2013) found an ABC transporter that probably functions as cardenolide efflux carrier, and protected the nervous tissues of lepidopterans. The other transporters of the major facilitator superfamily, which include many transporters of the SLC family, were downregulated in *M. sexta* larvae fed on wild tobacco mutants that were unable to produce JA (Govind et al., 2010). The other members of the same superfamily were upregulated in herbivore insects shifted from their preferred host to less suitable host (de la Paz Celorio-Mancera et al., 2013).

4.2. Pharmacodynamic Responses

The pharmacodynamic response comprises adaptations at the level of interactions between allelochemicals and their target-site(s) (Taylor and Feyereisen, 1996; Kennedy and Tierney, 2013). This is also known as mechanism of decreased sensitivity. While many herbivores become adapted to xenobiotics and developed mechanisms that directly or indirectly divert these substances from their target sites, or reduce the target-site sensitivity. Although these mechanism are associated with herbivore resistance to pesticides (Van Leeuwen et al., 2010b; Ffrench-Constant, 2013; Feyereisen et al., 2015), but few studies have linked these mechanisms to insect resistance to plant allelochemicals, e.g., The target-site insensitivity to phytotoxins that confer resistance to the plant toxin ouabain, a cardenolide produced by several members of the Apocynaceae, is that of amino acid substitutions in the subunit of Na^+/K^+- ATPase and reported from four different insect orders (Lepidoptera, Coleoptera, Heteroptera and Diptera) (Agrawal et al., 2012; Dobler et al., 2012; Zhen et al., 2012; Dalla et al., 2013). Another target site based resistance mechanism is tolerance to L-canavanine, a non-proteinogenic amino acid of leguminous plants and reported in many insect species. The toxicity of L-canavanine is caused by its incorporation into proteins and replacement of L-arginine, e.g., the bruchid beetle *Caryedes brasiliensis* have specialized in feeding on the L-canavanine-rich seeds of *Dioclea megacarpa* and have evolved an arginyl-tRNA synthetase that can discriminate between L-canavanine and L-arginine, and thereby effectively avoids the adverse biochemical effect of L-canavanine (Rosenthal et al., 1976; Leisinger et al., 2013). In addition to target-site insensitivity, mechanisms of decreased exposure to L-canavanine have also been reported for the tobacco budworm *Heliothis virescens* (Melangeli et al., 1997).

5. Other Metabolic Resistance Mechanisms

In insects, detoxification enzymes play a major role in the development of resistance against plant toxins, but other metabolic resistance mechanisms have also been identified in various plant-insect interactions.

For example, the soybean cysteine protease inhibitor soyacystatin N (scN) activate an array of counter defence genes in the larvae of the cowpea bruchid *Callosobruchus maculates* to adapt to this toxin. The over expression of cathepsin-like cysteine proteases saturate the protease inhibitor molecules and play a basic role in the adaptation to ingested scN (Moon et al., 2004). Another study reported that when the polyphagous arctiid moth *Estigmene acrea* feeds on plants of Asteraceae, it not only sequesters pyrrolizidine alkaloids (PAs) present in the plants, but also detoxifies them by a specific flavin-dependent monooxygenase (Hartmann et al., 2005). This enzyme usually has general functions in an insect's primary metabolism, but in the arctiid moth they have evolved after gene duplication and catalyse N-oxidation of pyrrolizidine alkaloids (Langel and Ober, 2011). The in vitro expression of a similar N-oxygenase from the arctiid moth *Tyria jacobaeae* confirmed its high metabolic activity toward PAs. This activity enables the insect to feed on plants containing toxic PAs and to accumulate PA metabolites as a predator deterrent (Naumann et al., 2002).

References

Agrawal, A. A., Petschenka, G., Bingham, R. A., Weber, M. G. and Rasmann, S. (2012). Toxic cardenolides: chemical ecology and coevolution of specialized plant–herbivore interactions. *New Phytologist*, 194: 28-45.

Ahmad, S. A. and Hopkins, T. L. (1993). *B*-glycosylation of plant phenolics by phenol *B*-glucosyltransferase in larval tissues of the tobacco hornworm *Manduca sexta* (L.). *Insect Biochemistry and Molecular Biology*, 23: 581-589.

Ahn, S. J., Dermauw, W., Wybouw, N., Heckel, D. G., Van Leeuwen, T. (2014). Bacterial origin of a diverse family of UDP-glycosyltransferase genes in the *Tetranychus urticae* genome. *Insect Biochemistry and Molecular Biology*, 50: 43-57.

Ahn, S-J., Badenes-Pérez, F. R., Reichelt, M., Svatoš, A., Schneider, B., Gershenzon, J. and Heckel, D. G. (2011). Metabolic detoxification of capsaicin by UDP-glycosyltransferase in three *Helicoverpa* species. *Archives of Insect Biochemistry and Physiology,* 78: 104-118.

Ananthakrishnan, T. N. and Raman, A. (1986). *Dynamics of insect-plant interaction* (New Delhi: Oxford and IBH).

Ananthakrishnan, T. N., Dhileepan, K. and Padmanaban, B. (1985). Behavioural responses in terms of feeding and reproduction in some grasshoppers (Orthoptera: Insecta). *Proceedings of Indian Academy of Sciences (Animal Sciences)*, 94: 443-461.

Baker, T. (2009). Representations of odour plume flux are accentuated deep within the moth brain. *Journal of Biology*, 8: 16.

Barbehenn, R. V. and Constabel, C. P. (2011). Tannins in plant-herbivore interactions. *Phytochemistry,* 72: 1551-1565.

Barbehenn, R. V., Niewiadomski, J. and Kochmanski, J. (2013). Importance of protein quality versus quantity in alternative host plants for a leaf-feeding insect herbivore. *Oecologia*, 173: 1-12.

Becerra, J. X. (2003). Synchronous coadaptation in an ancient case of herbivory. *Proceedings of Natural Academy of Sciences USA*, 100: 12804-12807.

Beddington, J. (2010). Food security: contributions from science to a new and greener revolution. *Philosophical Transactions of the Royal Society London B Biological Sciences*, 365: 61-71.

Bede, J. C., Musser, R. O., Felton, G. W. and Korth, K. L. (2006). Caterpillar herbivory and salivary enzymes decrease transcript levels of *Medicago truncatula* genes encoding early enzymes in terpenoid biosynthesis. *Plant Molecular Biology*, 60: 519-531.

Behmer, S. T., Simpson, S. J. and Raubenheimer, D. (2002). Herbivore foraging in chemically heterogeneous environments: nutrients and secondary metabolites. *Ecology*, 83: 2489-2501.

Bennett, R. N. and Wallsgrove, R. M. (1994). Secondary metabolites in plant defence mechanisms. *New Phytologist*, 127: 617-633.

Beran, F., Pauchet, Y., Kunert, G., Reichelt, M., Wielsch, N., Vogel, H., Reinecke, A., Svatoš, A., Mewis, I., Schmid, D., Ramasamy, S., Ulrichs, C., Hansson, B. S., Gershenzon, J. and Heckel, D. G. (2014). *Phyllotreta striolata* flea beetles use host plant defense compounds to create their own glucosinolate-myrosinase system. *Proceedings of the National Academy of Sciences of the USA*, 111: 7349-7354.

Berenbaum, M. (1982). Coumarins and caterpillars: a case for coevolution. *Evolution,* 37: 163-179.

Berenbaum, M. R. (2001) Chemical mediation of coevolution: Phylogenetic evidence for Apiaceae and associates. *Annals of the Missouri Botanical Garden*, 88:45-59.

Bernays, E. A. and Chapman, R. F. (1994). *Host Plant Selection by Phytophagous Insects*. New York: Chapman & Hall.

Bos, J. I. B, Prince, D., Pitino, M., Prince, D., Pitino, M., Maffei, M. E., Win, J., Hogenhout, S. A. (2010). A functional genomics approach identifies candidate effectors from the aphid species *Myzus persicae* (green peach aphid). *PLoS Genetics*, 6: e1001216.

Brattsten, L. B. (1992). *Metabolic defenses against plant allelochemicals. In: Herbivores: their interactions with secondary plant metabolitesII: ecologicaland evolutionary processes*, (ed. Rosenthal, G. A. and Berenbaum, M. R.), San Diego: Academic, pp.175-242.

Bruce, T. J. A. and Pickett, J. A. (2011). Perception of plant volatile blends by herbivorous insects - Finding the right mix. *Phytochemistry*, 72: 1605-1611.

Cardé, R. and Willis, M. (2008). Navigational strategies used by insects to find distant, wind-borne sources of odor. *Journal of Chemical Ecology*, 34: 854-s866.

Chapman, R. F. (2003). Contact chemoreception in feeding by phytohagous insects. *Annual Review of Entomology*, 48: 455-484.

Cianfrogna, J. A., Zangerl, A. R., Berenbaum, M. R. (2002). Dietary and developmental influences on induced detoxification in an oligophage. *Journal of Chemical Ecology*, 28: 1349-1364.

Cohen, M. B., Schuler, M. A. and Berenbaum, M. R. (1992). A host-inducible cytochrome P-450 from a host specific caterpillar: molecular cloning and evolution. *Proceedings of Natural Academy of Sciences USA*, 89: 10920-10924.

Cornell, H. and Hawkins, B. (2003). Herbivore responses to plant secondary compounds: a test of phytochemical coevolution theory. *American Naturalist*, 161: 507-522.

Cotter, S. C. and Edwards, O. R. (2006). Quantitative genetics of prefence and performance on chickpeas in the noctuid moth, *Helicoverpa armigera. Heredity*, 96: 396-402.

Dalla, S., Swarts, H. G. P., Koenderink, J. B. and Dobler, S. (2013). Amino acid substitutions of Na, K-ATPase conferring decreased sensitivity to cardenolides in insects compared to mammals. *Insect Biochemistry and Molecular Biology*, 43: 1109-1115.

de la Paz Celorio-Mancera, M., Wheat, C. W., Vogel, H., Soderlind, L., Janz, N. andNylin, S. (2013). Mechanisms of macroevolution: polyphagous plasticity in butterfly larvae revealed by RNA-Seq. *Molecular Ecology*, 22: 4884-4895.

Dermauw, W., Wybouw, N., Rombauts, S., Menten, B., Vontas, J., Grbić, M., Clark, R. M., Feyereisen, R. and Van Leeuwen, T. (2013). A link between host plant adaptation and pesticide resistance in the polyphagous spider mite *Tetranychus urticae. Proceedings of the National Academy of Sciences of the USA*, 110: E113-E122.

Despres, L., David, J-P. and Gallet, C. (2007). The evolutionary ecology of insect resistance to plant chemicals. *Trends in Ecology and Evolution*, 22: 298-307.

Dethier, V. G. (1982). Mechanism of Host-Plant Recognition. *Entomologia Experimentalis et Applicata,* 31: 49-56.

Devonshire, A. L. and Moores, G. D. (1982). A carboxylesterase with broad substrate specificity causes organophosphorus, carbamate and pyrethroid resistance in peach-potato aphids (*Myzuspersicae*). *Pesticide Biochemistry and Physiology*, 18: 235-246.

Devonshire, A. L. and Sawicki, R. M. (1979). Insecticide-resistant *Myzus persicae* as an example of evolution by gene duplication. *Nature,* 280: 140-141.

Dobler, S., Dalla, S., Wagschal, V. and Agrawal, A. A. (2012). Community-wide convergent evolution in insect adaptation to toxic cardenolides by substitutions in the Na,K-ATPase. *Proceedings of the National Academy of Sciences of the USA*, 109: 13040-13045.

Dussourd, D. E. (2003). Chemical stimulants of leaf-trenching by cabbage loopers: Natural products, neurotransmitters, insecticides and drugs. *Journal of Chemical Ecology*, 29: 2023-2047.

Eigenbrode, S. D. and Espelie, K. E. (1995). Effects of plant epicuticular lipids on insect herbivores. *Annual Review of Entomology*, 49: 171-194.

Elbaz, M., Halon, E., Malka, O., Malitsky, S., Blum, E., Aharoni, A. and Morin, S. (2012). Asymmetric adaptation to indolic and aliphatic glucosinolates in the B and Q sibling species of *Bemisia tabaci* (Hemiptera: Aleyrodidae). *Molecular Ecology*, 21: 4533-4546.

Elsayed, G. (1994). *Does chronic toxicity display biochemical coevolution in the desert locust, Schistocerca gregaria*? [Ph.D. thesis] [Egypt]: University of Cairo.

Enayati, A. A., Ranson, H. and Hemingway, J. (2005). Insect glutathione transferases and insecticide resistance. *Insect Molecular Biology*, 14: 3-18.

Feyereisen, R. (2005). Insect cytochrome P450. In: *Comprehensive Molecular Insect Science* (ed. Gilbert, L. I. and Gill, S. S.), Elsevier, Oxford, UK, pp 1-77.

Feyereisen, R. (2012). Insect CYP genes and P450 enzymes. In: (ed. Gilbert, L. I.) *Insect molecular biology and biochemistry.* Elsevier, Academic Press, London.

Feyereisen, R., Dermauw, W. and Van Leeuwen, T. (2015). Genotype to phenotype, the molecular and physiological dimensions of resistance in arthropods. *Pesticide Biochemistry and Physiology*, 121: 61-77.

Ffrench-Constant, R. H. (2013). The molecular genetics of insecticide resistance. *Genetics,* 194: 807-815.

Fiedler, K., Krug, E. and Proksch, P. (1993). Complete elimination of host-plant quinolizidine alkaloids by larvae of a polyphagous lycaenid butterfly *Callophrys rubi. Oecologia.* 94: 441-445.

Fox, C. W., Stillwell, R. C., Amarillo-S, A. R., Czesak, M. E. and Messina, F. J. (2004). Genetic architecture of population differences in oviposition behaviour of the seed beetle *Callosobruchus maculatus*. *Journal of Evolutionary Biology*, 17: 1141-1151.

Francis F., Vanhaelen, N. and Haubruge, E. (2005). Glutathione S-transferases in the adaptation to plant secondary metabolites in the *Myzus persicae* aphid. *Archives of Insect Biochemistry and physiology*, 58: 166-174.

Francis, F., Lognay, G., Wathelet, J-P. and Haubruge, E. (2001). Effects of allelochemicals from first (Brassicaceae) and second (*Myzus persicae* and *Brevicoryne brassicae*) trophic levels on *Adalia bipunctata*. *Journal of Chemical Ecology*, 27: 243-256.

Freeland, W. J. and Janzen, D. H. (1974). Strategies in herbivory by mammals: the role of plant secondary compounds. *American Naturalist*, 108: 269-289.

Gaertner, L. S., Murray, C. L. and Morris, C. E. (1998). Transepithelial transport of nicotine and vinblastine in isolated malpighian tubules of the tobacco hornworm (*Manduca sexta*) suggests a P-glycoprotein-like mechanism. *Journal of Experimental Biology*, 201: 2637-2645.

Glendinning, J. I. (2002). How do herbivorous insects cope with noxious secondary plant compounds in their diet? *Entomologia Experimentalis et Applicata*, 104: 15-25.

Glendinning, J. I., Davis, A. and Ramaswamy, S. (2002). Contribution of different taste cells and signaling pathways to the discrimination of "bitter" taste stimuli by an insect. *Journal of Neuroscience*, 22: 7281-7287.

Gloss, A. D., Vassão, D. G., Hailey, A. L., Dittrich, A. C. N., Schramm, K., Reichelt, M., Rast, T. J., Weichsel, A., Cravens, M. G., Gershenzon, J., Montfort, W. R. and Whiteman, N. K. (2014). Evolution in an ancient detoxification pathway is coupled with a transition to herbivory in the Drosophilidae. *Molecular Biology and Evolution*, 31: 2441-2456.

Govind, G., Mittapalli, O., Griebel, T., Allmann, S., Bocker, S. and Baldwin, I. T. (2010). Unbiased transcriptional comparisons of generalist and specialist herbivores feeding on progressively defenseless *Nicotiana attenuata* plants. *PLoS One*, 5: e8735.

Hartmann, T., Theuring, C., Beuerle, T., Klewer, N., Schulz, S., Singer, M. S. and Bernays, E. A. (2005). Specific recognition, detoxification and metabolism of pyrrolizidine alkaloids by the polyphagous arctiid *Estigmene acrea*. *Insect Biochemistry and Molecular Biology*, 35: 391-411.

Helmus, M. R. and Dussourd, D. E. (2005). Glues or poisons: which triggers vein cutting by monarch caterpillars? *Chemoecology,* 15: 45-49.

Houlihan, D. F. (1969). The structure and behaviour of *Notiphila riparia* and *Erioptera squalida*, two root-piercing insects. *Journal of Zoology*, 159: 249-267.

Kant, M. R., Jonckheere, W., Knegt, B., Lemos, F., Liu, J., Schimmel, B. C. J., Villarroel, C. A., Ataide, L. M. S., Dermauw, W., Glas, J. J., Egas, M., Janssen, A., Van Leeuwen, T., Schuurink, R. C., Sabelis, M. W. and Alba, J. M. (2015).

Mechanisms and ecological consequences of plant defence induction and suppression in herbivore communities. *Annals of Botany*, 115: 1015-1051.

Karban, R. and Agrawal, A. A. (2002). Herbivore offense. *Annual Review of Ecology, Evolution, and Systematics,* 33: 641-664.

Kennedy, C. and Tierney, K. (2013). Xenobiotic protection/resistance mechanisms in organisms. In: *Environmental toxicology* (ed. Laws, E. A.), New York: Springer.

Krieger, R. I., Feeny, P. P. and Wilkinson, C. F. (1971). Detoxication enzymes in the guts of caterpillars: an evolutionary answer to plant defenses? *Science,* 172: 579-580.

Kuhn, J., Petersson, E. M., Feld, B. K., Burse, A., Termonia, A., Pasteels, J. M. and Boland, W. (2004). Selective transport systems mediate sequestration of plant glucosides in leaf beetles: a molecular basis for adaptation and evolution. *Proceedings of Natural Academy of Sciences USA*, 89: 13808-13813.

Langel, D. and Ober, D. (2011). Evolutionary recruitment of a flavin-dependent monooxygenase for stabilization of sequestered pyrro-lizidine alkaloids in arctiids. *Phytochemistry,* 72: 1576-1584.

Leisinger, A-K., Janzen, D. H., Hallwachs, W. and Igloi, G. L. (2013). Amino acid discrimination by the nuclear encoded mitochondrial arginyl-tRNAsynthetase of the larva of a bruchid beetle (*Caryedes brasiliensis*) from northwestern Costa Rica. *Insect Biochemistry and Molecular Biology*, 43: 1172-1180.

Li, W., Schuler, M. and Berenbaum, M. (2001). Molecular analysis of multiple CYP6B genes from polyphagous *Pailio* Species. *Insect Biochemistry and Molecular Biology*, 31: 999-1011.

Li, X., Berenbaum, M. R. and Schuler, M. A. (2002a). Plant allelochemicals differentially regulate *Helicoverpa zea* cytochrome P450 genes. *Insect Molecular Biology*, 11: 343-351.

Li, X., Schuler, M. A. and Berenbaum, M. R. (2002b). Jasmonate and salicylate induce expression of herbivore cytochrome P450 genes. *Nature*, 419: 712-715.

Li, X., Schuler, M. and Berenbaum, M. (2007). Molecular mechanisms of metabolic resistance to synthetic and natural xenobiotics. *Annual Review of Entomology,* 52: 231-253.

Li, X., Zangerl, A. R., Schuler, M. A. and Berenbaum, M. R. (2000). Cross resistance to alpha-cypermethrin after xanthotoxin ingestion in *Helicoverpa zea* (Lepidoptera; Noctuidae). *Journal of Economic Entomology*, 93: 18-25.

Luque, T., Okano, K. and O'Reilly, D. R. (2002). Characterization of a novel silkworm (*Bombyx mori*) phenol UDP-glucosyltransferase. *European Journal of Biochemistry*, 269: 819-825.

Martin, J. P., Beyerlein, A., Dacks, A. M., Reisenman, C. E., Riffell, J. A., Hand, L. and Hildebrand, J. G. (2011). The neurobiology of insect olfaction: Sensory processing in a comparative context. *Progress in Neurobiology*, 95: 427-447.

Mattson, W. J. (1980). Herbivory in relation to plant nitrogen content. *Annual Review of Ecology and Systematics*, 11: 119-161.

Melangeli, C., Rosenthal, G. A. and Dalman, D. L. (1997). The biochemical basis for L-canavanine tolerance by the tobacco budworm *Heliothis virescens* (Noctuidae). *Proceedings of the National Academy of Sciences of the USA*, 94: 2255-2260.

Mitchell, C., Brennan, R. M., Graham, J. and Karley, A. J. (2016). Plant defense against herbivorous pests: exploiting resistance and tolerance traits for sustainable crop protection. *Frontiers in Plant Science*, 7: 1132.

Moon, J., Salzman, R. A., Ahn, J.-E., Koiwa, H. and Zhu-Salzman, K. (2004). Transcriptional regulation in cowpea bruchid guts during adaptation to a plant defence protease inhibitor. *Insect Molecular Biology*, 13: 283-291.

Musser, R. O., Hum-Musser, S. M., Eichenseer, H., Peiffer, M., Ervin, G., Murphy, J. B., Felton, G. W. (2002). Caterpillar saliva beats plant defences. *Nature,* 416: 599-600.

Naumann, C., Hartmann, T. and Ober, D. (2002). Evolutionary recruitment of a flavin-dependent monooxygenase for the detoxification of host plant-acquired pyrrolizidine alkaloids in the alkaloid-defended arctiid moth *Tyria jacobaeae*. *Proceedings of national Academy of Sciences USA*, 99: 11223-11228.

Nealis, V. G. and Nault, J. R. (2005). Seasonal changes in foliar terpenes indicate suitability of Douglas-fir buds for western spruce budworm. *Journal of Chemical Ecology*, 31: 683-696.

Nishida, R. (2002). Sequestration of defensive substances from plants by Lepidoptera. *Annual Review of Entomology,* 47: 57-92.

Nyman, T. and Julkunen-Tiitto, R. (2000). Manipulation of the phenolic chemistry of willows by gall-inducing sawflies. *Proceedings of the National Academy of Sciences of the USA*, 97: 13184-13187.

Ode, P. J. (2006). Plant chemistry and natural enemy fitness: effects on herbivore and natural enemy interactions. *Annual review of Entomology*, 47: 57-92.

Opitz, S. W. and Muller, C. (2009). Plant chemistry and insect sequestration. *Chemoecology,* 19: 117-154.

Pearse, I. S., Harris, D. J., Karban, R. and Sih, A. (2013). Predicting novel herbivore-plant interactions. *Oikos,* 122: 1554-1564.

Petersen, R. A., Zangerl, A. R., Berenbaum, M. R. and Schuler, M. A (2001). Expression of CYP6B1 and CYP6B3 cytochrome P450 monooxygenases and furanocoumarin metabolism in different tissues of *Papilio polyxenes* (Lepidoptera: *Papilionidae*). *Insect Biochemistry and Molecular Biology,* 31: 679-690.

Petschenka, G., Pick, C., Wagschal, V. and Dobler, S. (2013). Functional evidence for physiological mechanisms to circumvent neurotoxicity of cardenolides in an adapted and a non-adapted hawk-moth species. *Proceedings of the Royal Society B: Biological Sciences,* 280: 20123089.

Popp, J., Peto, K. and Nagy, J. (2013). Pesticide productivity and food security. A review. *Agronomy for Sustainable Development*, 33:243-255.

Pulliam, H. R. (1975). Diet optimization with nutrient constraints. *American Naturalist,* 109: 765-768.

Rahfeld, P., Kirsch, R., Kugel, S., Wielsch, N., Stock, M., Groth, M., Boland, W. and Burse, A. (2014). Independently recruited oxidases from the glucose-methanol-choline oxidoreductase family enabled chemical defences in leaf beetle larvae (subtribe *Chrysomelina*) to evolve. *Proceedings of the Royal Society B: Biological Sciences*, 281: 20140842.

Rapport, D. J. (1980). Optimal foraging for complementary resources. *American Naturalist*, 116: 324-346.

Ratzka, A., Vogel, H., Kliebenstein, D. J., Mitchell-Olds, T. and Kroymann, J. (2002). Disarming the mustard oil bomb. *Proceedings of the National Academy of Sciences of the USA*, 99: 11223-11228.

Riffell, J. A., Lei, H., Christensen, T. A. and Hildebrand, J. G. (2009). Characterization and coding of behaviorally significant odour mixtures. *Current Biology*, 19: 335-340.

Rosenthal, G., Dahlman, D. and Janzen, D. (1976). A novel means for dealing with L-canavanine, a toxic metabolite. *Science,* 192: 256-258.

Schoonhoven, L. M. and Dethier, V. G. (1966). Sensory aspects of host plant discrimination by lepidopterous larvae. *Archives Néerlandaises de Zoologie*, 16: 497-530.

Schoonhoven, L. M., van Loon, J. J. A., Dicke, M. (2005). *Insect-plant Biology.* Oxford University Press: Oxford, UK.

Schoonhoven, L. M., Jermey, T., van Loon, J. J. A. (1998). *Insect Plant-biology: from Physiology to Evolution.* Chapman and Hall: London, UK.

Shukle, R. H., Subramanyam, S., Salzmann, K. A. and Williams, C. E. (2010). Ultrastructural changes in the midguts of hessian fly larvae feeding on resistant wheat. *Journal of Insect Physiology*, 56: 754-760.

Silva, C. P., Terra, W. R., Grossi de Sá, M. F., Samuels, R. I., Isejima, E. M., Bifano, T. D. and Almeida, J. S. (2001). Induction of digestive α-amylases in larvae of *Zabrotessubfasciatus* (Coleoptera: Bruchidae) in response to ingestion of common bean α-amylase inhibitor 1. *Journal of Insect Physiology,* 47:1283-1290.

Simpson, S. J. and Simpson, C. L. (1989). The mechanisms of nutritional compensation by phytophagous insects. In: (ed. Bernays, E. A.) *Insect-Plant Interactions.* Boca Raton (FL), CRC Press, Pp 111-160.

Sobhy, I. S., Erb, M., Lou, Y. and Turlings, T. C. J. (2014). The prospect of applying chemical elicitors and plant strengtheners to enhance the biological control of crop pests. *Philosophical Transactions of the Royal Society London B Biological Sciences*, 369: 20120283.

Stevens, J. L., Snyder, M. J., Koener, J. F. and Feyereisen, R. (2000). Inducible P450s of the CYP9 family from larval *Manduca sexta* midgut. *Insect Biochemistry and Molecular Biology*, 30: 559-568.

Taylor, M. and Feyereisen, R. (1996). Molecular biology and evolution of resistance of toxicants. *Molecular Biology and Evolution*, 13: 719-734.

Termonia, A., Hsiao, T. H., Pasteels, J. M. and Milinkovitch, M. C. (2001). Feeding specialization and host derived chemical defense in Chrysomeline leaf beetles did not lead to an evolutionary dead end. *Proceedings of National Academy of Sciences USA*, 98: 3909-3914.

Van Leeuwen, T., Vontas, J., Tsagkarakou, A., Dermauw, W. and Tirry, L. (2010). Acaricide resistance mechanisms in the two-spotted spider mite *Tetranychus urticae* and other important Acari: a review. *Insect Biochemistry and Molecular Biology*, 40: 563-572.

Walling, L. L. (2009). Adaptive defense responses to pathogens and pests. *Advances in Botanical Research,* 51: 551-612.

War, A. R., Paulraj, M. G., Ahmad, T., Buhroo, A. A., Hussain, B., Ignacimuthu, S. and Sharma, H. C. (2012). Mechanisms of plant defense against insect herbivores. *Plant Signaling & Behavior*, 7:1306-1320.

Webster, B., Bruce, T., Pickett, J. and Hardie, J. (2010). Volatiles functioning as host cues in a blend become nonhost cues when presented alone to the black bean aphid. *Animal Behaviour*, 79: 451-457.

Wen, Z., Baudry, J., Berenbaum, M. R. and Schuler, M. A. (2005). Ile115Leu mutation in the SRS1 region of an insect cytochrome P450 (CYP6B1) comprises substrate turnover via changes in the predicted product release channel. *Protein Engineering, Design and Selection*, 18: 191-199.

Whittaker, R. H. and Feeny, P. P. (1971). Allelochemics: chemical interactions between species. *Science,* 171: 757-770.

Will, T., Tjallingii, W. F., Thonnessen, A., Van Bel, A. J. E. (2007). Molecular sabotage of plant defense by aphid saliva. *Proceedings of the National Academy of Sciences of the USA*, 104: 10536-10541.

Willinger, G. and Dobler, S. (2001). Selective sequestration of iridoid glycosides from their host plants in *Longi tarsus* flea beetles. *Biochemical Systematics and Ecology*, 29: 335-346.

Wink, M. and Schneider, D. (1990). Fate of plant-derived secondary metabolites in three moth species (*Syntomis mogadorensis, Syntomeida epilais,* and*Creatonotos transiens*). *Journal of Comparative Physio-logy*, 160: 389-400.

Wittstock, U., Agerbirk, N., Stauber, E. J., Olsen, C. E., Hippler, M., Mitchell-Olds, T., Gershenzon, J. and Vogel, H. (2004). Successful herbivore attack due to metabolic diversion of a plant chemical defense. *Proceedings of the National Academyof Sciences of the USA*, 101: 4859-4864.

Wouters, F. C., Reichelt, M., Glauser, G., Reichelt, M. Glauser, G., Bauer, E., Erb, M., Gershenzon, J. and Vassão, D. G. (2014). Reglucosylation of the benzoxazinoid DIMBOA with inversion of stereochemical configuration is a detoxification strategy in lepidopteran herbivores. *Angewandte Chemie*, 126: 11502-11506.

Yu, S. J. (1996). Insect glutathione S-transferases. *Zoological Studies*, 35: 9-19.

Zagrobelny, M., Bak, S., Rasmussen, A. V., Jorgensen, B., Naumann, C. M. and Møller, B. L. (2004). Cyanogenic glucosides and plant-insect interactions. *Phytochemistry,* 65: 293-306.

Zangerl, A. R. (1990). Furanocoumarin induction in wild parsnip: evidence for an induced defense against herbivores. *Ecology,* 71:1926-1932.

Zarate, S. I., Kempema, L. A. and Walling, L. L. (2007). Silverleaf whitefly induces salicylic acid defenses and suppresses effectual jasmonic acid defenses. *Plant Physiology*, 143: 866-875.

Zhang, P. J., Zheng, S. J., van Loon, J. J. A., Boland, W., David, A., Mumm, R. and Dickea, M. (2009). Whiteflies interfere with indirect plant defense against spider mites in Lima bean. *Proceedings of the National Academy of Sciences of the USA*, 106: 21202-21207.

Zhang, P., Zhu, X. and Huang, F. Liu, Y., Zhang, J., Lu, Y. and Ruan, Y. (2011). Suppression of jasmonic acid-dependent defense in cotton plant by the mealy bug *Phenacoccus solenopsis*. *PLoSOne*, 6: e22378.

Zhen, Y., Aardema, M. L., Medina, E. M., Schumer, M. and Andolfatto, P. (2012). Parallel molecular evolution in an herbivore community. *Science,* 337: 1634-1637.

In: Recent Advances in Plant Research
Editors: B. Welch and M. Wilkerson
ISBN: 978-1-53614-170-2

Chapter 6

MID-CRETACEOUS ANGIOSPERM FLOWERS IN MYANMAR AMBER

George Poinar, Jr.*

Department of Integrative Biology,
Oregon State University, Corvallis, OR, US

ABSTRACT

Amberization is a unique type of fossilization that preserves both animal and plant parts for millions of years through a unique fixation- dehydration process. Flowers in amber often show details of their reproductive parts that are not evident in blooms preserved in sedimentary deposits. The present chapter discusses angiosperm flowers that have been found in mid-Cretaceous Myanmar (Burmese) amber. Two new genera are described, *Chenocybus allodapus* gen et sp. nov. and *Diaphoranthus burmensis* gen et sp. nov. These two new genera, along with other flowers described from this amber source, namely, *Antiquifloris latifibris*, *Cascolaurus burmitis*, *Endobeuthos paleosum, Eoepigynia burmensis*, *Jamesrosea burmensis*, *Lachnociona terriae, Micropetasos burmensis, Palaeoanthella huangii, Programinis burmitis, Programinis laminatus, Tropidogyne pikei* and *Tropidogyne pentaptera*, are considered extinct at the generic level. However, the level of extinction of the various lineages varies and they can be arranged in roughly three groups. Group one consists of flowers with no obvious affinities to any extant lineages. Group 2 are flowers with possible affinities to extant lineages above the family level. Group 3 are flowers that can be placed in extant families. Comments on possible pollinators in the Myanmar amber forest and extinction events regarding Myanmar amber flowers are included.

1. INTRODUCTION

Fossil angiosperm flowers are very rare since most flowers found in subtropical-tropical forests do not have an opportunity to become fossilized in sedimentary deposits. Only plants growing near lowland basins where their remains are washed into depressions and rapidly covered by sediment have an opportunity to become fossilized. In such cases, only portions of the foliage are usually recovered since the floral parts are too fragile to persist. Not only does amber capture flowers from angiosperms growing in forests from sea level to high elevations but it has the ability to preserve the delicate floral parts as well. This is due to the abilities of plant resins to infiltrate, fix, and dehydrate embedded tissue. During these processes, oxygen is restricted and associated decay-causing microorganisms are destroyed. Terpenoids and resin acids are responsible for tissue fixation, while sugars and terpines replace water by inert dehydration (Poinar and Hess, 1985). Angiosperm flowers in Myanmar amber are significant in showing floral diversity that occurred at that period in the earth's history.

The Myanmar amber flowers discussed in the present work originated from mines excavated in the Hukawng Valley and located southwest of Maingkhwan in Kachin State (26º20´N, 96º36´E) in Myanmar. Based on paleontological evidence, these sites were dated to the late Albian of the Early Cretaceous (Cruickshank and Ko, 2003), placing the age at 97 to 110 Ma. A more recent study using U-Pb zircon dating determined the age to be 98.79 ± 0.62 Ma or at the Albian/Cenomanian boundary (Shi et al., 2012). Nuclear magnetic resonance (NMR) spectra and the presence of araucaroid wood fibers in amber samples from the Noije Bum 2001 Summit Site indicate an araucarian tree source for the amber (Poinar et al., 2007a).

The Myanmar amber forest was a tropical rainforest where the seasons were only marked by the amount of rain that fell each succeeding month (Poinar and Poinar, 2008). The dominant plants were mostly gymnosperms such as conifers, ginkgos, and cycads. Ancient ferns, liverworts, lycopods, equisetums, and mosses were common. The most important tree from our perspective was a member of the Araucariaceae that produced the resin that later polymerized into amber. Based on findings in Myanmar amber, there were numerous lineages of angiosperms present at that time. These angiosperm flowers can be roughly grouped into three groups.

Group one consists of flowers with no obvious affinities to any extant lineages. Group 2 are flowers with possible affinities to extant lineages above the family level. Group 3 are flowers that can be placed in extant families. Comments on possible pollinators in the Myanmar amber forest and extinction events regarding the Myanmar amber flowers are included.

1.2. Myanmar Amber Flowers with No Obvious Affinities to Extant Groups

1.2.1. Antiquifloris Latifibris Poinar, Buckley and Chen (2016)

This perfect flower has a perianth composed of 14 separate spirally arranged fleshy tepals, a central cluster of 12 stamens with tetrasporangiate anthers that possess extended apical connectives, and an ovary surrounded by the cluster of stamens. The style is single, short and narrow and bears a minute stigma. The bases of the 3-veined, short, broad filaments are composed of vacuolated parenchyma and surrounded by white, pulvinate tissue. No known extant plant family has these features.

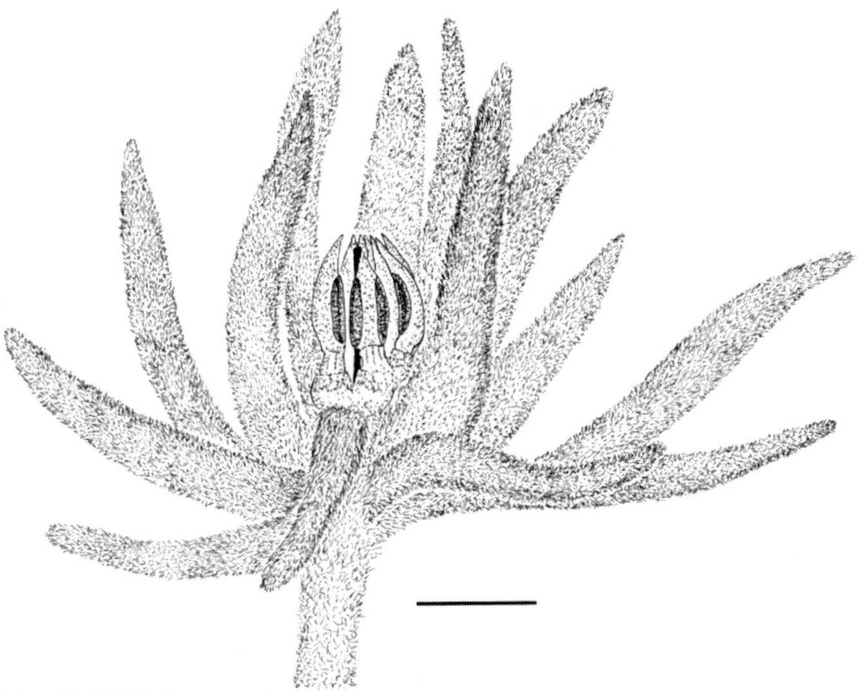

Figure 1. Flower of *Antiquifloris latifibris*. Scale bar = 3.8 mm.

1.2.2. Micropetasos Burmensis Poinar, K. L. Chambers and Wunderlich (2013)

It is unusual to find a single fossil inflorescence with 18 attached flowers, as in the case of *Micropetasos burmensis*. The flowers are hypogynous and about 1 mm in diameter. The perianth consists of 5 spreading, often unequal, basally connate sepals.

The petals are absent. The numerous stamens are tightly clustered around the pistil. The anthers are bisporangiate and vary in length. The gynoecium consists of a single carpel, with a short, curved style and inconspicuous stigma (Figure 2). The pollen is triaperturate. The flowers show no clear affinity with any modern family.

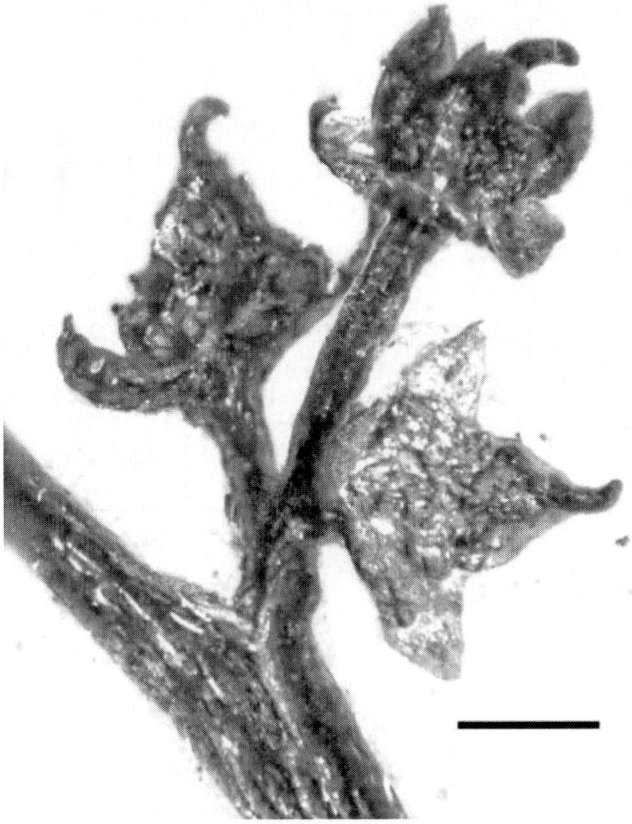

Figure 2. Three flowers of *Micropetasos burmensis*. Scale bar = 0.38 mm.

1.2.3. Chenocybus Allodapus Gen. Et Sp. Nov.

This is a very strange flower with the floral elements arranged in a trimerous, whorled condition. The petals are thick, fleshy and spirally arranged, the anthers are large and shield-like and bear spherical, inaperturate pollen grains. The superior ovary lacks a style, but has a stigmatic crest. The fruits are capsules containing indehiscent fruitlets.

- *Chenocybus* gen. n.
- Type species: *Chenocybus allodapus* gen. et sp. nov.
- *Chenocybus allodapus* gen. et sp. nov. (Figures 3-8)

1.2.3.1. Diagnosis

Flower asymmetric, trimerous with 3 connate sepals positioned at the base of the peduncle (Figures 3, 4). The first of two floral whorls consists of 6 petals and 3 stamens. A short peduncle extension leads to the second whorl that includes 6 petals, 6 stamens, 2

Figure 3. Holotype of *Chenocybus allodapus* gen. et sp. nov. in Myanmar amber. Arrow shows a damaged stamen on a separate twig. Scale bar = 2.0 mm.

ovaries and one dehisced capsule (Figure 4). The stamens have narrow, elongate filaments and broad, flat two-valved anthers containing spherical pollen grains (Figures 5, 6) The ovary with a sessile stigma is borne on an extended narrow stalk (Figures 5,7). The fruit is a capsule filled with small indehiscent fruitlets containing seeds (Figure 8).

1.2.3.2. Description

Length of flower (not including peduncle), 7.0 mm; width of flower, 7.2 mm; length of peduncle, 4.4 mm; perianth members in spiral sequence, length of sepals, 2.0 mm; length of petals, 1.0-7.3 mm; stamens club shaped, length of filaments, 2.3-3.6 mm; length of anthers, 0.9-1.3 mm; pollen round, 25-32 μm in diameter; ovaries superior, positioned on extended stalks 2.3-3.2 mm in length; stigmas 107-370 μm in length; single dehisced capsule 644 μm in length and 686 μm in width; fruitlets protruding from dehisced capsule 105- 140 μm in greatest diameter; seeds (inside fruitlets) oval, 48-60 μm in length, surface with a dark longitudinal ridge or furrow.

Figure 4. Floral parts of *Chenocybus allodapus* gen. et sp. nov. Solid back areas represent stamens. Stippled areas represent developing pistils. White structures represent petals. Arrow shows sepals positioned at base of peduncle. Scale bar = 1.6 mm.

The long slender filaments subtending the pistils and capsules, stamen structure and extended floral axis are characters unique to *Chenocybus allodapus* gen. et sp. nov. The anthers open by a longitudinal slit located between the two locules. The pollen grains (Figure 6) are spherical, inaperturate monads that superficially resemble those of the extant annonid, *Polyalthia laui* Merrill. Ornamentation of the pollen grains is verrucate and superficially resembles that of pollen of the extant annonid *Miliusa balansae* Finet and Gagnepain (Gan et al., 2015).

Holotype: accession # B-An-6 deposited in the Poinar amber collection maintained at Oregon State University, Corvallis, Oregon.

Type locality and horizon: Hukawng Valley southwest of Maingkhwan in Kachin State (26º20´N, 96º36´E), Myanmar, uppermost Albian-lowermost Cenomanian (mid-Cretaceous).

Etymology: Generic name from the Greek "chen"= goose and the Greek "kybe" = head in reference to the shape of the peduncles and capsules. The specific epithet is taken from the Greek "allodapos" = belonging to another kind.

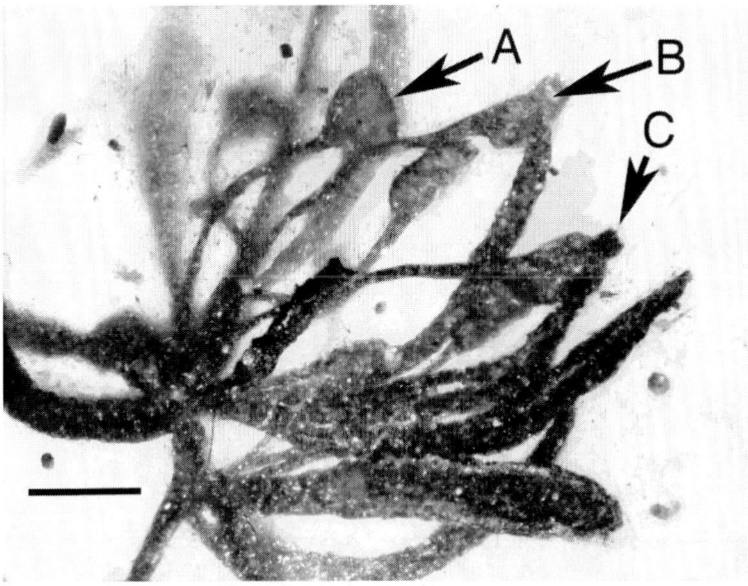

Figure 5. *Chenocybus allodapus* gen. et sp. nov. A = dehisced capsule. B = anther. C = stigma on developing ovary. Scale bar = 1.0 mm.

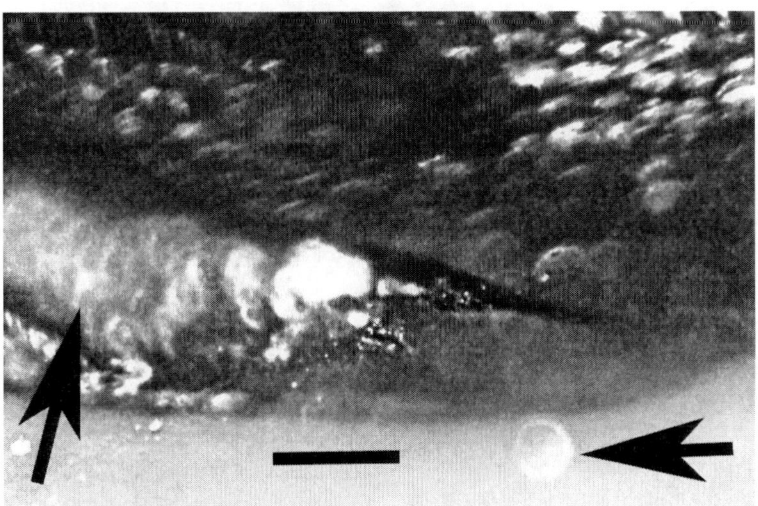

Figure 6. Exposed pollen grains (left arrow) in damaged anther of *Chenocybus allodapus* gen. et sp. nov. Right arrow shows a released pollen grain. Scale bar = 70 μm.

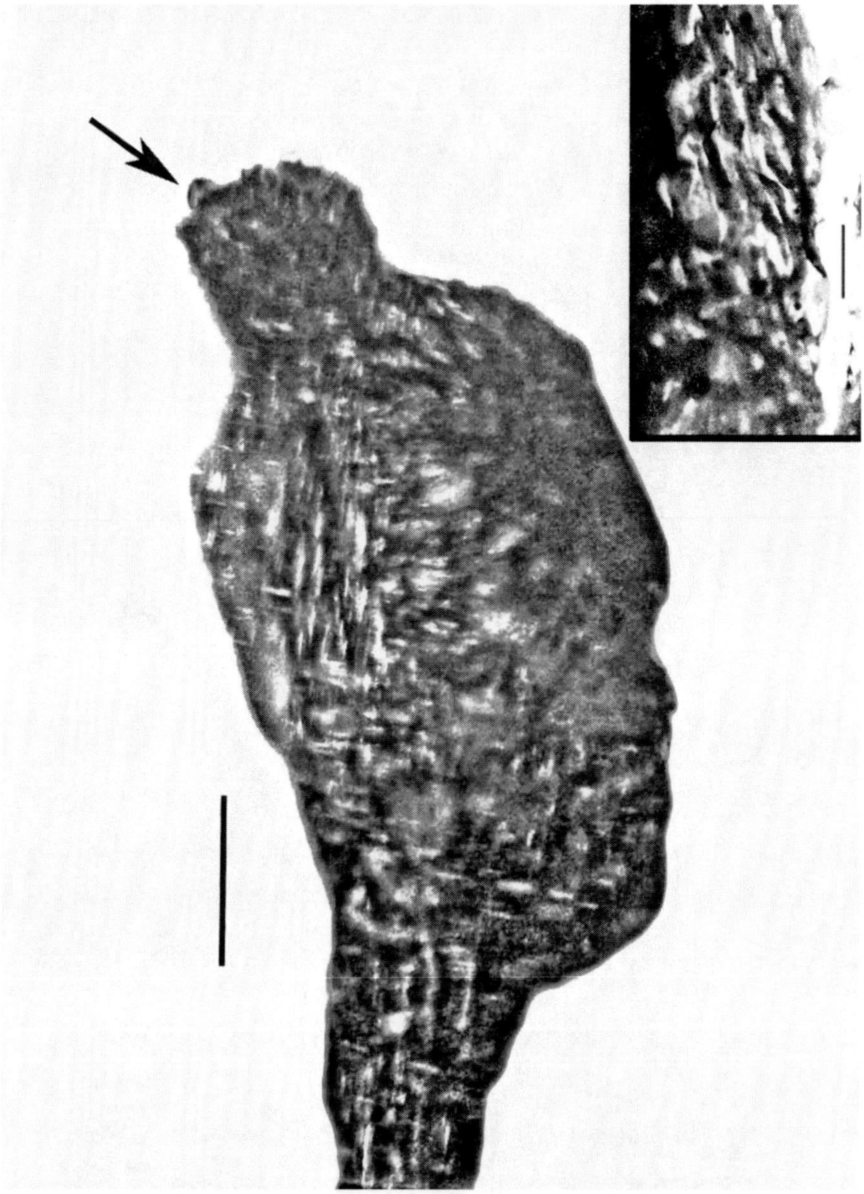

Figure 7. Mature ovary of *Chenocybus allodapus* gen. et sp. nov. Arrow shows pollen grain on stigma. Scale bar = 65 μm. Insert shows sclerid cells on surface of ovary. Scale bar = 10 μm.

1.2.3.3. Conclusion

The fossil cannot be assigned to an extant family, even though the trimerous and whorled state of the floral elements, thick, fleshy, spirally arranged petals, large shield-like anthers bearing spherical, inaperturate pollen grains, a superior ovary lacking a style, but with a stigmatic crest, and fruits with indehiscent fruitlets are features the fossil shares with some members of the Annonaceae (Chatrou et al., 2004, Croat, 1978, Heywood, 1993, Endress, 1994, Stebbins, 1974, Thorne, 1974). Thorne (1974) placed the Annonaceae in the suborder Annonineae, a large, diverse collection of generally relict families with unspecialized, primitive features that inhabit moist, tropical and warm-temperate forests. However, the long slender stalks of *Chenocybus allodapus* gen. et sp. nov. subtending the carpels, stamen structure and extended floral axis are unique characters of the fossil that show no clear affinity with any modern family.

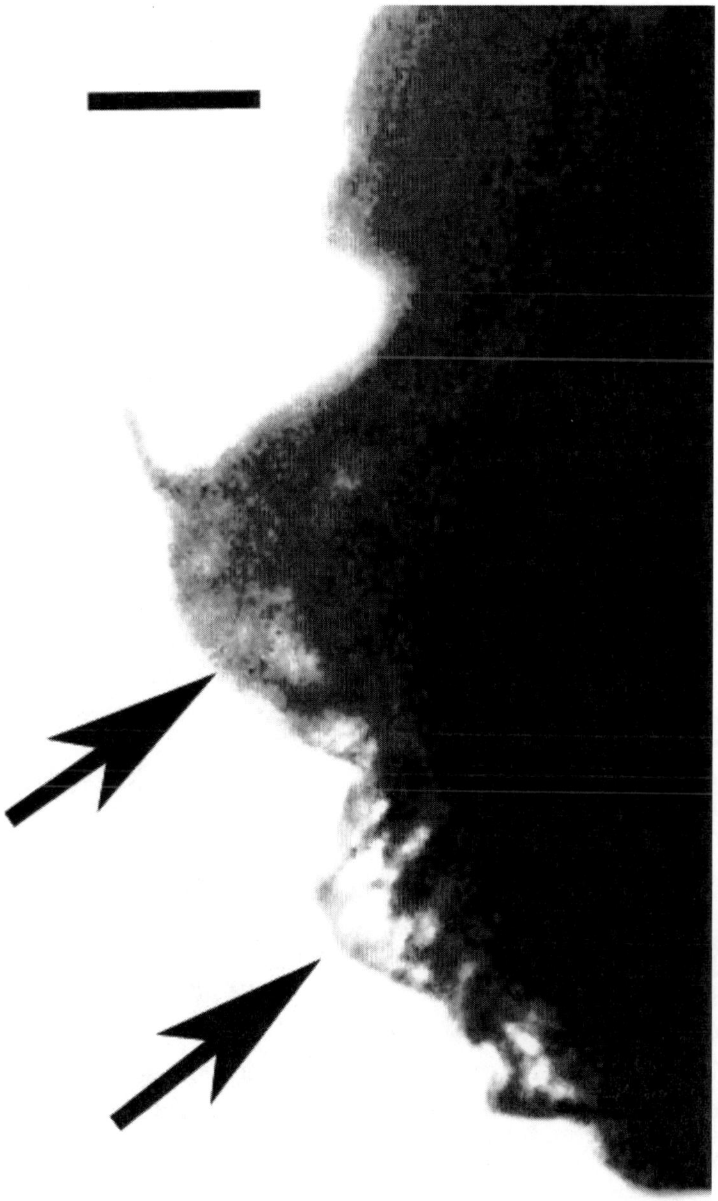

Figure 8. Two fruitlets (arrows) protruding from a dehisced capsule of *Chenocybus allodapus* gen. et sp. nov. Scale bar = 75 μm.

1.2.4. Diaphoranthus Burmensis Gen. Et Sp. Nov.

This unique Myanmar amber flower possesses primitive athecal anthers that are positioned along the tips of the filaments. Pollen grains are produced in sporogenous tissue in the center of the anthers and the oblong, monocolpate pollen grains are released through the anther walls. The anther structure and method of pollen formation and release distinguish this flower from all known extinct and extant angiosperm flowers. The fossil flower is in anthesis and pollen grains can be detected in the anthers positioned at the tips of the filaments and associated with the stigmas.

- Genus *Diaphoranthus* gen. nov.
- Type species: *Diaphoranthus burmensis* gen et sp. nov.
- *Diaphoranthus burmensis* sp. nov. (Figures 9-16).

1.2.4.1. Diagnosis

Flower small, pedicellate, bisexual, actinomorphic, 5-merous, with a single perianth whorl of five separate, spreading tepals with 3 main and a few indistinct secondary veins. Stamens 10 in two whorls, the first whorl alternate with the tepals and the second whorl opposite the tepals. Filaments long, inserted at base of narrow intrastaminal disc; anthers athecal, borne along the tips of the filaments; pollen grains oblong, monocolpate; gynoecium of 5 carpels; ovary syncarpous, superior, glabrous; style short, with 5 large stigmatic branches; fruit a capsule.

Holotype: accession # B-An-4 deposited in the Poinar amber collection maintained at Oregon State University, Corvallis, Oregon.

Type locality and horizon*:* Hukawng Valley southwest of Maingkhwan in Kachin State (26°20´N, 96°36´E), Myanmar, uppermost Albian-lowermost Cenomanian (mid-Cretaceous).

Etymology: Generic name from the Greek "diaphoros"= different and the Greek "anthos" = flower. The specific epithet indicates the geographic origin of the fossil.

1.2.4.2. Description

Total length flower (without pedicel), 2.1 mm; pedicel, 4.1 mm in length; 5 separate lanceolate-ovate tepals ~1.0 mm in length and 0.5 mm in greatest width; stamens 10, ~1.0 mm in length; anthers athecal, consisting of sporogenous tissue borne at the tips of the filaments; anther portions narrow to slightly swollen in shape, ranging from 45 μm to 300 μm in length; pollen grains oblong, monocolpate, smooth walled, length, 11μm-16 μm, width, 7μm- 9μm; ovary superior, 0.75 mm in length, 0.7 mm in width, surface smooth, glossy; style single, short, 130 μm in length, stigmas 5, all ~ 450 μm in length.

Comments: Adjacent to the flower is a large separate tepal that is 2.4 mm in length and 1.0 mm in width (Figure 9). It is possible that this larger tepal represents a caduceus petal. The pollen grains are small, but compare in size to pollen grains described from other angiosperm flowers in Myanmar amber (Poinar and Chambers, 2005, Poinar et al., 2007b, 2013, 2016).

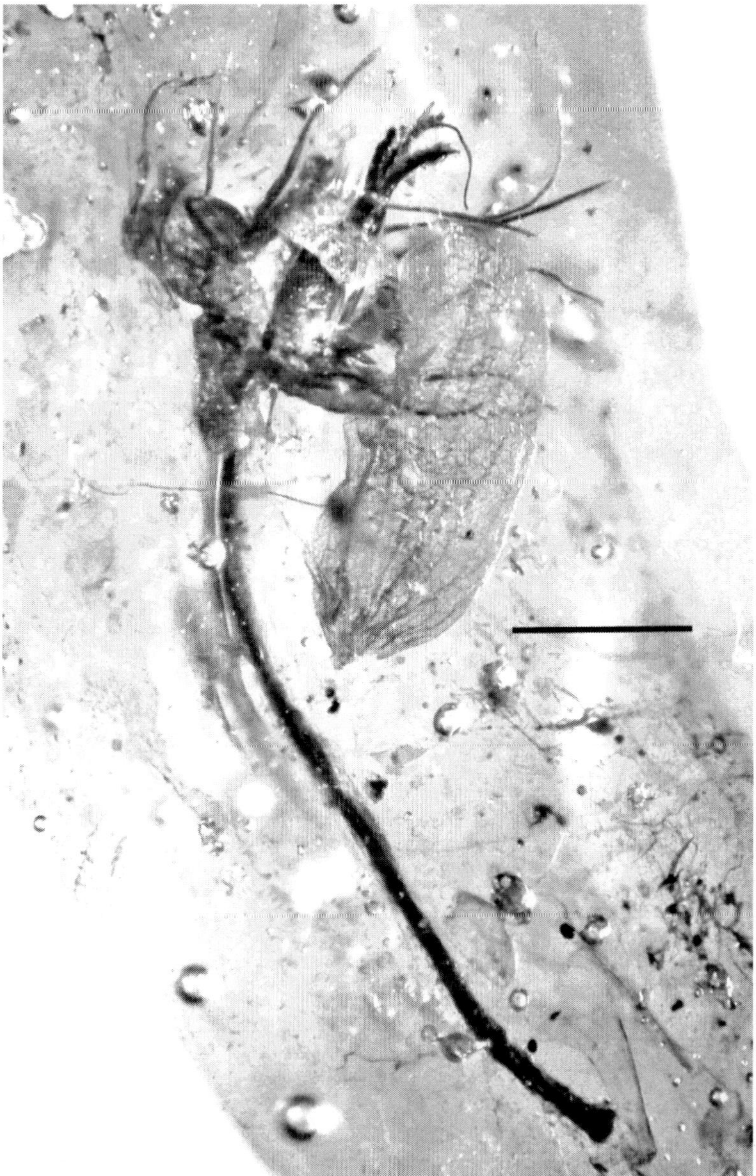

Figure 9. Holotype of *Diaphoranthus burmensis* gen. et sp. nov. with adjacent tepal. Scale bar = 0.9 mm.

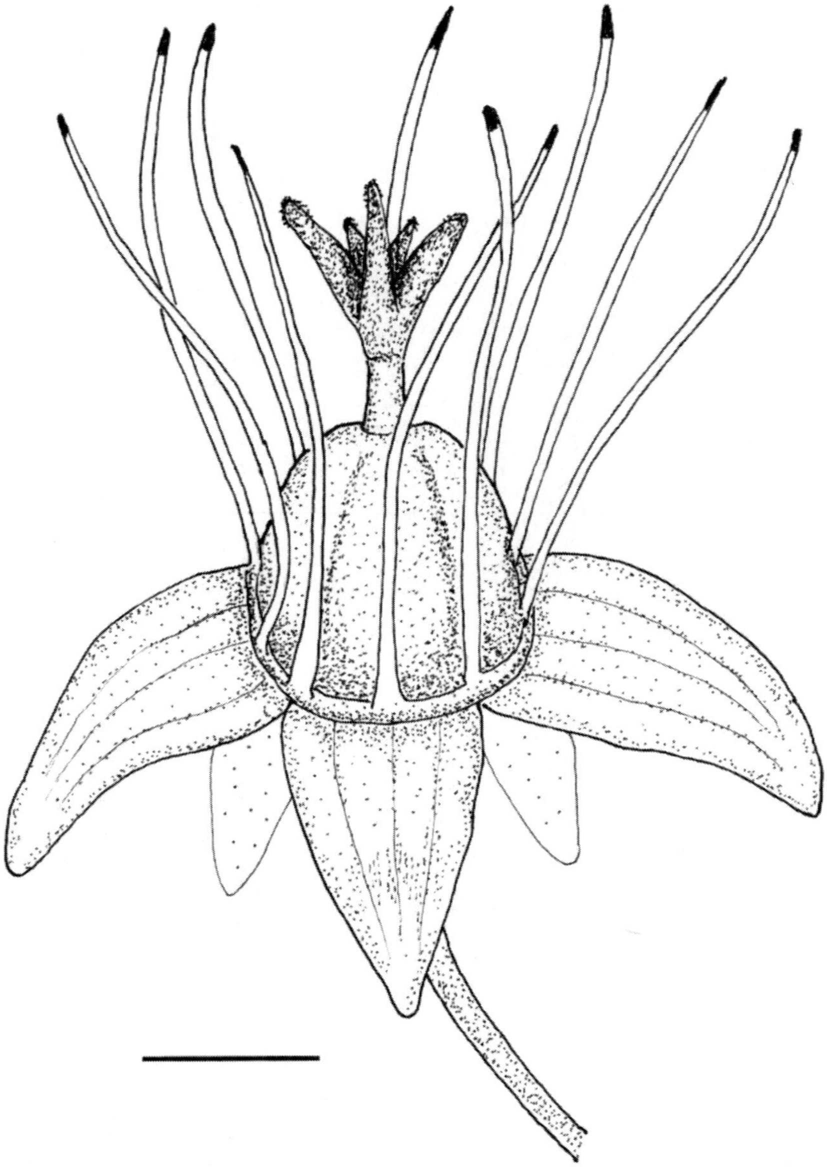

Figure 10. Reconstruction of *Diaphoranthus burmensis* gen. et sp. nov. Scale bar = 450 μm.

1.2.4.3. Conclusion

Angiosperm anthers are quite diverse in size, shape and dehiscence but for the most part are composed of a pair of thecae, each of which contains two pollen sacs (Stebbins, 1974, Endress, 1994, Endress and Stumpf, 1990, 1991). When mature, the 4 pollen sacs in the two lateral thecae usually open along a preformed line of dehiscence (the stomum) through which pollen is released. In *Diaphoranthus burmensis*, such typical anthers are replaced with athecal anthers bearing pollen grains that are released through the anther wall. Pollen mother cells developing in the center portion of the anthers produce mature pollen grains that accumulate along the parietal walls, eventually breaking through the anther wall for dispersal (Figures 12-15). At this stage, the pollen grains are often arranged in vertical rows along the walls of the anther (Figures 12, 15).

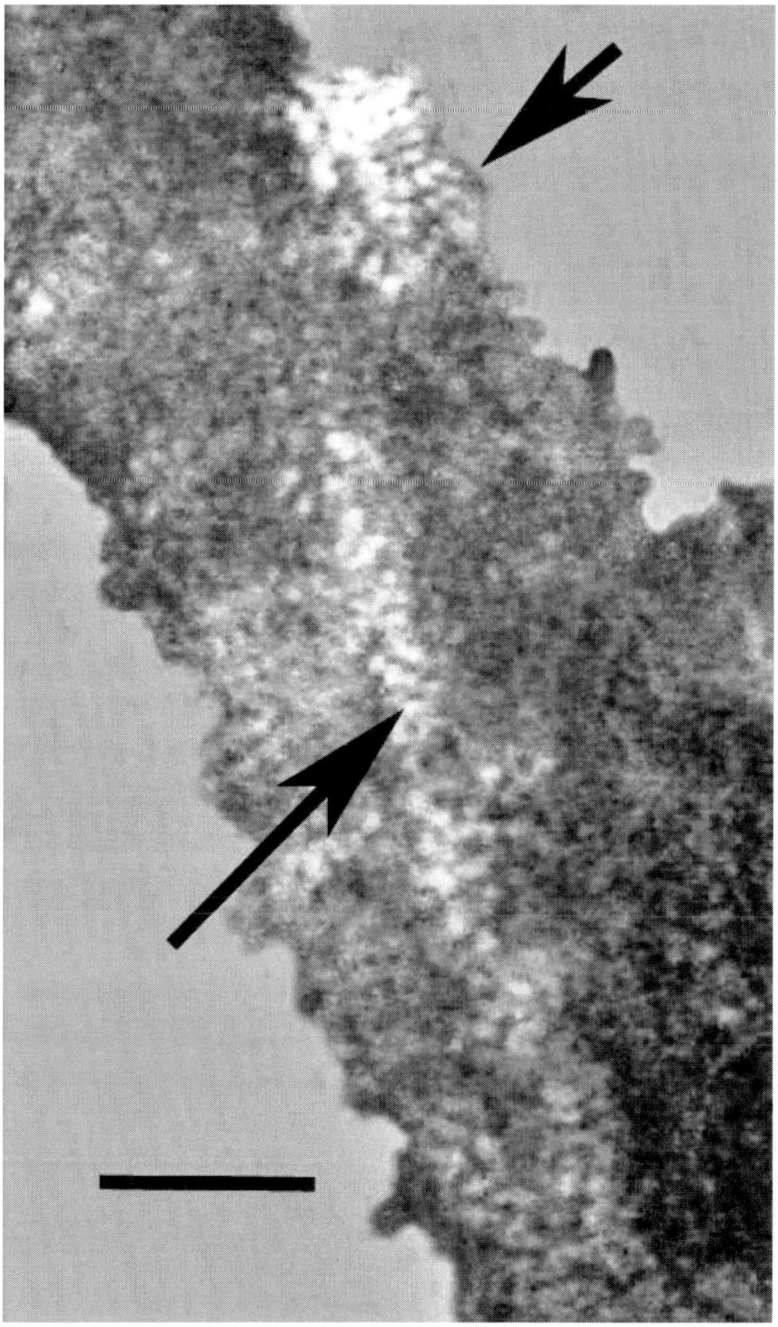

Figure 11. Portion of an anther of *Diaphoranthus burmensis* gen. et sp. nov. showing developing pollen grains in sporogenous tissue (lower arrow) and a cluster of pollen grains retained by the outer wall of the anther (top arrow). Scale bar = 65 μm.

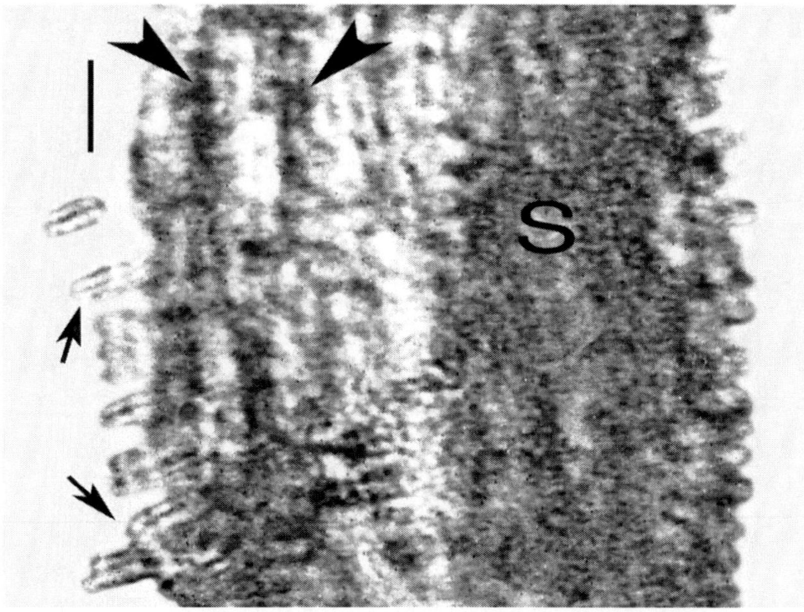

Figure 12. Section of an anther of of *Diaphoranthus burmensis* gen. et sp. nov. showing pollen grains on left side (arrows) exiting through the anther wall. Arrowheads show two rows of possible tapetal cells. S = sporogenous zone with pollen mother cells. Scale bar = 23 μm.

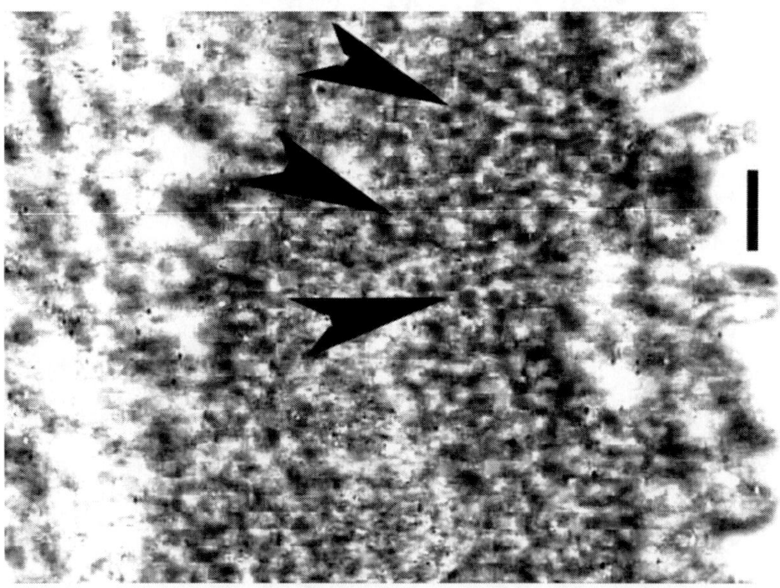

Figure 13. Detailed view of a portion of the anther of *Diaphoranthus burmensis* gen. et sp. nov. showing the sporogenous zone. Clusters of black spots (arrowheads) represent pollen mother cells undergoing meiosis. Scale bar = 8 μm.

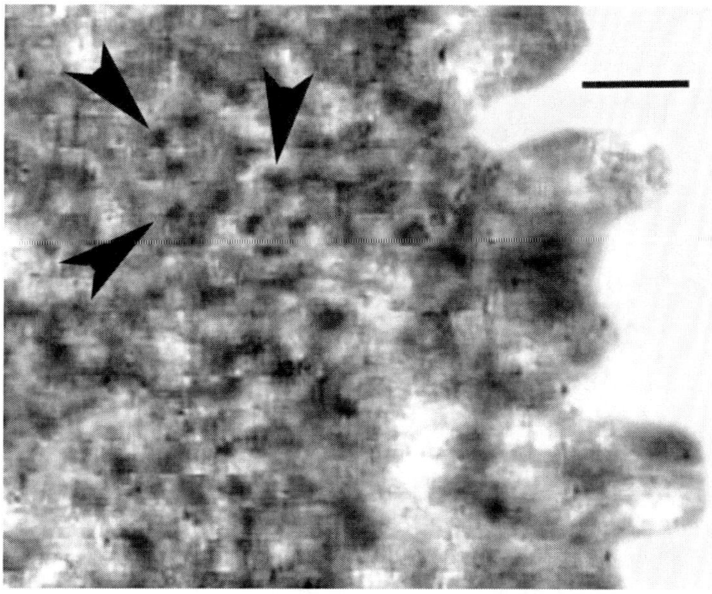

Figure 14. Detail of sporogenous zone of an anther of *Diaphoranthus burmensis* gen. et sp. nov. Black spots (arrowheads) represent nuclei of the embryonic pollen cells. Note pollen grains protruding from the anther wall on the right. Scale bar = 5 μm.

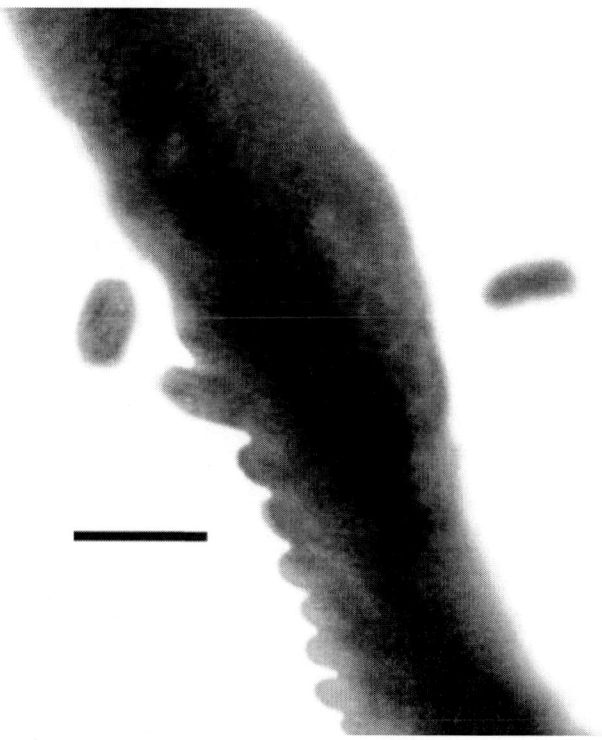

Figure 15. Elongate anther of *Diaphoranthus burmensis* gen. et sp. nov. releasing pollen grains. Scale bar = 18 μm.

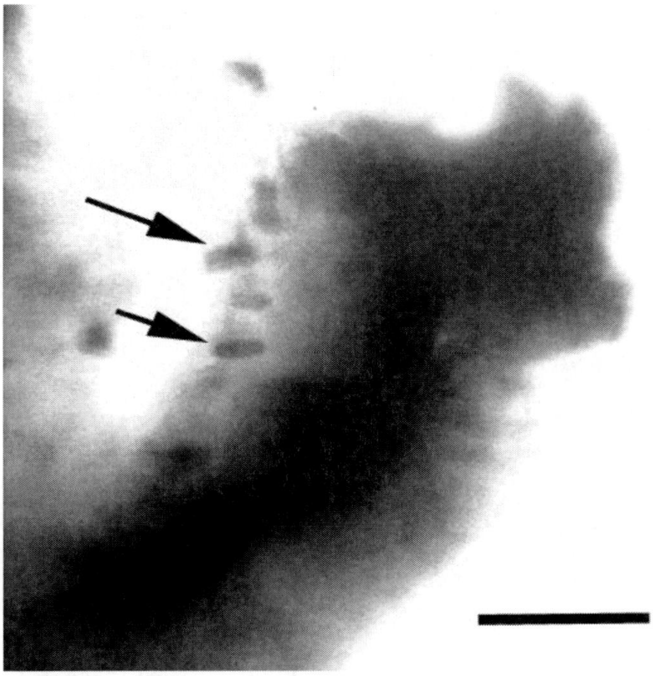

Figure 16. Pollen grains attached to one of the stigmatic lobes of *Diaphoranthus burmensis* gen. et sp. nov. Scale bar = 55 μm.

Originally, it was suspected that the anthers of *Diaphoranthus burmensis* were caduceus, however a more detailed examination showed that pollen was being released from sporogenous tissue positioned along the filament tips. These terminal tips, which vary little in shape from the subtending filaments, represent athecal anthers. The pollen-bearing tissue can be distinguished from the filaments by the absence of vascular bundles. A single vascular bundle in the filament that extends up to but not into the adjacent anther is the prevailing condition in filaments of extant angiosperms (Esau, 1953). Pollen grains are produced in sporogenous zones, which vary in length and width, along the center core of the anther (Figures 11-15). There is no evidence of connectives in any of the anthers.

Normally, pollen-bearing tissue in Angiosperms is surrounded by a tapetum that provides nourishment and mechanical tension during pollen release (Stebbins, 1974, Endress, 1994). In *Diaphoranthus burmensis*, it is difficult to determine if tapetal cells are present. However in some areas adjacent to sporogenous zones containing pollen mother cells, there are rows of rectangular cells that alternate with the developing grains (Figure 12). These rows of rectangular cells could be tapetal (or endothecial) cells. Since the cells maintain their individuality and position, they could represent the secretory type (Pacini, 1990).

The pollen grains of *Diaphoranthus burmensis* possess a single aperture with a single, straight furrow that extends the entire length of the grain. Such monosulcate pollen grains resemble those of some gymnosperms (Zamiaceae; Ginkoaceae), many monocots and a few primitive dicots including some members of the Magnoliales (Kapp et al., 2000). Dicots with pollen grains possessing a single furrow, like the *Clavatipollenites*- monocolpate type, are considered to be an evolutionary primitive condition (Kapp et al., 2000), which later led to triaperturate grains that are common in dicotyledons today.

Diaphoranthus burmensis has some features of wind-pollinated flowers. These include the small pollen size, anthers with abundant pollen, manner of pollen release and anthers positioned high above the stigmas. The type of pollen release from athecal anthers would have provided a higher probability that the grains would fall on the stigmas below. This self-pollination scenario is supported by the viscous nature of the broad stigmatic branches that retain the smooth-walled pollen grains (Figure 8D) (Stebbins, 1974).

While *Diaphoranthus burmensis* may have been wind pollinated, the expanded tepals along with a nectar disc could have attracted pollinators. The size of the flower suggests that any pollinators had to be very small, which would include minute insects. Based on the fossil record of extant insect pollinators of angiosperm flowers, beetles are considered to have pollinated the most primitive flowers (Stebbins, 1974, Bernhardt, 2000). Adult carrion beetles (Coleoptera: Dermestidae) are one possible group that could have visited the flowers of *Diaphoranthus burmensis.* These beetles are quite small, are known to visit a range of extant flowers for nectar and pollen and have been reported in Myanmar amber (Poinar and Poinar, 2016).

Some polysporangiate anthers in a few extant angiosperms, e.g., Viscaceae, Rafflesiaceae (*Mitrastemon*), and Icacinaceae (*Polyporandra*), also have athecal anthers (Endress, 1994) but they differ from those of *Diaphoranthus burmensis*. In the cases of these extant polysporangiate athecal anthers, each of the many sporangia opens individually by a pore through which pollen is released in slimy masses (Endress and Stumpf, 1990).

Aside from its athecal anthers, the fossil flower has some general features of extant members of the global families Ixonanthaceae, Linaceae and Hugoniaceae. Flowers in all three of these families are small, bisexual, mostly actinomorphic with 5-merous perianth members and 5- 15 stamens with slender filaments. They all have superior syncarpous ovaries with 1-5 styles and the fruit is a capsule or drupe (Steyermark and Luteyn, 1980, Kearns, 2004a, 2004b, Ramírez, 2004). However members of the above families have thecal anthers, the perianth is in two series and the pollen is tricolporate. *Diaphoranthus burmensis* shows no clear affinity with any modern family.

Anthers of extant Angiosperms are phylogenetically advanced structures (Esau, 1953). During their embryonic development, these advanced anthers pass through various stages, including one in which the sporogenous cells occur as simple primordia before any thecae have developed (Boke, 1949, Schewe et al., 2011). With sporangia developing in tissue located at the tips of the filaments, the anthers of *Diaphoranthus burmensis* mimic stages in the early embryonic development of angiosperm flowers. Such a primitive anther configuration was obviously a temporary evolutionary stage since it is not known to occur in any extant angiosperm flowers.

1.2.5. Endobeuthos Paleosum Poinar and K. L. Chambers (2018) (Figure 17)

The description of this angiosperm is based on six flowers from a single inflorescence. The flowers vary from 3.0 mm to 4.2 mm in length and from 3.2 mm to 4.6 mm in width. The small 50 plus helically arranged sepals, petals mostly hidden within the calyx, numerous stamens with apparent enlarged anther connectives with single bisporangiate, sessil thecae and apparent absence of pistils are characters that prevent an alignment of the fossil with any modern families.

Figure 17. Two flowers of *Endobeuthos paleosum.* Scale bar = 1.5 mm.

1.3. Flowers with Some Affinities to Extant Groups

1.3.1. Eoepigynia Burmensis Poinar, K. L. Chambers and Buckley (2007) (Figure 18)

Flowers of this species are small, perfect and actinomorphic. They possess a single series of basally connate sepals and four equal distinct petals. The four included stamens alternate with the petals and the inferior ovary bears a single style with a bilobed stigma. The pollen is triaperturate. The flowers show affinities to members of the order Cornales.

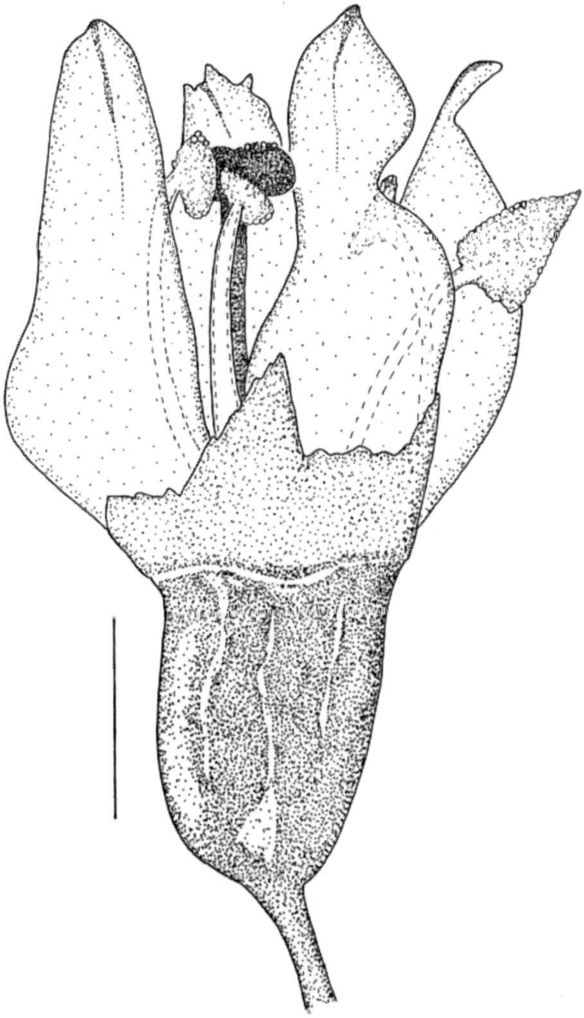

Figure 18. Flower of *Eoepigynia burmensis*. Scale bar = 0.34 mm.

1.3.2. Tropidogyne Pikei K. L. Chambers, Poinar and Buckley (2010) and T. Pentaptera Poinar and K. L. Chambers (2017) (Figure 19)

Flowers of *Tropidogyne* are small, bisexual, epigynous, apetalous with five spreading sepals containing one large central vein and lesser veins with branched, anastomosing veinlets. Anthers of the 10 stamens have two thecae that dehisce by longitudinal slits.

Tropidogyne pikei has a three-carpellate 10-ribbed inferior ovary surmounted by a conspicuous disc. There are three short, acute styles, and a possible secretory gland is located at the summit of each rib of the ovary. *Tropidogyne pentaptera* is very similar but differs in being bicarpellate and having only two elongate, slender styles. A relationship of these two species with the family Cunoniaceae was proposed.

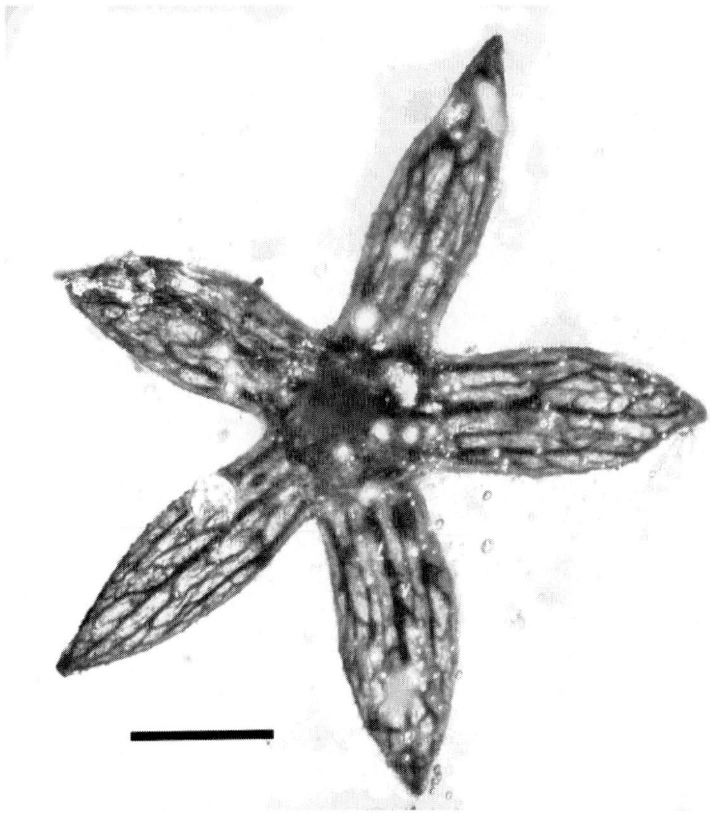

Figure 19. Flower of *Tropidogyne pentaptera*. Scale bar = 1.0 mm.

1.3.3. Palaeoanthella Huangii Poinar and K. L. Chambers (2005) (Figure 20)

Male flowers of this species are characterized by a cup-shaped perianth of 8 fused lobes (tepals) arranged in one series. There are 8 equal subsessile, unappendaged stamens in a single whorl, more or less alternating with the tepals. The 4-locular anthers open by longitudinal slits. Adjacent pollen is inaperturate with finely ridged and grooved exines. The fossil shows affinities with the family Monimiaceae.

1.3.4. Lachnociona Terriae Poinar, K. L. Chambers and Buckley (2008)

This is a small, hirsute, pistillate flower with 5 strongly recurved perianth parts, 10 staminodial filaments and a 5-carpellate pistil with 5 stout, erect, connivent or connate styles (Figure 21).

The sharply acute recurved tips of the styles are assumed to have decurrent ventral stigmas. The gynoecium is densely lanate-hirsute. Three of the filaments show thin pieces of vestigial anther tissue at their tips. The perianth is interpreted as a calyx, with the corolla absent. Possible relationships with families in the orders Oxalidales and Saxifragales were suggested for this lineage.

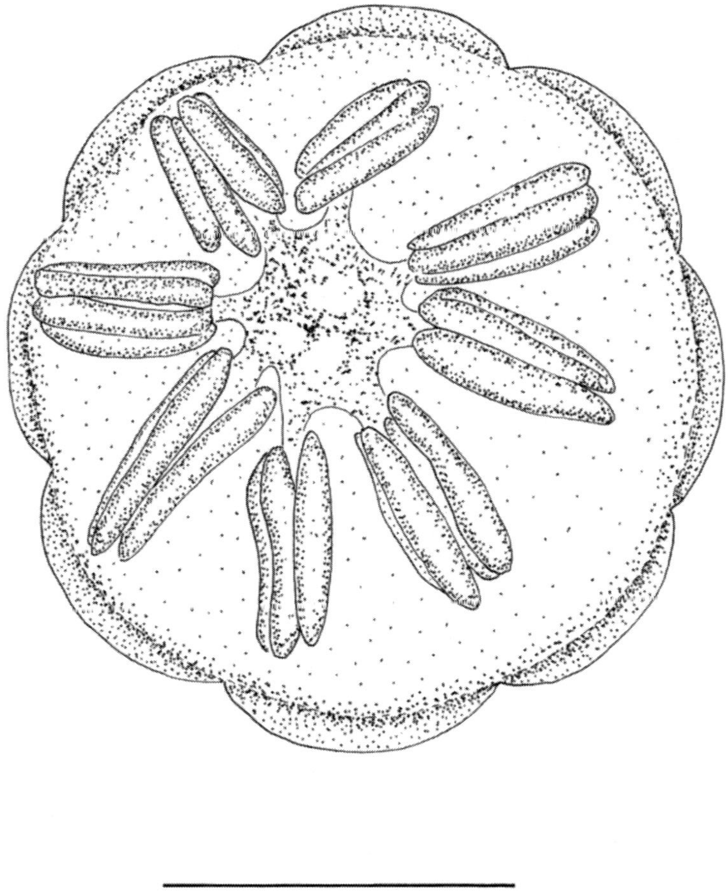

Figure 20. Flower of *Palaeoanthella huangii*. Scale bar = 0.5 mm.

1.4. Flowers Assigned to Extant Families

1.4.1. Cascolaurus Burmitis Poinar (2017a)

This male flower falls within the extant family Lauraceae and most closely resembles extant unisexual flowers of the recent genus *Litsea* Lam. in the Tribe Laureae. The size difference of the two perianth whorls, three whorls of tetrasporangiate anthers with apical valves arranged in 2 rows with introse dehiscence in whorls 1 and 2 and extrose dehiscence in whorl 3, plus the extremely large staminal glands, distinguish the fossil from extinct and extant genera of the family Lauraceae.

Another Lauralean flower described from Myanmar amber is *Jamesrosea burmensis* Crepet. Nixon, Grimaldi, Ricco (Crepet et al., 2016). Determination that *Jamesrosea* is an early representative of the Laurales was based on the "rather typical laurad hypanthium, lauralean bilocular, valvately dehiscing pollen sacs, and the diagnostic presence of paired basal stamen appendages typical of the Atherospermataceae-Gomortegaceae clade" (Crepet et al., 2016). *Jamesrosea* was placed within the Siparunaceae-Atherospermataceae clade and falls in category 2, namely lineages with possible affinities with some extant groups.

The authors claimed that modern family groups of this clade had not yet formed, however the presence of *Cascolaurus burmitis* in Myanmar amber shows that some lineages from that age fall within our modern concept of the Lauraceae. Obviously, primitive and advanced lauralean lineages were both present in the ancient Myanmar amber forest.

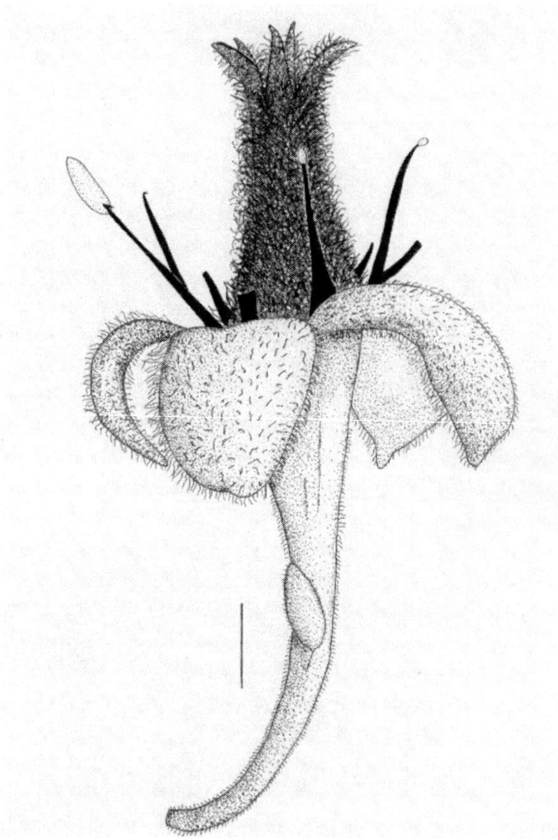

Figure 21. Flower of *Lachnociona terriae*. Scale bar = 1.0 mm.

Figure 22. Flower of *Cascolaurus burmitis*. Scale bar = 1.1 mm.

1.4.2. Programinis Burmitis Poinar (2004) and P. Laminatus Poinar (2004, 2011), (Poinar et al., 2015)

Species of *Programinis* are early members of the grass family (Poaceae) and *Programinis burmitis* possesses laterally compressed spikelets with two basal sterile glumes, a series of lemmas and paleas, stamens and gynoecia characteristic of modern grasses (Figure 23). Leaf fragments of *Programinis laminatus* show stomata with sausage-shaped guard cells and elongate nuclei (Figure 24), rows of epidermal cells with long and short cells and spherical and elliptical silica-like bodies in cuboid epidermal cells, which is consistent with features of modern grasses. Modern grasses are common hosts of ergot fungi and discovering such a fungus attacking a species of *Programinis*-like grass in Myanmar amber (Figure 25) is consistent with present day grass-fungal associations.

Figure 23. Spikelet of *Programinis burmitis*. Scale bar = 1.7 mm.

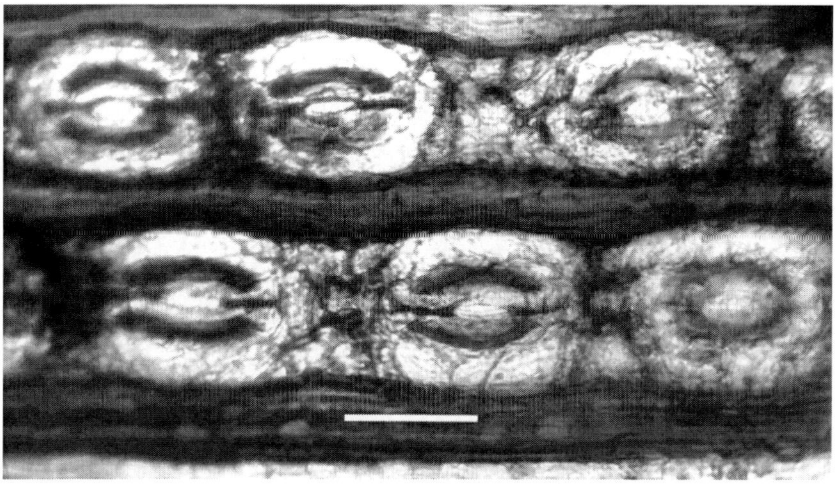

Figure 24. Stomata of *Programinis laminatus* with sausage-shaped guard cells and elongate dark guard cell nuclei. Scale bar = 46 μm.

Figure 25. An ergot fungus, *Palaeoclaviceps parasiticus* (arrowhead), infecting a *Programinis*-like spikelet in Myanmar amber. Scale bar = 1.6 mm.

2. Pollinators

The flowers thus far reported from Myanmar amber are small, suggesting that pollinators were probably very small insects. Based on the fossil record of extant insect pollinators of angiosperm flowers, beetles are considered to have pollinated the most primitive flowers (Stebbins, 1974, Bernhardt, 2000). Since beetles are guided more by odor than sight, many of the beetle pollinated flowers were probably white or dull colored but some probably produced nectar and attractive fragrances (Raven et al., 1970). However, most beetles visiting early angiosperm flowers were probably attracted to the floral elements as a food source and haphazardly picked up and transferred pollen to neighboring flowers.

Adult carrion beetles (Coleoptera: Dermestidae) are one possible group that could have served as pollinators. These beetles are quite small, are known to visit a range of extant flowers for nectar and pollen and have been reported in Myanmar amber (Poinar and Poinar, 2016). Another beetle group that could have been efficient pollinators were short-winged flower beetles (Coleoptera; Kateretidae). One species from Myanmar amber, *Furcalabratum burmanicum* Poinar and Brown (2017a) was just 2.6 mm in length. Both adults and larvae of extant Kateretidae are anthophagous and feed on angiosperm flowers of diverse families. Larval host families include Agavaceae, Caprifoliaceae, Cyperaceae, Juncaceae, Scrophulariaceae, Papaveriaceae and Urticaceae. The adults also feed on members of the Rosaceae, Melanthiaceae, Asteraceae and Apiaceae (Jelínek and Cline, 2010, Hisamtsu, 2011).

The most dependable pollinators were those that collected pollen to nourish their young. Bees, one of the most important groups of pollinating insects today, were just appearing 100 million years ago. Evolving from wasps, it is not surprising that some of the early bees retained a few wasp-like features. One such species was the tiny (body length of 2.95 mm) now extinct solitary bee, *Melittosphex burmensis* Poinar and Danforth (2006), (Danforth and Poinar, 2011) that carried pollen in branched hairs on its hind legs. This bee could have pollinated many of the flowers discussed above.

3. Plant Extinctions

Why did some Myanmar amber angiosperm lineages, such as those representing the families Lauraceae and Poaceae survive to the present, while others like *Antiquifloris*, *Micropetasos, Diaphoranthus, Endobeuthos* and *Chenocybus* apparently became extinct? Were the latter five genera victims of mass extinctions, which can abruptly destroy half of the species in certain regions, allowing no time for survival through adaptation? Or did they disappear by randomly determined background extinctions? It would appear that the lineages of *Eoepigynia*, *Tropidogyne*, *Jamesrosea, Palaeoanthella* and *Lachnociona* underwent pseudoextinctions, leaving modified descendants. The only definite extant family lineages found so far in Myanmar amber are *Cascolaurus* in the Lauraceae and *Programinis* in the Poaceae.

We can only speculate on the crucial determinants that resulted in these survival or extinction patterns. All species have some traits that make them either susceptible or resistant to extinctions and there are many factors that determine whether a plant will survive in a particular habitat. Some abiotic factors are loss of habitat resulting from changes in temperature, moisture, soil nutrients or exposure to sunlight. Biotic factors include competition with other plant lineages for space and availability of pollinators. The presence of vertebrate and invertebrate herbivores, especially those that can transmit disease causing agents, could have been significant. Viruses and their insect vectors are thought to have co-evolved 200 million years ago (Harris and Maramorosch, 1980). Important vectors of extant viral diseases of angiosperms are aphids and leafhoppers, both of which have been recovered from Myanmar amber (Poinar, 2017b, Poinar and Brown, 2017). The devastating effects of fungal diseases, especially those carried by bark beetles (Coleoptera: Scolytidae) were discussed earlier, using the Dutch elm disease and chestnut blight as examples (Poinar and Poinar, 2011). A combination of both biotic and abiotic factors probably caused the disappearance of the Myanmar amber angiosperms lineages.

CONCLUSION

An analysis of angiosperm flowers currently known from Myanmar amber reveals 12 distinct generic lineages at various levels of evolutionary development. Five of these lineages cannot be aligned with any contemporary angiosperm groups while five others show possible affinities with some extant groups and two lineages can be placed in modern families. Some of the more unusual features of Myanmar amber flowers that are unknown in present day angiosperms are the athecal anthers of *Diaphoranthus burmensis*, the extended floral axis, long slender stalks subtending the carpels and stamen structure of *Chenocybus allodapus* and the numerous helically arranged sepals and multiple stamens with anthers possessing single bisporangiate abaxial thecae in *Endobeuthos paleosum*.

ACKNOWLEDGMENTS

I thank Roberta Poinar for reviewing the manuscript and Barney Lipscomb of the Botanical Research Institute of Texas (BRIT), editor of *Sida* and the *Journal of the*

Botanical Research Institute of Texas, for granting permission to reuse the drawings that appear in Figures 1, 18, 20, and 21.

REFERENCES

Bernhardt, P. 2000. Convergent evolution and adaptive radiation of beetle-pollinated angiosperms. *Plant Systematics and Evolution,* 222, 293-320.

Boke, N. H. 1949. Development of the stamens and carpels in *Vinca rosea* L. *American Journal of Botany,* 36, 535-547.

Chambers, K. L., Poinar, G. O., Jr. and Buckley, R. 2010. *Tropidogyne,* a new genus of Early Cretaceous Eudicots (*Angiospermae*) from Burmese amber. *Novon*, 20, 23- 29.

Chatrou, L.W., Rainer, H. and Maas, P. J. M. 2004. Annonaceae (Soursop Family). In: Smith, N., Mori, S.A., Henderson A., Stevenson D. Wm., Heald, S.V. (Eds.) *Flowering plants of the Neotropics*. pp. 18-20. Princeton, NJ: Princeton University Press.

Crepet, W. L, Nixon, K. C., Grimaldi, D. and Ricco, M. 2016. A mosaic lauralean flower from the Early Cretaceous of Myanmar. *American Journal of Botany,* 103, 290-297.

Croat, T. B. 1978. *Flora of Barro Colorado Island.* Stanford, CA: Stanford University Press.

Cruickshank, R. D. and Ko K. 2003. Geology of an amber locality in the Hukawng Valley, northern Myanmar. *Journal of Asian Earth Sciences,* 21, 441-455.

Danforth, B. D. and Poinar, Jr., G. O. 2011. Morphology, Classification, and Antiquity of *Melittosphex burmensis* (Apoiodea: Melitto-sphecidae) and implications for early bee evolution. *Journal of Paleontology*, 85, 882- 891.

Endress, P. K. 1994. *Diversity and evolutionary biology of tropical flowers.* Cambridge, Cambridge University Press.

Endress, P. K. and Stumpf S. 1990. Non-tetrasporangiate stamens in the angiosperms: structure, systematic distribution and evolutionary aspects. *Botanische Jahrbücher,* 112, 193-240.

Endress, P. K., and Stumpf, S. 1991. The diversity of stamen structures in 'Lower' Rosidae (Rosales, Fabales, Proteales, Sapindales). *Botanical Journal of the Linnean Society,* 107, 217-293.

Esau, K. 1953. *Plant Anatomy*. New York, NY: John Wiley & Sons, Inc.

Gan, Y., Liu, Y. and Xu, F. 2015. Pollen morphology of selected species from Annonaceae. *Grana,* 54, 271-281.

Harris, K. F. and Maramorosch, K. 1980. *Vectors of plant pathogens*, New York, NY: Academic Press.

Heywood, V. H., 1993. *Flowering plants of the world.* Oxford University Press, Oxford.

Hisamtsu, S. 2011. A review of the Japanese Kateretidae fauna (Coleoptera: Cucujoidea). *Acta Entomologica Musei Nationalis Pragae*, 51, 551-585.

Holmes, F. W. 1980. Bark Beetles, *C. ulmi* and Dutch Elm Disease. In: Harris, K. F., Maramorosch, K. (Eds.). *Vectors of Plant Pathogens.* pp. 133-147. New York. NY: Academic Press.

Jelínek, J. and Cline, A. R. 2010. Kateretidae Erichson in Agassiz, 1846. In: Leschen, R. A. B., Beutel, R. G. and Lawrence, J. F. (Eds.). *Handbook of Zoology, Arthropoda: Insecta, Coleoptera, Beetles Morphology and Systematics.* pp. 386- 390. Berlin: De Gruyter.

Kapp, R. O., Davis, O. K. and King, J. E. 2000. *Pollen and spores.* College Station, TX: The American Association of Stratigraphic Palynologists.

Kearns, C. A. 2004a. Linaceae (Flax family). In: Smith, N., Mori, S. A., Henderson, A., Stevenson, D. Wm. and Heald, S. V. (Eds.). *Flowering plants of the Neotropics.* pp. 215-216. Princeton NJ: Princeton University Press.

Kearns, C. A. 2004b. Hugoniaceae (Hugonia family). In: Smith, N., Mori, S. A., Henderson, A., Stevenson, D. Wm. and Heald, S. V. (Eds.). *Flowering plants of the Neotropics.* pp. 184-185. Princeton NJ: Princeton University Press.

Pacini, E. 1990. Tapetum and microspore function. In: Blackmore, S. and Knox, R. B. (Eds.) *Microspores, evolution and ontogeny.* pp. 213-237. New York, NY: Harcourt Brace Jovanovich.

Poinar, Jr., G. O. 2004. *Programinis burmitis* gen. et sp. nov., and *P. laminatus* sp. nov., Early Cretaceous grass-like monocots in Burmese amber. *Australian Systematic Botany,* 17, 497-504.

Poinar, Jr., G. O. 2011. Silica bodies in the Early Cretaceous *Programinis laminatus* Angiospermae: Poales). *Palaeodiversity,* 4, 1-6.

Poinar, Jr., G. 2017a. A mid-Cretaceous Lauraceae flower, *Cascolaurus burmitis* gen. et sp. nov., in Myanmar amber. *Cretaceous Research,* 71, 96-101.

Poinar, Jr. G. O. 2017b. A new genus and species of aphids, *Tanyaulus caudisetula* gen. & sp. nov. (Hemiptera: Aphidoidea: Burmitaphidae) in mid-Cretaceous Myanmar amber. *Cretaceous Research,* 82, 36-39.

Poinar, Jr., G. O., Alderman, S., and Wunderlich, J. 2015. One hundred million year old ergot: psychotropic compounds in the Cretaceous? *Palaeodiversity,* 8, 13-19.

Poinar, Jr., G. O. and Brown, A. 2017. A new genus of leafhoppers (Hemiptera: Cicadellidae) in mid-Cretaceous Myanmar amber. *Historical Biology,* http://dx.doi.org/10.1080/08912963.2017.1384472.

Poinar, Jr., G. O. and Brown, A. E. 2018. *Furcalabratum burmanicum* gen. et sp. nov., a Short-winged Flower Beetle (Coleoptera: Kateretidae) in mid-Cretaceous Myanmar amber. *Cretaceous Research*, 84: 240-244.

Poinar, Jr., G. O., Buckley, R., and Chen, H. 2016. A primitive mid-Cretaceous angiosperm flower, *Antiquifloris latifibris* gen. & sp. nov., in Myanmar amber. *Journal of the Botanical Research Institute of Texas,* 10, 55-162.

Poinar, Jr., G. O. and Chambers K. L. 2005. *Palaeoanthella huangii* gen. and sp. nov., an early Cretaceous flower (Angiospermae) in Burmese amber. *Sida,* 21, 2087-2092.

Poinar, Jr., G. O. and Chambers, K. L. 2017. *Tropidogyne pentaptera*, sp. nov., a new mid-Cretaceous fossil angiosperm flower in Burmese amber. *Palaeodiversity,* 10, 135-140.

Poinar, Jr., G. O. and Chambers, K. L. 2018. *Endobeuthos paleosum* gen. et sp. nov., fossil flowers of uncertain affinity from mid-Cretaceous Myanmar amber. *Journal of the Botanical Research Institute of Texas* (in Press).

Poinar, Jr., G. O., Chambers, K. L., and Buckley, R. 2007b. Eoëpigynia burmensis gen. and sp. nov., an Early Cretaceous eudicot flower (Angiospermae) in Burmese amber. *Journal of the Botanical Research* Institute of Texas, 1, 91-96.

Poinar, Jr., G. O., Chambers, K., and Buckley, R. 2008. An early Cretaceous angiosperm fossil of possible significance in rosid floral diversification. *Journal of the Botanical Research Institute of Texas,* 2, 1183-1192.

Poinar, Jr., G. O., Chambers, K. L., and Wunderlich, J. 2013. *Micropetasos*, a new genus of Angiosperms from mid-Cretaceous Burmese amber. *Journal of the Botanical Research Institute of Texas,* 7, 745-750.

Poinar, Jr., G. and Danforth, B. N. 2006. A fossil bee from Early Cretaceous Burmese amber. *Science,* 314, 614.

Poinar, Jr., G. O. and Hess, R. 1985. Preservative qualities of recent and fossil resins: Electron micrograph studies on tissue preserved in Baltic amber. *Journal of Baltic Studies,* 16, 222-230.

Poinar, Jr., G., Lambert, J. B., and Wu, Y. 2007a. Araucarian source of fossiliferous Burmese amber: spectroscopic and anatomical evidence. *Journal of the Botanical Research Institute of Texas,* 1, 449-455.

Poinar, Jr., G. and Poinar, R. 2008. *What Bugged the Dinosaurs?* Princeton, NJ: Princeton University Press.

Poinar, Jr., G. and Poinar, R. 2016. Ancient hastisetae of Cretaceous carrion beetles (Coleoptera: Dermestidae) in Myanmar amber. *Arthropod Structure and Development,* 45, 642-645.

Ramírez N. 2004. Ixonanthaceae (Ixonanthes family). In: Smith, N., Mori, S. A., Henderson, A., Stevenson, D. Wm. and Heald, S. V. (Eds.). *Flowering plants of the Neotropics.* pp. 195-196. Princeton: Princeton University Press, NJ.

Raven, P., Evert, R. F. and Curtis, H. 1981. *Biology of Plants,* 3rd ed. New York, NY: Worth Publishers, Inc.

Russin, J. S., Shain, L. and Nordin, G. L. 1984. Insects as carriers of virulent and cytoplasmic hypovirulent isolates of the chestnut blight fungus. *Journal of Economic Entomology,* 77, 838-846.

Schewe, L. C., Sawhney, V. K. and Davis, A. R. 2011. Ontogeny of floral organs in Flax. (*Linum usitatissiumum*; Linaceae). *American Journal of Botany,* 98, 1077-1085.

Shi, G., Grimaldi, D. A., Harlow, G. E., Wang, J., Wang, J. Y. M., Lei, W., Li, Q., Li, X. 2012. Age constraint on Burmese amber based on U-Pb dating of zircons. *Cretaceous Research*, 37, 155-163.

Stebbins, G. L. 1974. *Flowering plants: Evolution above the species level.* Cambridge MA, Harvard University Press.

Steyermark, J. A. and Luteyn, J. L. 1980. Revision of the genus *Ochthocosmus* (Linaceae). *Brittonia,* 32, 128-143.

Thorne, R. F. 1974. A phylogenetic classification of the Annoniflorae. *Aliso* 8, 147-209.

In: Recent Advances in Plant Research
Editors: B. Welch and M. Wilkerson

ISBN: 978-1-53614-170-2

Chapter 7

PRODUCTION, CULTIVATION, AND CONSERVATION OF DOMESTIC AROMATIC PLANTS IN JAPAN

Maiko Inoue**, *Shinichiro Hayashi and Lyle E. Craker

[1]Specified Nonprofit Corporation of Horticulture Therapy and Zen, Hamamatsu, Japan

[2]GREEN FLASK Inc., Tokyo, Japan

[3]Stockbridge School of Agriculture, University of Massachusetts, Amherst, MA, US

ABSTRACT

Japan, which extends north and south, has four clearly defined seasons that provide a variety of unique habitats for the growth of various aromatic and medicinal plants. Recent focus on these traditional plants has begun to revitalize local communities, providing diversification to alternative medicines and the development of other products. Aromatic plants are widely used in Japanese foods such as Washoku, the cuisine of Japan, which is recognized on the UNESCO Heritage list. The traditional Japanese "Ko-do" (an elegant incense ceremony) uses several aromatic plants that provide graceful aromas. Indigenous aromatic and medicinal plants continue the connection of plants with life and with spirit for the Japanese. Due to global warming, a change may limit plant growth and alter constituents, endangering both domestic and wild species. Thus, conservation of genetic resources is necessary to insure that the unique aromatic plant species of Japan, such as *Eutrema japonicum* (Wasabi) and *Cryptomeria japonica* (Sugi), are maintained. While universities, companies, institutes, and homes own a number of gardens and conservatories, the most important concerns are the use, conservation, and return of natural resources in a daily life. Various locations in Japan practice a circulation-type forestry. This method creates new jobs that eventually lead to regional revitalization.

* Corresponding Author Email: den8mai@yahoo.co.jp.

Recently, the introduction of profiles and advantages of domestic Japanese medicinal and aromatic plants have developed markets with remarkable growth in Japan, suggesting increased interest and use of plants for better health, education, and environment in Japan.

Keywords: indigenous, Japanese products, medicinal plants, regional vitalization, and social contributions

1. Introduction

Japan, which is 3,000 kilometers long from north to south and east to west, has subarctic to subtropical climates in the land of Japanese long islands. Almost 70 percent of the national land of Japan is mountainous with a mean annual precipitation of approximately 1,700 mm, twice the world average (Ministry of Land, Infrastructure, Transport, and Tourism, 2008). Except for the southern tropical islands, Japan has four seasons that are clearly prevalent throughout the country. The landform and climate of Japan provide various and unique habits for native aromatic and medicinal plants (Table 1). In addition, various plants that came via the sea, such as those plants from China and Korea, have become wide spread in Japan. At the present time, abundant plant resources exist in Japan.

Since ancient times, Japanese people have used aromatic plants for housing, food, and clothing. *Thujopsis dolabrata* and *Chamaecyparis obtusa* were used to build the historic temples and castles, such as the Chusonji Golden Hall, the Five-storied pagoda in Houryuji temple, and the Hirosaki castle, all of which have a remarkable durability and resistance to insect damage and moisture because of a repelling aroma from the plants used in construction.

Before traditional Chinese medicine was introduced into Japan, an older Japanese medicine known as "*Waho*" was prevalent. A book written in the sixth century AD records the use of indigenous plants in Japan, such as *Geranium thunbergii* (Gennosyoko), *Houttuynia cordata* (Dokudami), and *Swertia japonica* (Senburi). During this time period, Buddhism was introduced from China and scented woods were burned in Buddhist rituals. The traditional Japanese medicine (*Kampo*) developed from Chinese medicine in the 15th century with many aromatic and medicinal plants being used as crude drugs. The famous *Kampo* drug, "*Kakkontou,*" contains several aromatic plants, such as cinnamon (*Cinnamomum verum*), ginger (*Zingiber officinale)*, licorice (*Glycyrrhiza glabra*), and Chinese jujube (*Ziziphus jujuba*).

The traditional incense culture known as "*Ko-do,*" began in the 16th century (Inoue et al., 2017), formed three natural incense wood constituents: Kyara, Jinko, and Byakudan. These wood incenses developed over time through the formation and ripeness of the aromatic resin in the plant. In the Edo era, a distillation method for extraction of essential oil was introduced with the essential oils of Japanese mint (*Mentha arvensis*) being exported in the Meiji era. Subsequently, the oil production significantly decreased due to the appearance of synthetic scent and foreign manufacture of scents. In the 1970s, potpourri became popular in Japan with the attention focused on dry herbs (Ozawa, 2015). The term "Aromatherapy," introduced in the 1980s and 1990s, initiated a large increase in gardening. Various scientific studies, such as scent constituents and interactions between human beings and aromas, were begun by researchers. Thereafter, various societies began, such as The Japanese Society for Complementary and Alternative Medicine in 1990, The Japanese Society of Aromatherapy in 1997, The Society for Integrative Medicine Japan in 2008, The Aroma Environment Association of Japan and The Japanese Society of Phytotherapy, both in 2012.

Table 1. Indigenous aromatic plants and applications

Scientific name	Japanese name	Tissue used	Applications
Prunus jamasakura	Yamazakura	Leaf, flower, xylem	Tea, food, crude drugs
Chamaecyparis obtusa	Hinoki	Leaf, xylem, resin, root	Essential oil, soap, cosmetics, crude drugs
Zanthoxylum piperitum	Sanshō	Leaf, peel	Essential oil, food, crude drugs
Magnolia kobus	Kobushi	Bud, flower, xylem	Essential oil, food, crude drugs
Lindera umbellata	Kuromoji	Branch, bark, root	Toothpicks, essential oil, food, crude drugs
Citrus unshiu	Unshumikan	Fruit	Food, soap, crude drugs
Eutrema japonica	Wasabi	Whole plant	Food, spice, crude drugs
Cryptotaenia japonica	Mitsuba	Whole plant	Food, crude drugs
Arelia elata	Taranoki	Sprout, bark, root skin	Food, crude drugs
Oenanthe javanica	Seri	Whole plant	Food, crude drugs
Aralia cordata	Udo	Sprout, shoot, stem, root	Food, crude drugs
Houttuynia cordata	Dokudami	Whole plant	Food, tea, crude drugs
Thymus quimquecostatus	Ibukijyakousou	Upper part of ground	Scent, crude drugs
Magnolia obovata	Hoonoki	Leaf, xylem, bark, seed	Kitchen, crude drugs

The Japanese aromatic plants sanshō (*Zantboxylum piperitum*), shiso (*Perilla frutescens*), and yuzu (*Citrus junos*) have delightful scents that are excellent in supporting the flavor of the traditional Japanese cuisine *"Washoku"* that is recognized on the UNESCO Heritage List. Wasabi (*Eutrema japonicum*) became especially popular for the spicy flavor and function as a necessary disinfectant for sushi, a most well-known

Japanese cuisine in the world. Yuzu was used as seasonal item of the winter solstice bath, a custom for good blood circulation begun in the Edo era.

The recent popularity of aromatic plants in Japan is remarkable. As proof, the 2015 market scale for aroma in Japan was 333.7 billion yen (2.930 billion dollars). This was a market increase of 126% from the 2011 market and is expected to keep increasing for a period of time (aromatopia, 2016). Recently, diverse industry areas are bringing domestic aromatic and medicinal plants to practical use, such as construction, gardening, foods, pharmaceuticals, cosmetics, relaxations, and groceries. In addition, domestically produced dry herbs and essential oils are recently becoming popular. This popularity of herbs and oils most likely comes from an idea known as, *Shindofuji*, which means eating foods that were produced in your native land are good for your health, that is things produced in Japan are well matched to the body of Japanese people. Although unproven scientifically, a sense of safety and familiarity recognize "*Shindofuji*" as a fact. The manufacturing of Japanese products using domestic aromatic plants creates new jobs and greater levels of employment, eventually leading regional vitality.

Conservation of plant resources is an urgent issue in Japan due to the global warming that becomes more serious yearly. Global warming changes the habitat and thus the population of aromatic plants. In Japan, the trend for germinating seeds of aromatic plants is probably slightly further north due to the progress of global warming creating and unfavorable environment for plant growth. A large number of botanical gardens are present in various parts of Japan for conservation and research. Additionally, the Genebank in Tsukuba is active in hunting, collection, characteristic evaluation, conservation, distribution, and publication of plant information. Recently, other countries started to cultivate unique, aromatic, Japanese plants because of the aging producers and the climate change causing a decline in production.

Recently, businesses have begun to use aromas to design public spaces for customers. Everything from a small and intimate setting to a large hotel, mall, or airport has been designed with characteristic aromas as a favorable location for people to gather. These aromatic areas are planned by providing a special aroma exclusively for a particular space, improving the value of spaces and sales. Evidently, the potential for practical uses of aromatic plants are ever-expanding in the future world.

2. Production

2.1. Indigenous and Domestic Aromatic Plants

Because Japan is a humid island country with four seasons, many indigenous and domestic aromatic plants exist (Figure 1, 2, Table 2). The indigenous aromatic plants have a variety of uses, in construction, Japanese cuisine, crude drugs, incense, cosmetics, soap, chopsticks, soy sauce barrels, and craft work. The cultivation and processing of indigenous and domestic aromatic plants is a traditional handwork in Japan. For example, *Echinacea purpurea* is cultivated at Saitama is finished using a manufacturing methodology for making green tea and *Crocus sativus* is cultivated at Oita inside buildings. In recent years, the tendency of using aroma for health support has grown, especially as people are using aroma for prevention of dementia and stress relief. As an example, two blended essential oils, one for daytime and one for nighttime dementia prevention, are well known in Japan. Additionally, the use of essential oils for body massages and reflexology is popular in Japan, especially in hand massages for senior citizens. To those interested in antiaging, beauty, and health, the functionality of phytochemicals is just as essential as the aromatic properties of plants. Many kinds of herbal teas, spices and supplements are sold in various shops.

Most growers of cultivated, domestic, aromatic plants manage processing with the manufactured products delivered to retail stores for sale (Figure 3). The traceability of the plant material is clear and easy to trace from the end user. Sales companies become involved in business cooperation with developers as fair-trade.

2.2. Sakhalin fir (*Abies sachalinensis*) in Hokkaido

Fir trees have been used historically for treatment of pneumonia and bronchitis Sebastian Kneipp (1821-1897), the forefather of the naturopathic "*Kneipp Cure*," recommended placing a small pot of fir indoors. Sakhalin fir (*Todo-matsu* in Japanese) is a representative tree of the northern forest vegetation, although observed only in the Hokkaido area of Japan. The scent of balsam from the fir leaves and branches along with the essential oil have the effect of purifying the air.

In 2000, a project started by the owner of the Shimokawa Town Forest Cooperative in Shimokawa, Hokkaido (Figure 4). The project was based repeatedly on a circular forestry that consisted of four parts, planting, trimming, cutting, and effective use of the forest. The forest resources are used for building materials, biomass, compost, charcoal, pyroligneous acid, smoking materials, bark, and ornaments. The cooperative has received a forest certification for national and private forests in Shimokawa that are administered by a cooperative from the Forest Stewardship Council (FSC, Figure 5) in 2003. The forest certification can be awarded only for forests and forest products that comply with the FSC standards for managed environmental, social, and economical responsibility (Forest Stewardship Council, 2017). To place the FSC logo on final products, the forest certification for all of parties participating in the manufacturing process is necessary. This local project involves the whole community giving rise to employment that leads the society to become active. The national government has designated Shimokawa is designated as an "eco-future city."

Table 2. Applications and properties of domesticated aromatic plants in Japan (GREEN FLASK, 2017)

Map	Scientific name	Common name	Growing district	Uses	Applications	Properties
1	*Betula pendula*	Silver birch	Bifuka, Hokkaido	Leaf (roasted)	Tea	Diuresis
2	*Abies sachalinensis*	Sakhalin fir	Northern Hokkaido	Leaf & branch	Essential oil	Respiratory system
3	*Mentha arvensis*	Japanese mint	Northern Hokkaido	Leaf (roasted)	Tea & essential oil	Flatulence & diarrhea
4	*Cucurbita pepo*	Pumpkin	Wassamu, Hokkaido	Seed	Food	Prostatomegaly & enuresis
5	*Eleutherococcus senticosus* (*Acanthopanax senticosus*)	Free pips	Hokkaido	Rhizome (roasted)	Tea	Depression & nutritional enhancement
6	*Thujopsis dolabrata*	Hiba	Shimokita peninsula, Aomori	Trunk	Essential oil	Antimicrobial, antibacterial, & insect-repellent
7	*Houttuynia cordata*	Dokudami	Gunma	Whole	Tea	Diuresis, aperient, antimicrobial, & antibacterial
8	*Vitis vinifera*	Grape vine	Obuse, Nagano	Leaf (roasted)	Tea	Antioxidation & venous vessel protection
9	*Echinacea purpurea*	Purple coneflower	Saitama	Leaf	Tea	Increase in immunity
10	*Lindera umbellata*	Kuromoji	Chichibu, Saitama	Leaf & branch	Essential oil & aromatic distilled water	Anxiety, inflammation, analgesic, & blood circulation
11	*Citrus tachibana*	Tachibana orange	Numazu, Shizuoka	Peel	Essential oil	Depression, blood circulation, & brain function
12	*Cymbopogon citratus*	Lemon grass	Izu, Shizuoka	Leaf	Tea, spice	Indigestion & insect-repellent
13	*Camellia japonica*	Japanese camellia, Tsubaki	Izu Ōshima, Tokyo	Seed	Oil	Moisturizing & bad cholesterol
14	*Chamaecyparis obtusa*	Japanese cypress, Hinoki	Kiso, Nagano	Branch	Essential oil	Antimicrobial & antibacterial

Map	Scientific name	Common name	Growing district	Uses	Applications	Properties
15	*Sciadopitys verticillata*	Japanese umbrella-pine, Kouyamaki	Wakayama	Branch & leaf	Essential oil	Tranquility
16	*Zingiber officinale*	Ginger	Shimanto, Kochi	Root	Essential oil	Inappetence, indigestion, & blood circulation
17	*Citrus tamurana*	Hyuganatsu, New Summer Orange	Kochi	Peel	Essential oil	Blood circulation & refreshment
18	*Citrus junos*	Yuzu	Kochi	Peel	Essential oil	Blood circulation & tranquility
19	*Mentha piperita*	Peppermint	Fukuoka	Leaf	Tea	Irritable bowel syndrome
20	*Cinnamomum camphora*	Camphor tree	Fukuoka	Branch & trunk	Essential oil	Stress, tension, & insomnia
21	*Rhus succedanea*	Japanese wax tree	Chikugo, Fukuoka	Fruit	Wax	100% botanical
22	*Crocus sativus*	Saffron crocus	Oita	Stigma	Tea & spice	Menopausal symptoms
23	*Morus alba*	White mulberry	Kagoshima	Leaf (roasted)	Tea	Suppress sugar absorption & blood sugar levels after meals
24	*Curcuma longa*	Turmeric	Okinawa	Root	Tea	Detoxification of liver & depression
25	*Momordica charantia*	Bitter melon	Okinawa	Fruit	Tea & food	Insulin support & diabetes
26	*Orthosiphon stamineus*	Misai Kucing, cat whiskers (Java tea)	Okinawa	Leaf & stem	Tea	Hypertension & gout
27	*Alpinia speciosa*	Shell-flower	Okinawa	Leaf	Tea	Relaxation & support digestion
28	*Citrus depressa*	Shequasar	Okinawa	Fruit & peel	Food & essential oil	Refresh & summer fatigue
29	*Alpinia zerumbet.excelsa*	Getto	Kitadaitojima, Okinawa	Leaf	Essential oil & aromatic distilled water	Relaxation & digestion support
30	*Perilla frutescens*	Shiso, perilla	Whole Japan, Hokkaido (essential oil)	Leaf	Food & essential oil	Antimicrobial & antibacterial
31	*Cryptomeria japonica*	Sugi	Yakushima, Fukushima	Leaf, branch, trunk	Essential oil	Air purification

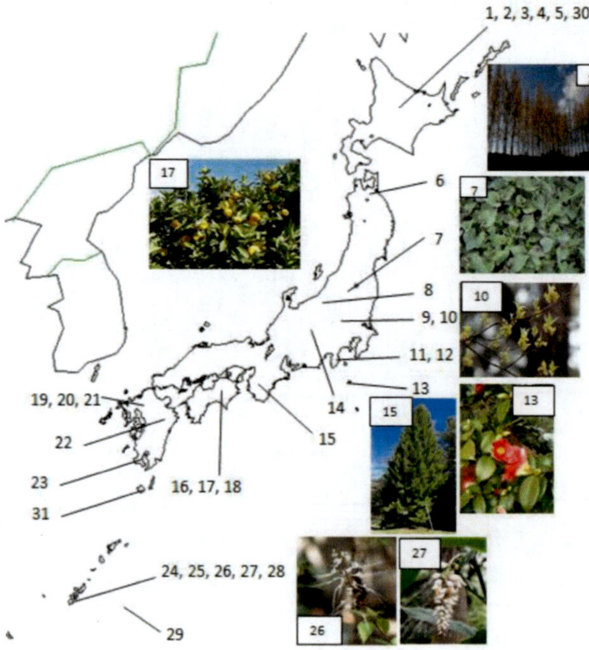

Figure 1. A Map of producing area of indigenous and domestic aromatic plants in Japan.

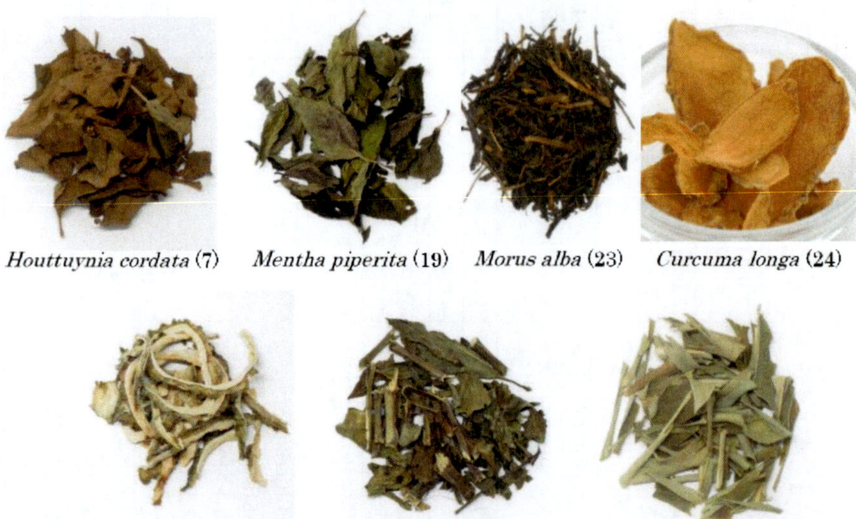

Houttuynia cordata (7) *Mentha piperita* (19) *Morus alba* (23) *Curcuma longa* (24)

Momordica charantia (25) *Orthosiphon stamineus* (26) *Alpinia speciose* (27)

Figure 2. Domestic aromatic plant products. Numbers in parenthesis indicate the numbers in Table 2 and map location in Figure 1.

Figure 3. Manufactured products in a retail store. From left: *Mentha piperita*, *Houttuynia cordata*, *Curcuma longa*.

The essential oil business of Sakhalin fir was a part of the project started in 2000. Primarily, the concept was to collect and use the leaves and branches that had been trimmed from the forest trees to develop new industries for the town. Fortunately, the Sakhalin fir produces more essential oil than other tree species that inhabit Hokkaido.

As of 2017, a total of 10,384 ha (national forest 7,702 ha, private forest 2,682 ha) of Sakhalin forest are controlled by the town of Shimokawa. Fresh leaves and branches of Sakhalin fir are gathered from the certificated forests as materials for extraction of the essential oil (Figure 6.1, 2). A total of 80 kg of plant material is used for one cycle of distillation (Figure 7). After distillation for two hours, 0.5 to 0.8 percent of tissue weight is extracted as essential oil that is subsequently bottled and labeled (Figure 8.1, 2). Aromatic distilled water of outgrowth is bottled as an alternative product as well. Additionally, some of the leaves and branches distilled were used for the inside of pillows. In this way, all tree resources are effectively used.

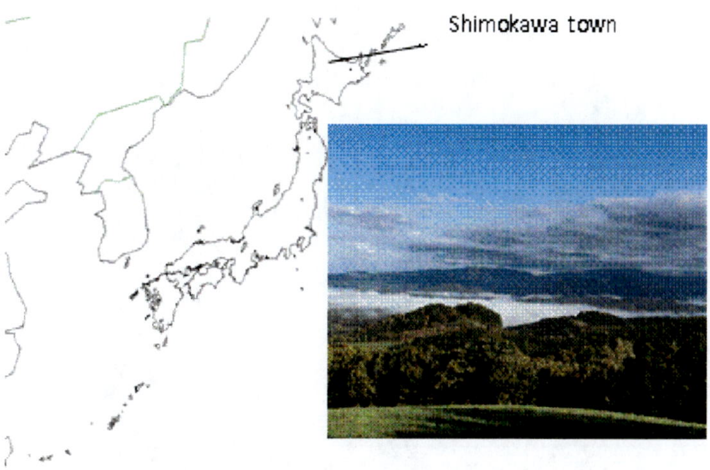

Figure 4. The location and photo of Shimokawa, Hokkaido.

FSC FOREST STEWARDSHIP COUNCIL

Figure 5. The logo of FSC.

Figure 6.1. Carried out fresh leaves and branches of Sakhalin fir from certificated forests.

Figure 6.2. Fresh leaves and branches of Sakhalin fir for extraction of the essential oil.

Figure 7. 80 kg of plant material is placed in a distillation still.

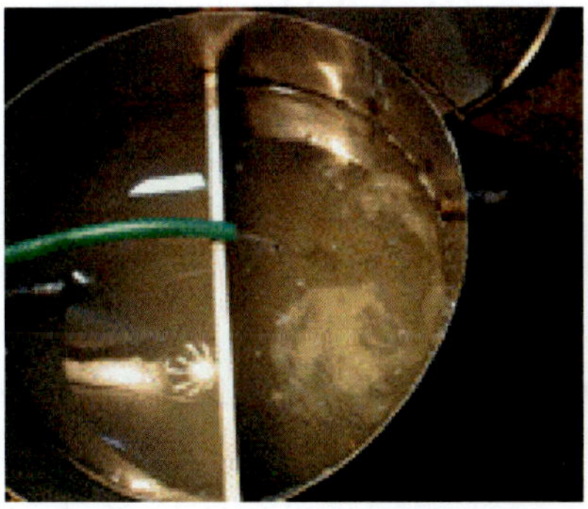

Figure 8.1. An essential oil extracted by 2 hours of distillation.

Figure 8.2. Collected and bottled essential oil is labeled for sale.

2.3. Sugi (*Cryptomeria japonica*), Hinoki (*Chamaecyparis obtuse*), Yuzu (*Citrus junos*), and Shequasar (*Citrus depressa*) in Japan

Untrimmed forests are a common problem throughout Japan, causing various problems in places with a local forestry association or NPO. Yakushima, a southern island famous for Sugi, has natural and reforested Sugi from which a company has started making aromatic products using the reforested Sugi circulating resources with a local NPO. Trimmed branches and leaves have been used for essential oil extraction. After distillation the leaves and branches are put on a nature trail. The forests are trimmed so sunlight can reach the ground, sustaining the forests in good condition. The Sugi of Yakushima has a slightly different aroma and constituents from the Sugi of the main island (Yatagai, 2010). The differences in the Sugi essential oils add value to the Sugi of Yakushima.

Kumano, Wakayama became famous due to the World heritage *"Kumano kodo."* Since ancient times, a forest of Kumano was sacred and the trees were used in rituals. The forests, Hinoki, Sugi, and Kuromoji (*Lindera umbellate*) are seen by many people walking the forest pilgrimage route throughout the year. A company that started to extract essential oil from Hinoki (*Chamaecyparis obtuse*) discovered an acetylcholinesterase (AChe) inhibitor constituent that is known for the prevention of dementia (McGleenon, 1999). The company has obtained a patent for the essential oil because currently the AChe inhibitor has only been detected within forest of Kumano Hinoki. The Japanese population is aging, resulting in a heightened awareness and interest in the anti-aging and prevention of dementia. As a result, products related to maintaining good health in old age are in high demand.

Yuzu (*Citrus junos*) is one of the most popular citrus fruits used in Japanese cuisine, such as hot pot dishes, jams, desserts, and flavorings. Most of the yuzu manufacturers use the juice and the fresh and refreshing scent familiar to Japanese people.

Shimanto, Kochi is the largest manufacture of Yuzu in Japan, therefore local residents, companies, and research institutions cooperated to find an effective use of local resources and to develop many kinds of products using Yuzu. One of the products was an essential oil derived from leftover peels from a manufacture. A method of extraction of essential oil was innovated through the use of impression ultrasonic waves. As extraction efficiency improved, a patent was obtained for the newest method. In addition to food, cosmetics, fresheners, and soaps are currently in production.

Shequasar (*Citrus depressa*) is a native species of Okinawa, where lots of manufacturing industries related to Shequasar can be seen. The *nobiletin*, a polymethoxylated flavone exhibiting anti-dementia properties, was identified within the peel of Shequasar (Nakajima et al., 2014). Since then, a long term study has taken place yielding evidence to suggest that a daily intake of citrus fruits could reduce the risk developing dementia by almost 15% (Zhang et al., 2017). Depending on the citrus consumption, ≤ 2 times/week, 3-4 times/week, and almost every day, the multivariate hazard ratios were different but the lowest score was indicated in the group that consumed citrus almost every day.

3. Cultivation and Conservation

3.1. Field Work

3.1.1. A Circular Forestry Business of Aromatic Tree

Forests account for 70% of the land in Japan. Domestic wood supply and demand was lowest in 2002 (18.8%) and had been decreasing since 1955 (96%) as a result of cheaper wood being imported from other countries. The number has been increasing since 2002, however, eventually amounting to 34.8% in 2016 (Forestry Agency, 2017). During the high-growth period from 1955 to 1973, large quantities of trees were planted in reforestation efforts meaning that most of the trees have reached appropriate age for cutting by now.

A circular forestry business is conducted in Shimokawa, Hokkaido. About 90% of the Shimokawa area is covered by forest, and several aromatic trees such as Sakhalin fir, Sakhalin spruce (*Picea glehnii*), and Japanese larch (*Larix kaempferi*) are grown in the forest. In 1953, the Shimokawa Town Forest Owner's Cooperative had purchased 1221 ha of national forest and thereafter continued to purchase 50 ha of national forest every year. Each 50 ha is afforested and after 60 years could be ready to cut. Thus, a total of 3,000 ha was used for circular forestry business (Hisayama, 2015). In Shimokawa, relatively young people have been engaging in the forestry business for the past five years. Trimming and putting cut brunches and boughs out are a necessary task in order to maintain a high quality forest. Moreover, the managed forests could sustain an environment and local people.

3.1.2. Japanese mint (Mentha arvensis)

Japanese mint grows wild in Hokkaido. Actively cultivated in Kitami, Hokkaido for its essential oil, at the height of prosperity Kitami mint accounted for 70% of the world product share. After World War II, a rivalry of Brazilian and Chinese mint, synthesized oil, and cold-weather damage caused a decline in price and the planted area gradually dwindled. In Takinoue, Hokkaido, approximately 10 ha of cultivated land remained, where the mint accounts for 95% of domestic product share (Takinoue town, 2017). The mint harvested in early September is dried and essential oil is extracted, which are mainly used for pharmaceuticals. After extraction, the mint is reused as an organic fertilizer.

3.2. Conservation of Domestic Aromatic Plants

In Japan, the cultivation of some domestic aromatic plants is being pressed for environmental changes. Global warming is a serious problem all over the world, a disturbance in rainfall, unstable weather, pests, high temperature, and desertification are observed. Recently, Japanese aromatic plants are starting to be cultivated abroad due to difficulties in a delicate ecology. For example, Wasabi (*Eutrema japonicum*) grows only in a clear stream at 10 to 14^0C stream throughout the year. Global warming causes a reduction of the places that slightly fulfill the needed environmental conditions, decreasing production with the production scale downsizing. At the same time, demand for Wasabi is increasing due to the recent popularity of Wasabi. Currently, England, Australia, and Canada have successfully cultivated Wasabi through a trial and error, and production has expanded.

A large number of public and private botanical gardens are spread throughout Japan and universities and pharmaceutical companies have developed their own gardens for materials, education, commerce, and conservation. Some historically important and large-scale botanical gardens have developed in Japan (Table 3). In the middle of the 16th century, the Toyama area was a prosperous apothecary region, where many medicinal gardens that have existed continue to thrive in the present. Several large pharmaceutical companies are operating in the Kyoto area and accordingly have medicinal gardens.

Table 3. The list of aromatic and medicinal botanical garden in Japan

	Name	URL	Location	Founded in	Species
1	Faculty of pharmaceutical sciences at Hokkaido University	https://www.pharm.hokudai.ac.jp/en/garden.html	Sapporo, Hokkaido	April, 1956	1,246
2	School of pharmaceutical sciences at Tohoku university	http://www.pharm.tohoku.ac.jp/~yakusoen/annai/	Sendai, Miyagi	April, 1974	1,200
3	Botanic Gardens of Toyama	http://www.bgtym.org/	Toyama, Toyama	Oct, 1993	4,900
4	Experimental Station for Medicinal Plant Research at Toyama University	http://www.pha.u-toyama.ac.jp/plant/garden_1.html	Toyama, Toyama	1923	2,000
5	Training Center of Medicinal plants at Toyama	http://www.pref.toyama.jp/branches/1285/center/	Nakaniikawa, Toyama	April, 1967	1,000
6	Faculty of pharmaceutical sciences at Hokuriku University	http://www.hokuriku-u.ac.jp/about/campus/y_garden.html	Kanazawa, Ishikawa	1976	1,200
7	Medicinal Plant Garden at Kanazawa University	http://www.p.kanazawa-u.ac.jp/~yakusou/	Kanazawa, Ishikawa	1953	
8	Fukui Botanical Garden	http://www.fukui-bot.jp/	Nyugun, Fukui	1994	3,000
9	Niigata Prefectural Botanical Garden	http://botanical.greenery-niigata.or.jp/	Niigata, Niigata	1998	
10	Koishikawa Botanical Garden	http://www.bg.s.u-tokyo.ac.jp/koishikawa/eng/	Bunkyo-ku, Tokyo	1684	100
11	Tokyo Metropolitan Park Association	https://www.tokyo-park.or.jp/english/	Chofu-shi, Tokyo	Oct, 1961	4,800
12	Higashiyama Zoo & Botanical Gardens	http://www.higashiyama.city.nagoya.jp/14_english/index.html	Nagoya, Aichi	1937	7,000
13	Hamamatsu Flower Park	http://www.inhamamatsu.com/activity/flower-park.php	Hamamatsu, Shizuoka	Sep, 1970	3,000
14	Yamashina Botanical Garden in Nippon Shinyaku CO., LTD.	http://www.nippon-shinyaku.co.jp/english/	Kyoto-shi, Kyoto	1934	3,000
15	Takeda Garden for Medicinal Plant Conservation, Kyoto	https://www.takeda.co.jp/kyoto/english/	Kyoto-shi, Kyoto	1933	2,800

	Name	URL	Location	Founded in	Species
16	Botanical Gardens at Osaka City University	http://www.sci.osaka-cu.ac.jp/biol/botan/	Katano-shi, Osaka	Mar, 1941	6,000
17	Nagai Botanical Garden	http://www.nagai-park.jp.e.aax.hp.transer.com/n-syoku/index.html	Osaka-shi, Osaka		1,200
18	Kobe City Arboretum	https://www.kobe-park.or.jp.e.abn.hp.transer.com/shinrin/	Kobe-shi, Hyogo	1940	1,200
19	Hyogo Prefectural Flower Center	http://www.hyogo-park.or.jp/flower-center/	Kasai-shi, Hyogo	Sep, 1974	4,546
20	Okayama City Handayama Botanical Garden	http://www.okayama-park.or.jp/facility/handayama	Okayama-shi, Okayama	May, 1964	3,200
21	The Hiroshima Botanical Garden	http://www.hiroshima-bot.jp/english/index.htm	Hiroshima-shi, Hiroshima	Nov, 1976	10,000
22	Wakunaga Garden	http://www.wakunaga.co.jp/english/garden.html	Akitakata-shi, Hiroshima	Jun, 1993	
23	The Kochi Prefectral Makino Botanical Garden	http://www.makino.or.jp/index_e.html	Kochi-shi, Kochi	Apri, 1958	3,000
24	Fukuoka City Botanical Garden	http://botanical-garden.city.fukuoka.lg.jp/	Fukuoka-shi, Fukuoka	Aug, 1933	2,600
25	Medicinal Garden in Pharmaceutical Sciences Kyusyu University	http://www.phar.kyushu-u.ac.jp/eng/index.php	Kasuya-gun, Fukuoka	Apri, 1974	1,000
26	Eco-Frontier Center of Medicinal Sesources	http://www.pharm.kumamoto-u.ac.jp/Labs/eco-frontier/en/	Kumamoto-shi, Kumamoto	1927	1,300
27	Florante Miyazaki	http://www.florante.or.jp/	Miyazaki-shi, Miyazaki	2000	
28	Flower Park Kagoshima	http://www.fp-k.org/english/index.html	Ibusuki-shi, Kagoshima	May, 1996	2,400
29	Medicinal Plant Garden in School of Pharmacy, Nippon University	http://mpgarden.pha.nihon-u.ac.jp/	Funabashi-shi, Chiba	1953	1,000
30	Tokyo University of pharmacy and Life Sciencee	https://www.toyaku.ac.jp/english-page	Hachiouji-shi, Tokyo		1,500
31	Medicinal Plant Garden, Kitasato University	http://www.pharm.kitasato-u.ac.jp/bio-garden/	Sagamihara-shi, Kanagawa	1965	2,000
32	Nikko Botanical Garden	http://www.bg.s.u-tokyo.ac.jp/nikko/eng/	Nikko-shi, Tochigi	1902	2,200

Koishikawa botanical garden is the oldest and with a prominent and long history in Japan (University of Tokyo, 2017). Originally, Koishikawa Medicinal Herb Garden was established in 1684. Presently, the garden has a total of 4,000 species of plants in the living collection, including hardy woody, hardy herbaceous, and tropical and subtropical species. The main purposes of this garden are propagation of extremely endangered species, the returning of propagules to original localities, and cultivation of endangered species for plant conservation biology.

In the Japanese Islands, about 7,000 plant species grow in the wild, with about 2,900 of these plants (40%) being indigenous flora distributed only in Japan (Tsukuba Botanical Garden, 2009). The ratio of indigenous flora is quite high compared with other countries. GeneBank, established in Tsukuba, Ibaraki in 1985, holds a great number of seeds and plants along with a huge collection of information related to genetic resources that is deposited and distributed for conservation, research, and education (GeneBank Project, NARO, 1997).

As with conservation, cultivation is very important as the Japanese used to live within the forest long ago. The forest of life named "*Tinjunomori*" would save people from various disasters such as Tsunami, fire, and flood. Now, Tinjunomori is rapidly decreasing year by year. Cultivating Tinjunomori takes a long time, however, once the nursery trees are planted, no care is needed after the first three years. Because many aromatic trees are growing inside of Tinjunomori, it is important for people to keep conserving and cultivating for the next generation.

4. Recent Researches Related an Interaction between Aroma and People

In the 1980s, the term "aromatherapy" was introduced in Japan. The circulation of publications relating to aromatherapy was extremely high during the late 2000s and the Japanese Society of Aromatherapy (JSA) and Aroma Environment Association of Japan (AEAJ) were established in 1997 and 2012, respectively. Various studies regarding the effect of scents on human beings and the functionality of aromas were performed. The essential oil of *Alpinia zerumbet* (Gettou), found in Okinawa in Japan, has a sweet flavor and was used to study the mental behavior of mice (Murakami et al., 2008). A mouse was exposed to Gettou essential oil and observed for 90 minutes, after which it was found that the anxiety level of the mouse had diminished and the level demonstrated a positive correlation with the dosage of essential oil.

Hanawa et al. used Japanese fragrances such as cherry blossom (Sakura), Japanese apricot (Ume), and Japanese wisteria (Fuji) to observe a sedative effecton sympathetic nervous activity (Hanawa et al., 2008). A total of 17 healthy female subjects participated and self-reported on their feelings while being exposed to Japanese flagrances, while the salivary amylase activities during a brain work were evaluated. Consequently, a decline of salivary amylase activity was indicated, and the sedation of sympathetic nervous activity was proved.

Extracts from the leaf and timber of Japanese cedar (*Cryptomeria japonica*) are effective for stress alleviation and the improvement of sleep quality (Sakamaki et al., 2013). Kawai and Matsubara (2017) reported the purification effect of wooden materials from Japanese cedar on the indoor air quality and the evaluation of volatiles emitted from wood on human health conditions. As a result, Japanese cedar timber as a building material absorbed air pollutants such as nitrogen dioxide, ozone, and formaldehyde. Inhaling air that contained volatiles emitted, suppressed the increase of sympathetic nervous activity and the emitted volatiles were pleasant and natural for the participants.

An essential oil of Ōyomogi (*Artemisia montana*) is effective as a traditional medicine for treatment of malaria, hepatitis, inflammation, hypertension, and cough (Jung et al., 2011). The aromatic components of essential oil of Ōyomogi were identified and the sedative effect was evaluated using sensory test and salivary α-amylase activity (Myoda et al., 2017). As a result, the stress-reducing effects of the Ōyomogi oil were the same as using lavender oil. An additional study suggests that 1,8-cineole is likely responsible for the sedative effects of Ōyomogi.

Recent studies indicate the biological activity of seeds from the spice known as the Grains of Paradise (GOP) (*Aframomum melegueta*) increase whole body energy when the extract is ingested. The GOP also have antimicrobial actions of against various bacteria (Sugita et al., 2013 and Ngwoke et al., 2014). In a study where an extract of GOP was fed to mice, Mitsunaga and Hattori (2017) noted that hepatic triglyceride (TG) concentration and total cholesterol (TC) were reduced. Consequently, the significant reductions of TG and TC suggest potential anti-obesity effects of GOP extracts.

CONCLUSION

In Japan, aromatic plants have undoubtedly been an integral part of people's lives, beginning when settlers first landed on the Japanese islands. At first, new generations of Japanese may not have fully recognized the value of medicinal and aromatic plants, but new health products and expanded applications of domestic and wild plant tissues are now important in Japanese culture.

Rural areas in Japan are becoming revived due to the value of medicinal and aromatic plants. In pursuit of health, food flavoring additives, and cosmetics, the Japanese people have become familiar with the importance of Japanese plants and plant extracts for improved health and quality of life due to various personal reasons, such as longevity, disease prevention, reduced life stress, and elderly care.

Increases in market sales of aromatic and medicinal plant extracts indicate that people are concerned with medical expenses in an aging society along with a tendency of people to be nature-conscious with expanded interest in complementary and alternative medicine. In recent years, the Japanese government has encouraged medical-agriculture cooperation and agriculture-welfare cooperation evidenced by the development and diffusion of functional and nursing-care food, an expansion of domestic medicinal plant production, and the employment of senior and disabled people in the processing and selling fields of agriculture.

A new form of health, education, and environment connected by plants was suggested as a prospective developing industry. Learning the functions of plants as food and self-treatment, having natural experiences for life and sense of body, protecting nature and cultivating therapist, and using of nature for health and medication practically are all necessary connections for the growing industries of health, education, and environment.

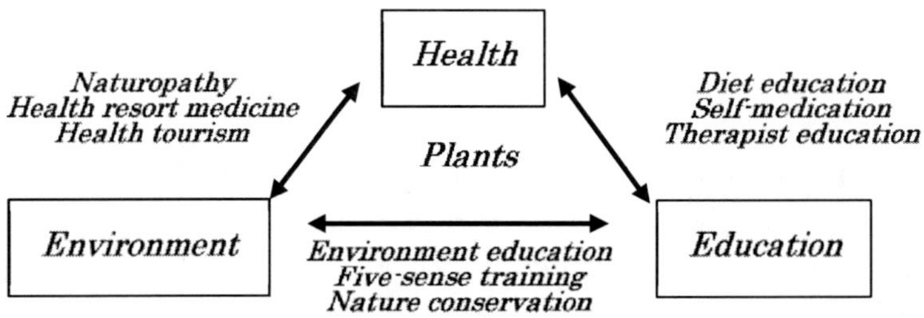

Figure 9. A new form of growing industry prospect.

The Health Resort Medicine (Figure 9) is a therapy based on scientific evidence performed in health resorts and indicates the medical activations, including promotion of health, prevention, treatment, and rehabilitation. The health promotion for a medium or long-term stay in the full of natural environment may well be effective in an upscale modern society. To solve serious problems, such as a medical expenses in an aging society and mature deficit disorders, using aromatic and medicinal plants are very important. The value of aromatic and medicinal plants will continually be firm in the future.

REFERENCES

Aromatopia. (2016). *Results of market research of aroma.* (in Japanese) *aromatopia. 139*, 1.

Forestry Agency. (2017). *Wood supply and demand table.* (in Japanese). Available from: http://www.rinya.maff.go.jp/j/riyou/ yusyutu/mokuzai-yusyutsu.html

Forest Stewardship Council. (2017 October 13). *Forest for all forever*. Available from: https://ic.fsc.org/en.

GeneBank Project, NARO. (1997). Available from: https://www.gene.affrc.go.jp/index_en.php.

GREEN FLASK. (2017). *GREEN FLASK catalogue*. (in Japanese) Tokyo: GREEN FLASK.

Hanawa, N., Saiki, Y. and Yamaguchi, M. (2008). Sedative effect of Japanese fragrance on Japanese women to sympathetic nervous activity. *Japan Society of Physiological Anthropology*. 13 (1), 49-56.

Hisayama, S. (2015). Live in comfort with forest based circular forestry business. (in Japanese) *Shoko Journal July 21-23*.

Inoue, M., Hayashi, S. and Craker L. (2017). Culture, history and applications of medicinal and aromatic plants in Japan. In E. H. El-Shemy (Ed.), *Aromatic and Medicinal Plants –Back to Nature* (pp. 95-110. Rijeka, Janeza Trdine: INTEC.

Jung, H. A., Islam, M. D., Kwon, Y. S., Jin, S. E., Son, Y. K., Park, J. J., Sohn, H. S. and Choi, J. S. (2011). Extraction and identification of three major aldose reductase inhibitors from *Artemisia montana. Food and Chemical Toxicology, 49(2)*: 376-384.

Kawai, S. and Matsubara, E. (2017). Effects of volatiles emitted from wooden materials on humans. (in Japanese) *Aroma Research, 18(3)*: 204-209.

McGleenon, B. M., Dynan, K. B. and Passmore, A. P. (1999). Acetylcholinesterase inhibitors in Alzheimer's disease. *Blackwell Science Ltd Br J Clin Pharmacol, 48*, 471–480.

Ministry of Land, Infrastructure, Transport, and Tourism. *Climate Change Adaptation Strategies to Cope with Water-related Disasters due to Global Warming* (Policy Report). (2008). Available from: http://www. mlit.go.jp/river/basic_info/english/pdf/policy_report.pdf.

Mitsunaga, T. and Hattori, H. (2017). Effects of the constituents of Grains of Paradise on anti-obesity and sympathetic nerve activity. (in Japanese) *Aroma Research, 18(3)*:229-234.

Murakami, S., Matsuura, M., Sato, T., Hayashi, S. and Koike, K. (2008). Study of the essential oil of Okinawa *Alpinia zerumbet* affects a mental behavior in mice. (in Japanese) *The pharmaceutical society of Japan*, 128, 26PE-am086.

Myoda, T., Toeda, K., Fujimori, T. and Nishizawa, M. (2017). Sedative effect of essential oil of *Artemisia montana*. (in Japanese) *Aroma research, 18(3)*: 225-228.

Nakajima, A., Ohizumi, Y. and Yamada, K. (2014). Anti-dementia Activity of *Nobiletin*, a Citrus Flavonoid: A Review of Animal Studies. *Clinical Psychopharmacology and Neuroscience, 12(2)*: 75-82.

Ngwoke, K. G., Chevallier, O., Wirkom, V. K., Stevenson, P., Elliott, C. T. and Situ, C. (2014). *In vitro* bactericidal activity of diterpenoids isolated from *Aframomum melegueta* K. Schum against strains of *Escherichia coli*, *Listeria monocytogenes* and *Staphylococcus aureus*. *Journal of Ethnopharmacology*, *151 (3)*:1147-1154.

Ozawa, N. (2015). *A study of a palatability and actual use of aromatic plants and the impression as foodstuffs.* (in Japanese) A thesis for a degree. Matsudo: Chiba University.

Sakamaki, S., Kofujita, H. and Sugawara, M. (2013). Effect of odors from *Cryptomeria japonica* essential oils on human electro-encephalogram. (in Japanese) *Aroma Research*, *14(1)*:64-71.

Sugita, J., Yoneshiro, T., Hatano, T., Aita, S., Ikemoto, T., Uchiwa, H., Iwanaga, T., Kameya, T., Kawai, Y. and Saito, M. (2013). Grains of paradise (*Aframomum melegueta*) extract activates brown adipose tissue and increases whole-body energy expenditure in men. *British Journal of Nutrition*, *110*: 733–738.

Takinoue Town. *Commerce and Tourism Section* (2017). Available from: http://town.takinoue.hokkaido.jp/kanko/hakkahakubutsukan/tenji/tenji2.html.

Tsukuba Botanical Garden. (2009). *Diversity conservation.* (in Japanese) Available from: http://www.tbg.kahaku.go.jp/diversity/protect/endan gered/conditions/index.html.

University of Tokyo. *Koishikawa Botanical Garden*. (2017 October 21). Available from: http://www.bg.s.u-tokyo.ac.jp/koishikawa/ eng/.

Yatagai, M. (2010). Scent and biological activity of plants. (in Japanese) *Flagrance Journal*, pp. 227.

Zhang, S., Tomata, Y., Sugiyama, K., Sugawara, Y. and Tsuji, I. (2017). Citrus consumption and incident dementia in elderly Japanese: the Ohsaki Cohort 2006 Study. *British Journal of Nutrition*, *117(8)*: 1-7.

In: Recent Advances in Plant Research
Editors: B. Welch and M. Wilkerson
ISBN: 978-1-53614-170-2

Chapter 8

A Survey of Angiospermic Flora and the Economic Potential of Species in "Caatinga-Cerrado" Transition Vegetation in the Piauí State, Northeastern Brazil

Lucas Santos Araújo[1], Elizabete Cristina de Cerqueira[1] and Jesus Rodrigues Lemos[2]

[1] Biological Sciences Course, Federal University of Piauí, Campus Ministro Reis Velloso-CMRV, Parnaíba, Piauí, Brazil

[2]Teacher's Degree in Biological Sciences, Federal University of Piauí/CMRV, Parnaíba, Piauí, Brasil

Abstract

Aiming to better understand the phytodiversity of a vegetation area in the Piauí state, Brazil, in addition to aspects such as the geographical distribution of species and their uses, this study was carried out in municipality of Bom Princípio, northern Piauí. The diverse flora was represented by trees, shrubs and herbaceous species, distributed in 54 families, 114 generas, and 146 species. The families of richest generas were *Fabaceae* (22 generas), followed by Malvaceae and Rubiaceae (seven generas) and Poaceae (five generas). Approximately 64,81% (35) of families were represented by a single genre at this location. It was found, according to the literature, that 29 species belonging to 16 families have economic potential. Most species have medicinal potential, with 22 species; followed by 15 timbers species, with 14 honey producer species; 12 ornamental species, food with eight species, forage nine and use for fuel, eight species. Of the species recorded in area 97 (66,4%) are not endemic in Brazil and 47 (32,2%) are endemic in this country. As consultations held in the collections of several Virtual Herbarium and Herbarium "HDelta," checked that 17 species did not have occurrence records in Piauí state. Analyzing the geographical distribution of species recorded in the study area and comparing with other surveys, it was found that 13 species found in survey's "cerrado" vegetation and 18 species in "caatinga" vegetation areas. The floristic survey in this semideciduous vegetation area in northern Piauí showed that in the area there are a number of species belonging to the plant formations of "cerrado" and "caatinga" vegetation, according to data of floristic lists made these vegetational formations, thus suggesting, treat itself to a transition area between these two types of vegetation.

Keywords: floristic diversity, xeric vegetation, Brazil's, Mid-North

INTRODUCTION

Brazil boasts the largest area of tropical forest in the world and is considered as the most abundant biodiversity in the country, ranking first in a number of plant species (Barbosa et al., 1996; Ribeiro & Walter, 1998). It is estimated that flora of Flowering Plants is 85,400 species as sampled by Brazil Flora Group (BFG, 2015).

However, this number is certainly apparent due to a shortage of floristic studies, both national and regional and local scale. Taking as an example, we have the flora of Brazil's northeastern is estimated that about 25,805 species of angiosperm (BFG, 2015).

Despite Brazil having poorly referenced vegetation, Giulietti et al. (2009) in order to ascertain the most important regions for the preservation of species and map the endemic species of Brazilian plants, mentions that there are 2,256 rare species in the country and 752 important areas for conservation of your biodiversity.

Souza et al. (1991) comments that the Northeast has an area of 154,867, 200 hectare, which corresponds to the most extent semiarid, with a predominance of "caatinga" vegetation and the remaining area, equivalent to 76,060,800 hectares, corresponding to "cerrado" and other vegetation types, such as forest babaçu, "carrasco" vegetation, forest, fields, floodplains and riparian forest. The distribution of vegetation in northeastern Brazil is deeply influenced by the strong climate gradient that goes from the humid east coast to the hinterland vast semi-arid (Andrade-Lima, 1981).

Initial studies indicated in the northeast region from the precursor Dárdano Andrade Lima, even in the 1950s (Santos, 2009). Although it is observed advantaged floristic diversity in plant communities of the Northeast (Zickel et al., 2004), there is a death of published works, with only isolated floristic surveys.

Due to the fact of geographical location, Piauí has a vegetation with an influence of different areas: Amazon, Plateau Central and Northeast. This is due to its distinctive climate and its geological formations (Farias & Castro, 2004).

However, this range has obtained in the past visiting naturalistic influential as Martius (1819), Gardner (1839), Schwacke (1878), Taubert (1895), Luetzelburg (1912/1914) and Assis Iglesias (1914 to 1919), vegetation in Piauí does not have a detailed study, envisioning an overall rating (Baptista, 1981).

Santos et al., (2009) commented that in the 70s, scientists have made extensive tours with scientific and exploratory method, giving origin to the first attempts of Piauí vegetation classification. Fernandes (1982) seeing the geographic layout, organized the Piauí vegetation into seven groups: forests, palm grove (carnaúba palm, babaçu and buritizais), "cerrado," "caatinga," "carrasco," grassland and coastal vegetation.

Santos Filho et al. (2009) avow that few bibliographic works and technical reports without scientific character summarized the Piauí Flora and data that emerged in the 80s the first academic research (master's essay and doctoral thesis) disseminating information on the formation and organization of Piauí´s vegetation.

From studies arising from this period have been the following: Oliveira et al. (1997), Farias (2003), Lemos (2004), Oliveira et al. (2007), Matos (2009), Santos Filho (2009), Costa (2010), Silva (2014), Amaral & Lemos (2015), Sousa et al. (2015). Still, there are many areas in the state of Piauí which does not know its floristic diversity. The main objective of this study is to know phytodiversity of green area in this state, as well as aspects such as the geographical distribution of species and their uses.

Materials and Methods

This research conducted a floristic inventory for a semideciduous vegetation stretch from the Bom Princípio´s municipality, locality belonging to the micro-region "Coast Piauiense." This micro-region has an estimated population in 2010 by IBGE in 303.511 inhabitants divided into fourteen municipalities: Bom Princípio do Piauí, Buriti dos Lopes, Cajueiro da Praia, Caraúbas do Piauí, Caxingó, Cocal, Cocal dos Alves, Ilha Grande, Luís Correia, Murici dos Portelas, Parnaíba, Piracuruca, São João da Fronteira e São José do Divino. This micro-region, with a total area of 9.658.107 km^2, in turn, belongs to the mesoregion "North Piauiense."

Bom Princípio city (03°11'27" S and 41°38'39" W) has an area of 521.572 square kilometers and is located 260 km from the state capital, at an altitude of 70 meters (Figure 1). Its estimated population in 2015 is of 5,497 inhabitants (IBGE, 2010).

The objected area of this study refers to the "Sítio da Gurita" private property located in Bom Princípio with the future idea, by the owners, the transformation of the same in a Biological Reserve. The area has flora still relatively preserved compared with the general surroundings. Presents tourism potential to possess, inside, stones with cave paintings, water fountains for baths and a representative floristic diversity of "caatinga" vegetation.

The field excursions took place twice a month, taking place in October 2016 to April 2018, following routine procedure field (Lawrence, 1973; Mori et al., 1989; Vaz et al., 1992). The collection of material followed the method of random walks held throughout the possible extension of the study area. The collected specimens were pressed with their respective field tokens and recorded according to standard methodology in Botany (Silva et al., 1989).

As the samples were being torn, some specimens were assessed and evaluated for their usefulness (timber, medicinal, forage, oil producers, food, honey, ornamental plants, etc.) and are based on literature (Lewington, 1990; Rizzini & Mors, 1995; Simpson & Ogorzaly, 1995; Agra, 1996; Costa et al., 2002; Lorenzi et al., 2003; Maia, 2004).

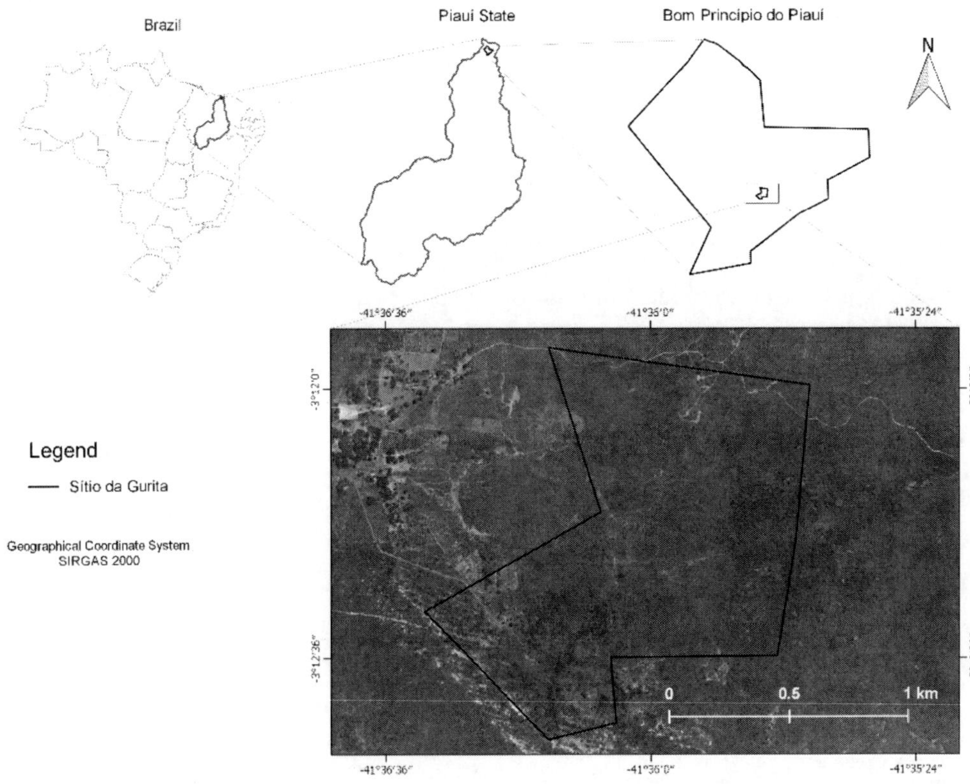

Figure 1. Location of the municipality Bom Princípio, northern Piauí.

Linked to species identification, research has been verified the geographical distribution of the same by searching in multiple virtual Herbarium collection, including in the database of "Flora of the Brazil 2020 under construction," available on the internet (florado-brasil.jbrj.gov.br),in Herbarium "HDelta" (Herbarium Delta of Parnaíba) of the Federal University of Piauí-UFPI/*Campus* Ministro Reis Velloso and yet in the specialized literature. The organization of botanical families followed the Angiosperm Phylogeny Group System III (APG III, 2009).

RESULTS AND DISCUSSION

The diversity of the flora "Sítio da of Gurita" in Bom Princípio, was represented by arboreal, shrubs and herbaceous species, distributed in 54 families, 114 genera and 146 species (Table 1). The families of greatest wealth of genres in Bom Princípio were *Fabaceae* (22 genera), followed by Malvaceae and Rubiaceae (seven genera) and Poaceae (five genera). Approximately 64.81% (35) of families were represented by a single genre at this location, 18.51% (10) for two, 3.7% (02) for three and 5.55% (03) for four genera. Genera with high richness were *Cyperus* (Cyperaceae) with five species; *Ipomoea* (Convolvulaceae) with four species, *Aeschynomene, Mimosa* (*Fabaceae*), *Turnera* (Turneraceae) and *Waltheria* (Malva-ceae), with of three species each one. Seventeen genera (14.9% of total) were represented by two especies and 91 genera (79.8%) of a single species. The families with the highest number of species were *Fabaceae*, 19.86% (29 spp.); Malvaceae, 6.16% (nine spp.); Rubiaceae, 6.16% (nine spp.); Convolvulaceae, Cyperaceae, Euphorbiaceae and Poaceae 3.4% (five spp. each one); Asteraceae and Lamiaceae 2.74% (four spp. each one). These nine families was represented by 51.38% of the overall species number and the other nine families are represented by three species (18.49%); eight by two species (10.95%) and 28 (19.17%) by only a single species (Figure 2). *Fabaceae* represented with 19.86% of the number of species, according to data from Brazil Flora Group (BFG, 2015) as the most family numerous in almost all ecosystems and biomes. Also according to BFG (2015), Rubiaceae, second family with the largest number of species in this study, ranks fourth in diversity in the country, among the Angiosperms.

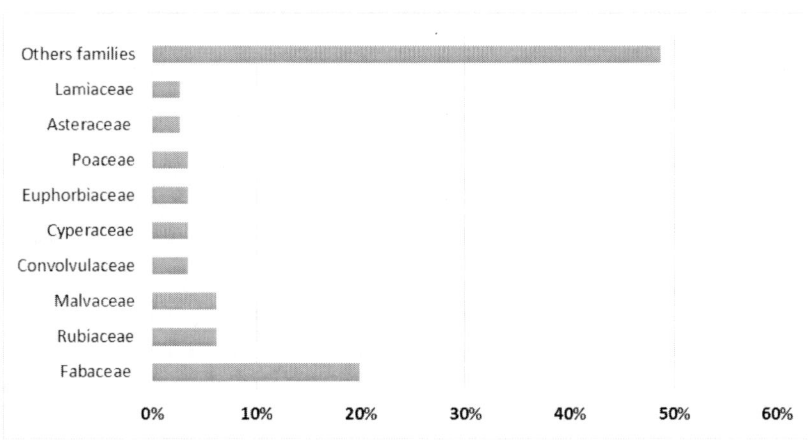

Source: Author's data.

Figure 2. Families with higher representativeness, in percentage of number of species in the floristic survey in Caatinga-Cerrado area.

Table 1. Families and species collected in "Sitio da Gurita", Bom Princípio, Piauí state, with their Common names, Habit and Collector/Collector Number (CN)

Family/Species		Common Name	Habit	Collector/CN
1	ACANTHACEAE			
1	*Dicliptera mucronifolia* Nees	-	Herb	Araújo, L. S./Lima, G. A./92
2	*Ruellia asperula* (Mart. ex Ness) Lindau	Melosa	Herb	Cerqueira, E. C. /114
3	*Ruellia paniculata* L.	-	Herb	Cerqueira, E. C./30
2	AIZOACEAE			
4	*Sesuvium portulacastrum* (L.) L.	-	Herb	Cerqueira, E. C./57
3	AMARANTHACEAE			
5	*Alternanthera brasiliana* (L.) Kuntze var. *brasiliana*	-	Herb	Nascimento, F. E. L./20; Cerqueira, E. C./69
6	*Alternanthera tenella* Colla	-	Sub-shrub	Cerqueira, E. C./ 198/ 49
7	*Froelichia humboldtiana* (Roem. & Schult.) Seub.	-	Herb	Cerqueira, E. C./56; Nascimento, F. E. L./16
4	AMARYLLIDACEAE			
8	*Griffinia* sp.	-	Herb	Cerqueira, E. C./242
5	ANACARDIACEAE			
9	*Anacardium humile* A. St.-Hil.	Cajuí	Tree	Cerqueira, E. C. /32
10	*Myracrodruon urundeuva* Allemão	Aroeira	Tree	Cerqueira, E. C./27
11	*Astronium fraxinifolium* Schott	Aroeira Branca	Tree	Cerqueira, E. C./142
6	APOCYNACEAE			
12	*Aspidosperma multiflorum* A. DC	-	Tree	Araújo, L. S./87
13	*Aspidosperma pyrifolium* Mart. & Zucc	Pereiro	Tree	Araújo, L. S./86; Nascimento, F. E. N./28
14	*Ditassa hastata* Decne.	-	Climb	Cerqueira, E. C./23
7	ARISTOLOCHIACEAE			
15	*Aristolochia labiata* Willd.	-	Climb	Araújo, L. S./88

Family/Species		Common Name	Habit	Collector/CN
8	ASTERACEAE			
16	*Acmella uliginosa* (Sw.) Cass.	-	Herb	Cerqueira, E. C./231
17	*Bidens pilosa* L.	-	Herb	Cerqueira, E. C./190
18	*Chresta pacourinoides* (Mart. ex DC.) Siniscalchi & Loeuille	-	Herb	Cerqueira, E. C./124; Araújo, L. S./71
19	*Emilia sonchifolia* (L.) DC. ex Wight	-	Herb	Araújo, L. S./74
9	BIGNONIACEAE			
20	*Fridericia dispar* (Bureau ex K. Schum.) L. G. Lohmann	-	Climb	Nascimento, F. E. L./25; Cerqueira, E. C./48; Cerqueira, E. C. /245
21	*Fridericia subverticillata* (Bureau & K. Schum.) L. G. Lohmann	Cipó Branco	Climb	Cerqueira, E. C./115 Cerqueira, E. C./230
22	*Handroanthus impetiginosus* (Mart. ex DC.) Mattos	Pau d`arco roxo	Tree	Cerqueira, E. C./128
10	BIXACEAE			
23	*Cochlospermum vitifolium* (Willd.) Spreng	Pacoté	Shrub	Cerqueira, E. C./127
11	BROMELIACEAE			
24	*Encholirium spectabile* Mart. ex Schult. & Schult. f.	-	Shrub	Nascimento, F. E. L./24
12	BURSERACEAE			
25	*Commiphora leptophloeos* (Mart.) J. B. Gillett	Amburana de espinho	Tree	Cerqueira, E. C./79
13	CAPPARACEAE			
26	*Cynophalla hastata* (Jacq.) J. Presl	Feijão Bravo	Shrub	Cerqueira, E. C./241
27	*Crateva tapia* L.	Trapiá	Shrub	Cerqueira, E. C./26
14	CHRYSOBALANACEAE			
28	*Licania rigida* Benth.	Oiticica	Tree	Cerqueira, E. C./28
29	*Licania tomentosa* (Benth.) Fritsch	Oiti Bravo	Tree	Cerqueira, E. C. /34
15	COMBRETACEAE			
30	*Combretum duarteanum* Cambess.	Mumfumbo de Xerém	Shrub	Nascimento, F. E. L./01
31	*Combretum leprosum* Mart.	Mofumbo	Tree	Lopes, J. R. S./04

Table 1. (Continued)

Family/Species		Common Name	Habit	Collector/CN
16	COMMELINACEAE			
32	*Aneilema brasiliense* C. B. Clarke	-	Herb	Cerqueira, E. C./75
33	*Commelina erecta* L.	-	Herb	Cerqueira, E. C./50
34	*Commelina obliqua* Vahl.	-	Herb	Nascimento, F. E. L./04
17	CONVOLVULACEAE			
35	*Ipomoea bahiensis* Willd. ex Roem. & Schult.	-	Climb	Nascimento, F. E. L./23
36	*Ipomoea megapotamica* Choisy	-	Climb	Araújo, L. S./66
37	*Ipomoea regnellii* Meisn.	-	Climb	Nascimento, F. E. L./15
38	*Ipomoea sericophylla* Meisn.	-	Climb	Cerqueira, E. C./129
39	*Turbina cordata* (Choisy) D.F. Austin & Staples	-	Climb	Cerqueira, E. C./184
18	CYPERACEAE			
40	*Cyperus aggregatus* (Willd.) Endl.	-	Herb	Cerqueira, E. C./62
41	*Cyperus brumadoi* D.A.Simpson	-	Herb	Cerqueira, E. C./58
42	*Cyperus schomburgkianus* Nees	-	Herb	Cerqueira, E. C./60
43	*Cyperus simplex* Kunth	-	Herb	Cerqueira, E. C./76
44	*Cyperus surinamensis* Rottb.	-	Herb	Cerqueira, E. C./59
19	DILLENIACEAE			
45	*Davilla rugosa* Poir.	Simbaíba	Tree	Cerqueira, E. C./37
20	DIOSCOREACEAE			
46	*Dioscorea ovata* Vell.	-	Climb	Araújo, L. S./67
21	ERIOCAULACEAE			
47	*Paepalanthus lamarckii* Kunthv	-	Herb	Cerqueira, E. C./227
22	EUPHORBIACEAE			
48	*Alchornea discolor* Poepp	-	Shrub	Pinheiro, Y. V. F./23
49	*Croton adenocalyx* Baill.	Marmeleiro Cedro	Shrub	Cerqueira, E. C./70 Pinheiro, Y. V. F./19

Family/Species		Common Name	Habit	Collector/CN
50	*Croton sonderianus* Müll. Arg.	Marmeleiro	Shrub	Cerqueira, E. C./01
51	*Jatropha mollissima* (Pohl) Baill.	Pinhão Bravo	Shrub	Cerqueira, E. C./171
52	*Sebastiania jacobinensis* (Müll. Arg.) Müll.Arg.	-	Shrub	Cerqueira, E. C./78
23	*FABACEAE*			
53	*Aeschynomene brevipes* Benth.	-	Herb	Cerqueira, E. C./53
54	*Aeschynomene paniculata* Willd. ex Vogel	-	Herb	Nascimento, F. E. L./18
55	*Aeschynomene viscidula* Michx.	-	Herb	Cerqueira, E. C./73
56	*Albizia niopoides* (Spruce ex Benth.) Burkart	AngicoBranco	Tree	Cerqueira, E. C./31
57	*Albizia polycephala* (Benth.) Killipex Record	Angico Preto	Tree	Araújo, L. S./83
58	*Amburana cearensis* (Allemão) A. C. Sm	Amburana	Tree	Cerqueira, E. C./228
59	*Anadenanthera colubrina* var. *cebil* (Griseb.) Altschul	Angico Preto	Tree	Cerqueira, E. C./112
60	*Andira humilis* Mart. ex Benth.	Cascudo	Tree	Lopes, J. R. S./05
61	*Bauhinia ungulata* L.	Mororó	Shrub	Nascimento, F. E. L./39; Cerqueira, E. C./25
62	*Dalbergia cearensis* Ducke	Violeta	Tree	Lopes, J. R. S. /09
63	*Centrosema brasilianum* (L.) Benth.	Feijão de Rolinha	Herb	Cerqueira, E. C./77
64	*Chamaecrista diphylla* (L.) Greene	-	Herb	Cerqueira, E. C./ 55; Nascimento, F. E. L./19
65	*Copaifera langsdorffii* Desf.	Podói	Tree	Araújo, L. S./84; Lopes, J. R. S./03
66	*Desmodium glabrum* (Mill.) DC.	-	Shrub	Cerqueira, E. C. /130
67	*Hymenaea* sp.	Jatobá	Tree	Cerqueira, E. C./66
68	*Libidibia ferrea* (Mart. ex Tul.) L.P. Queiroz	Jucá (Pau Ferro)	Tree	Lopes, J. R. S./01
69	*Macroptilium atropurpureum* (Sessé & Moc. ex DC.) Urb.	-	Climb	Cerqueira, E. C. /118
70	*Macroptilium lathyroides* (L.) Urb	-	Herb	Araújo, L. S./79
71	*Mimosa caesalpiniifolia* Benth.	Sabiá	Shrub	Nascimento, F. E. L./12

Table 1. (Continued)

Family/Species		Common Name	Habit	Collector/CN
72	*Mimosa tenuiflora* (Willd.) Poir.	Jurema Preta	Tree	Cerqueira, E. C./119
73	*Mimosa verrucosa* Benth.	Jurema	Shrub	Nascimento, F. E. L./24; Nascimento, F. E. L./38
74	*Piptadenia stipulacea* (Benth.) Ducke	Jurema Branca	Shrub	Cerqueira, E. C./14
75	*Pityrocarpa moniliformis* (Benth.) Luckow & R. W. Jobson	Catanduva	Tree	Araújo, L. S./85
76	*Poincianella gardneriana* (Benth.) L. P. Queiroz	Catingueira	Shrub	Cerqueira, E. C./15
77	*Senna gardneri* (Benth.) H. S. Irwin & Barneby	-	Shrub	Cerqueira, E. C./125; Nascimento, F. E. L./29
78	*Stylosanthes guianensis* (Aubl.) Sw.	-	Herb	Cerqueira, E. C./22; Araújo, L. S./77
79	*Stylosanthes macrocephala* M. B. Ferreira & Sousa Costa	-	Herb	Nascimento, F. E. L./22
80	*Tephrosia purpurea* (L.) Pers.	-	Shrub	Cerqueira, E. C./16
81	*Vigna halophila* (Piper) Maréchal et al.	Retirana	Climb	Cerqueira, E. C./113
24	IRIDACEAE			
82	*Cipura paludosa* Aubl.	-	Herb	Araújo, L. S./70
25	LAMIACEAE			
83	*Amasonia campestris* (Aubl.) Moldenke	-	Herb	Nascimento, F. E. L./17
84	*Gymneia platanifolia* (Mart. ex Benth.) Harley & J. F. B. Pastor	-	Herb	Araújo, L. S /Lima, G. A./875
85	*Mesosphaerum suaveolens* (L.) Kuntze	-	Herb	Araújo, L. S./82
86	*Rhaphiodon echinus* Schauer	-	Herb	Cerqueira, E. C./197
26	LENTIBULARIACEAE			
87	*Utricularia fimbriata* Kunth	-	Herb	Araújo, L. S./75
88	*Utricularia subulata* L.	-	Herb	Araújo, L. S./73
27	LORANTHACEAE			

Family/Species		Common Name	Habit	Collector/CN
89	*Struthanthus andrastylus* Eichler	Fava de Passarinho	Herb	Nascimento, F. E. L./34
90	*Struthanthus marginatus* (Desr.) Blume	Enxerto	Herb	Cerqueira, E. C./29
28	LYTHRACEAE			
91	*Cuphea campestris* Koehne	Orquídea de Xerém	Herb	Nascimento, F. E. L./11; Cerqueira, E. C./54; Araújo, L. S./68
92	*Cuphea teleandra* Lourteig	-	Shrub	Cerqueira, E. C./ 67
29	MALPIGHIACEAE			
93	*Byrsonima correifolia* A. Juss.	Murici	Shrub	Nascimento, F. E. L./24
94	*Diplopterys lutea* (Griseb.) W. R. Anderson & C. C. Davis	-	Shrub	Cerqueira, E. C./36
95	*Stigmaphyllon paralias* A. Juss	-	Shrub	Cerqueira, E. C./126
30	MALVACEAE			
96	*Helicteres muscosa* Mart.	Pimentinha	Shrub	Cerqueira, E. C./02
97	*Pavonia cancellata* (L.) Cav.	Malva	Herb	Cerqueira, E. C./202
98	*Pseudobombax marginatum* (A. St.-Hil., Juss. & Cambess.) A. Robyns	Embiratanha	Tree	Cerqueira, E. C./229
99	*Sida cordifolia* L.	-	Shrub	Cerqueira, E. C./03
100	*Sterculia striata* A. St.-Hil. & Naudin	Chichá	Tree	Nascimento, F. E. L./33
101	*Waltheria brachypetala* Turcz.	-	Shrub	Cerqueira, E. C./24
102	*Waltheria bracteosa* A. St.-Hil. & Naudin	-	Sub-shrub	Cerqueira, E. C./186
103	*Waltheria indica* L.	-	Herb	Cerqueira, E. C./131
104	*Wissadula hernandioides* (L. Hér.) Garcke	-	Shrub	Cerqueira, E. C./116
31	MELIACEAE			
105	*Cedrela odorata* L.	-	Shrub	Cerqueira, E. C./21
32	MYRTACEAE			
106	*Campomanesia aromatica* (Aubl.) Griseb.	Guabiraba	Shrub	Cerqueira, E. C./248
107	*Myrcia guianensis* (Aubl.) DC.	-	Tree	Nascimento, F. E. L./26
108	*Myrcia polyantha* DC.	Maria Preta	Tree	Lopes, J. R. S./02

Table 1. (Continued)

Family/Species		Common Name	Habit	Collector/CN
33	MORACEAE			
109	*Dorstenia tubicina* Ruiz & Pav.	-	Herb	Araújo, L. S./225
34	NYCTAGINACEAE			
110	*Guapira graciliflora* (Mart. ex Schmidt) Lundell	-	Tree	Pinheiro, Y. V. F./16
35	OCHNACEAE			
111	*Ouratea hexasperma* (A. St.-Hil.) Baill.	-	Tree	Cerqueira, E. C./20
36	ONAGRACEAE			
112	*Ludwigia leptocarpa* (Nutt.) H. Hara	-	Shrub	Cerqueira E. C./234
37	OPILIACEAE			
113	*Agonandra brasiliensis* Miers ex Benth. & Hook.f.	Pau Marfim	Shrub	Nascimento, F. E. L./37
38	OLACACEAE			
114	*Ximenia americana* L.	Ameixa	Tree	Cerqueira, E. C./141
39	OXALIDACEAE			
115	*Oxalis frutescens* L.	-	Herb	Cerqueira, E. C./74
40	PASSIFLORACEAE			
116	*Passiflora foetida* L.	-	Climb	Cerqueira, E. C./05
41	POACEAE			
117	*Cenchrus ciliares* L.	-	Herb	Cerqueira, E. C./200
118	*Panicum* sp.	-	Herb	Nascimento, F. E. L./27
119	*Paspalum conspersum* Schrad.	-	Herb	Cerqueira, E. C./61; Pinheiro, Y. V. F./21
120	*Streptostachys asperifolia* Desv.	-	Herb	Cerqueira, E. C./72
121	*Urochloa fusca* (Sw.) B. F. Hansen & Wunderlin	-	Herb	Pinheiro, Y. V. F./17
42	POLYGALACEAE			
122	*Asemeia violacea* (Aubl.) J. F. B. Pastore & J. R. Abbott	-	Herb	Araújo, L. S./81

Family/Species		Common Name	Habit	Collector/CN
123	*Polygala appendiculata* Vell.	-	Herb	Araújo, L. S./72
43	POLYGONACEAE			
124	*Triplaris gardneriana* Wedd.	Pajeú	Tree	Nascimento, F. E. L./32
44	PLUMBAGINACEAE			
125	*Plumbago scandens* L.	-	Sub-shrub	Araújo, L. S /Lima, G. A./93
45	RHAMNACEAE			
126	*Ziziphus joazeiro* Mart.	Juazeiro	Tree	Nascimento, F. E. L./40
46	RUBIACEAE			
127	*Amaioua intermedia* Mart. ex Schult. & Schult.f.	-	Shrub	Lopes, J. R. S./07
128	*Chomelia anisomeris* Müll. Arg.	-	Shrub	Pinheiro, Y. V. F./22
129	*Guettarda viburnoides* Cham. & Schltdl.	-	Tree	Araújo, L. S./78; Cerqueira, E. C./244
130	*Mitracarpus baturitensis* Sucre	-	Herb	Cerqueira, E. C./19/201
131	*Randia armata* (Sw.) DC.	-	Shrub	Cerqueira, E. C./144; Nascimento, F. E. L./35
132	*Randia calycina* Cham.	-	Shrub	Cerqueira, E. C./04
133	*Richardia grandiflora* (Cham. & Schltdl.) Steud	-	Herb	Araújo, L. S./94
134	*Tocoyena formosa* (Cham. & Schltdl.) K. Schum.	Jenipapinho Bravo	Shrub	Cerqueira, E. C./68; Lopes, J. R. S./08; Araújo, L. S./65
135	*Tocoyena bullata* (Vell.) Mart.	-	Shrub	Pinheiro, Y. V. F./20
47	RUTACEAE			
136	*Zanthoxylum gardneri* Engl.	Limãozinho	Tree	Cerqueira, E. C. 64; Pinheiro,Y. V. F.18
48	SALICACEAE			
137	*Casearia ulmifolia* Vahlex Vent.	-	Tree	Cerqueira, E. C./65
49	SAPINDACEAE			
138	*Toulicia guianensis* Aubl.	-	Tree	Cerqueira, E. C./117
50	SAPOTACEAE			

Table 1. (Continued)

Family/Species		Common Name	Habit	Collector/CN
139	*Pouteria ramiflora* (Mart.) Radlk.	Pitomba-de-leite	Shrub	Cerqueira, E. C./33/143
51	SIMAROUBACEAE			
140	*Homalolepis cedron* (Planch.) Devecchi & Pirani	-	Tree	Cerqueira, E. C./35
141	*Simarouba versicolor* A. St.-Hil.	Paraíba	Tree	Nascimento, F. E. L./36
52	TURNERACEAE			
142	*Turnera pumilea* L.	-	Shrub	Cerqueira, E. C./52
143	*Turnera pumilea* var. *piauhyensis* Urb.	-	Herb	Cerqueira, E. C./51; Nascimento, F. E. L./14
144	*Turnera coerulea* var. *surinamensis* (Urb.) Arbo & Fernández	-	Herb	Cerqueira, E. C./17 Cerqueira, E. C./247
53	VIOLACEAE			
145	*Pombalia calceolaria* (L.) Paula-Souza	-	Herb	Araújo, L. S./ 226
54	VITACEAE			
146	*Cissus verticillata* (L.) Nicolson & C. E. Jarvis	Guardião	Climb	Cerqueira, E. C./13

Source: Authors' data.

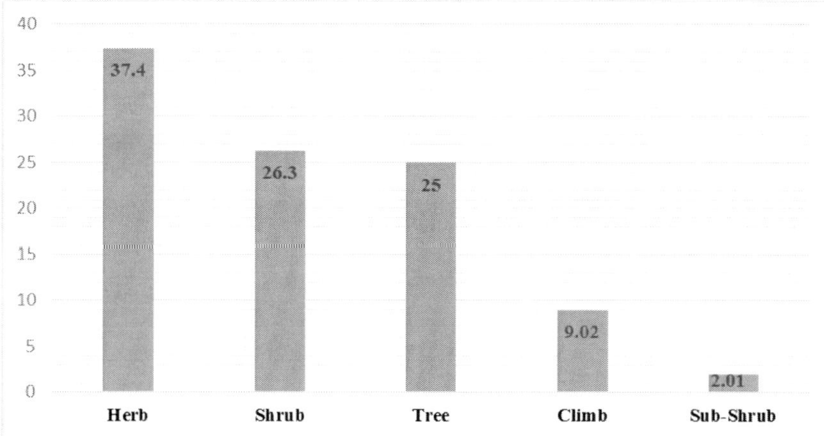

Source: Authors' data.

Figure 3. Distribution of species by habit.

So far, there were no rare species in the study area, as the Red Book of Flora of Brazil (Martinelli, 2013) and Rare Plants in Brazil (Giulietti, 2009).

Concerning the 145 species, 54 (37.0%) are represented by herbaceous plant habit, 38 (26.0%) is represented by bushes, 37 (25.3%) by trees, climbs by 14 (9.6%) and 3 (2%) are considered subshrubs plants (Figure 3). For the study area, the species of the *Fabaceae* and Rubiaceae family, in most cases, have a wide range of habit, being shrub, herbaceous and woody.

Checking the geographic distribution of species were recorded in the study area, the species as unique to the Northeast, such as *Croton adenocalyx*, *Cuphea campestris*, *Piptadenia stipulacea* and *Ziziphus joazeiro*.

From the total of the species, 97 (66.4%) are not endemic in Brazil, 47 (32.2%) are endemic and 2 (1.3%) have unknown endemism. As consultations on the Brazil 2020 Flora website, 17 species were found to occur in Piauí: *Alchornea discolor*, *Amaioua intermedia*, *Andira humilis*, *Aristolochia labiata*, *Campomanesia aromatica*, *Cedrela odorata*, *Cuphea teleandra*, *Cyperus brumadoi*, *Myrcia guianensis*, *Paspalum conspersum*, *Randia calycina*, *Ruellia asperula*, *Sebastiania jacobinensis*, *Sesuvium portulacastrum*, *Struthanthus andrastylus*, *Tocoyena bullata*, *Wissadula hernandioides*. However, carrying out consultations in another data base (species Link) the number of species not recorded in Piauí decreased to eight.

Checking the geographic species distribution in the study area in other vegetation types occurring in Piauí, it was found that there was record of species in the "cerrado" formations such as *Aspidosperma multiflorum*, *Aspidosperma pyrifolium*, *Bauhinia ungulata*, *Byrsonima correifolia*, *Cochlospermum vitifolium*, *Guettarda viburnoides*, *Mimosa caesal-piniifolia*, *Mimosa verrucosa*, *Myrcia guianensis*, *Pouteria ramiflora*, *Pseudobombax marginatum*, *Randia armata*, *Simarouba versicolor* and; the "caatinga," such as: *Amburana cearensis*, *Anadenanthera colubrina*, *Aspidosperma multiflorum*,

Aspidosperma pyrifolium, *Cochlospermum vitifolium*, *Commiphora leptophloeos*, *Copaifera langsdorffii*, *Croton sonderianus*, *Helicteres muscosa*, *Mimosa tenuiflora*, *Myracrodruon urundeuva*, *Passiflora foetida*, *Pavonia cancellata*, *Piptadenia stipulacea*, *Pseudobombax marginatum*, *Ruellia asperula*, *Sida cordifolia* and *Simarouba versicolor*.

Table 2. Species recorded in the study area with economic potential, "Sítio da Gurita", Bom Princípio, Piauí state

Families/Species		**Common name**	**Use categories**
1	AMARANTHACEAE		
1	*Alternanthera brasiliana* (L.) Kuntze var. *brasiliana*	-	a
2	*Alternanthera tenella* Colla	-	a, g
2	ANACARDIACEAE		
3	*Myracrodruon urundeuva* Allemão	Aroeira	a, b, c, e, f, g
3	APOCYNACEAE		
4	*Aspidosperma pyrifolium* Mart. & Zucc	Pereiro	a, b, c, e, f
4	ASTERACEAE		
5	*Bidens pilosa* L.	-	A
5	BIGNONIACEAE		
6	*Handroanthus impetiginosus* (Mart. ex DC.) Mattos	Pau d`arco roxo	a, c, e
6	BIXACEAE		
7	*Cochlospermum vitifolium* (Willd.) Spreng	Pacoté	a, c, e, g
7	BURSERACEAE		
8	*Commiphora leptophloeos* (Mart.) J. B. Gillett	Amburana de espinho	a, c, d, e, g
8	COMBRETACEAE		
9	*Combretum leprosum* Mart.	Mofumbo	a, c, e, g
9	EUPHORBIACEAE		
10	*Croton sonderianus* Müll. Arg.	Marmeleiro	a, b, c, d, g
Families/Species		Common name	Use categories
10	*FABACEAE*		
11	*Amburana cearensis* (Allemão) A. C. Sm	Amburana	a, c, e, f, g
12	*Anadenanthera colubrina* (Vell.) Brenan	Angico Preto	a, b, c, d, e, f, g
13	*Andira humilis* Mart. ex Benth.	Cascudo	a, e
14	*Bauhinia ungulata* L.	Mororó	a, b
15	*Centrosema brasilianum* (L.) Benth.	Feijão de Rolinha	G
16	*Copaifera langsdorffii* Desf.	Podói	D
17	*Mimosa caesalpiniifolia* Benth.	Sabiá	a, b, c, e, g
18	*Mimosa tenuiflora* (Willd.) Poir.	Jurema Preta	a, c, f
19	*Piptadenia stipulacea* (Benth.) Ducke.	Jurema Branca	a, b, c, f

Families/Species		Common name	Use categories
20	*Pityrocarpa moniliformis* (Benth.) Luckow& R. W. Jobson	Catanduva	G
11	LAMIACEAE		
21	*Mesosphaerum suaveolens* (L.) Kuntze		A
12	MALPIGHIACEAE		
22	*Byrsonima correifolia* A. Juss.	Murici	d
13	MALVACEAE		
23	*Pavonia cancellata* (L.) Cav.	Malva	g
24	*Pseudobombax marginatum* (A. St.-Hil., Juss. &Cambess.) A. Robyns	Embiratanha	a, b, c, d, e, f
25	*Sida cordifolia* L.	-	a
14	RHAMNACEAE		
26	*Ziziphus joazeiro*Mart.	Juazeiro	a, b, c, d, e, f, g
15	RUBIACEAE		
27	*Richardia grandiflora* (Cham. &Schltdl.) Steud	-	g
16	SAPOTACEAE		
28	*Pouteria ramiflora* (Mart.) Radlk.	Pitomba-de-leite	d
17	SIMAROUBACEAE		
29	*Simarouba versicolor* A. St.-Hil.	Paraíba	a, c

Source: Authors data.

Use categories: a = Medicinal, b = Forage, c = timber, d = Food, e = ornamental, f = Fuel, g = honey producer

Among the species collected in the study area was found in the literature, which has the same economic potential in different categories, such as medical purposes, forage, wood, food, ornamental fuel and honey producer (Table 2).

Thus, it was found that 29 species belonging to 17 families have economic potential. Most species have medicinal potential, with 22 species; followed by timber, 15 species; honey producer species, 14; 12 ornamental species; 8 species with food, feed with 9 and; use for fuel, eight species.

The species had a higher number of records with versatile uses in the literature were *Anadenanthera colubrina* and *Ziziphus joazeiro* with seven utilities each. On the other hand, *Myracrodruon urundeuva* and *Pseudobombax marginatum* presented six types of uses each one, and *Amburana cearensis*, *Aspidosperma pyrifolium*, *Commiphora leptophloeos*, *Croton sonderianus* and *Mimosa caesalpiniifolia* had five records of uses each one.

FINAL CONSIDERATIONS

The floristic survey this semideciduous vegetation area in northern Piauí showed that there is a representative number of species belonging to the plant formations of savanna and savanna, according to data of floristic lists held and has published these formations, suggesting thus treat yourself to a transition area between these two types of vegetation.

In this sense, this survey contributed to the greater knowledge of another area phytodiversity in the Piauí state, providing, in the near future, tools for a better understanding of plant formations as a whole, including with regard to ecological aspects and use of their species.

Acknowledgments

The authors thank Nicolas Chanove for the receptivity and permission for this research in the area studied and students who assisted in fieldwork.

References

Agra, M. D. F. (1996). *Plantas da medicina popular dos* [*Plants of folk medicine*] Cariris Velhos, Paraíba, Brasil. JoãoPessoa, Editora União.

Amaral, M. C., & Lemos, J. R. (2015). Floristic Survey of a Portion of the Vegetation Complex of the Coastal Zone in Piauí State, Brazil. *American Journal of Life Sciences*, 3(3), 213-218.

Andrade-Lima, D. (1951). A flora da praia de Boa Viagem. *Boletim da Secretaria de Agricultura de Pernambuco, Recife* [The flora of Boa Viagem beach. *Bulletin of the Secretariat of Agriculture of Pernambuco, Recife*], 18(1,2), 121-125.

Angiosperm Phylogeny Group. (2009) An update of the Angiosperm Phylogeny Group classification for the orders and families of flowering plants: APG III. *Botanical Journal of the Linnean Society*, 161,105-121.

Baptista, J. G. (1981). *Geografia física do Piauí.* Teresina: COMEPI. [*Physical geography of Piauí*]

Barbosa, M. D. V., Mayo, S. J., Castro, A. D., Freitas, G. D., Pereira, M. D. S., Neto, P. D. C., Gadelha Neto, P. C. & Moreira, H. M. (1996). Checklist preliminar das angiospermas. In: Sampaio, E.V.S.B.; Mayo, S. J. & Barbosa, M. R. (Eds.). *Pesquisa Botânica Nordestina: Progresso e Perspectivas* (pp. 253-415). Recife: Sociedade Botânica do Brasil, SeçãoRegional de Pernambuco [Recife: Botanical Society of Brazil, Regional Section of Pernambuco].

Brazil Flora Group- BFG (2015). *Growing knowledge: an overview of Seed Plant diversity in Brazil.* Rodriguésia, 66(4), 1085-1113.

Costa, J. A. S., Nunes, T. S., Ferreira, A. P. L., Stradmann, M. T. S., & de Queiroz, L. P. (2002). *Leguminosas forrageiras da caatinga: espécies importantes para as comunidades rurais do sertão da Bahia.* Feira de Santana: Universidade Estadual de

Feira de Santana [*Leguminous forage of the caatinga: species important for the rural communities of the sertão of Bahia.* Feira de Santana: State University of Feira de Santana], SASOP.

Costa, J. L. P. de O., Cavalcanti, A. P. B. (2010). Fitogeografia da planície deltaica do Rio Paranaíba, Piauí/Maranhão – Brasil: Análise da distribuição das espécies e interferência antrópica [Phytogeography of the Paranaíba River Delta Plain, Piauí / Maranhão - Brazil: Analysis of species distribution and anthropogenic interference]. *Observatorium: Revista Eletrônica de Geografia*, 2 (4),84-92.

Farias, R. R. S. (2003). *Florística e fitossociologia em trechos de vegetação do complexo de Campo Maior, Campo Maior, Piauí.* Dissertação de mestrado, Programa de pós graduação em biologia vegetal, Universidade Federal de Pernambuco, Recife, PE, Brasil [*Floristic and phytosociology in vegetation stretches of the Campo Maior complex, Campo Maior, Piauí.* Master's Dissertation, Graduate Program in Plant Biology, Federal University of Pernambuco, Recife, PE, Brazil].

Farias, R. R. S., & Castro, A. A. J. F. (2004). Fitossociologia de trechos da vegetação do Complexo de Campo Maior [Phytosociology of vegetation stretches of the Campo Maior Complex], Campo Maior, PI, Brasil. *Acta bot. bras*, 18(4), 949-963.

Fernandes, A. (1982). A vegetação do Piauí. [The vegetation of Piauí] In *Congresso nacional de botânica* (Vol. 32, pp. 313-3). Teresina: Sociedade Botânica do Brasil.

Flora do Brasil 2020 em construção. (2017). Instituto de Pesquisa Jardim Botânico do Rio de janeiro. Recuperado em 10 de maio, 2018, de Brazilian Botany Society. [*Flora of Brazil 2020 under construction.* (2017). Botanical Garden Research Institute of Rio de Janeiro. Retrieved on May 10, 2018, from] http://floradobrasil.jbrj.gov.br.

Giulietti, A. M., Rapini, A., Andrade, M. J. G., Queiroz, L. P. & Silva, J. M. C. (2009). *Plantas raras do Brasil* [*Rare plants of Brazil*]. Belo Horizonte: Conservação Internacional.

Instituto Brasileiro de Geografia e Estatística- IBGE. (2010). Censo *Demográfico 2010 Retratos do Brasil e do Piauí. Recuperado em 20 de janeiro* [*Demographic Census 2010 Portraits of Brazil and Piauí.* Retrieved on January 20], 2018, de https://ww2.ibge.gov.br/ home/presidencia/noticias/pdf/censo_2010_piaui.pdf.

Lawrence, G. H. M. (1973). *Taxonomia das plantas vasculares* [*Taxonomy of vascular plants*] (vol. 2, p. 256). Lisboa: Fundação Calouste Gulbenkian.

Lemos, J. R. (2004). Composição florística do parque nacional Serra da Capivara, Piauí, Brasil. *Rodriguésia* [Floristic composition of Serra da Capivara national park, Piauí, Brazil], 5, 55-66.

Lewington, A. (1990). *Plants for people*. London. The Natural History Museum.

Lorenzi, H., Souza, H. M. Torres, M. A. V. & Bacher, L. B. (2003). *Árvores Exóticas no Brasil: madeireiras, ornamentais e aromáticas.* Nova Odessa: Instituto Plantarum

[*Exotic trees in Brazil: logging, ornamental and aromatic.* Nova Odessa: Instituto Plantarum].

Maia, G. N. (2004). *Caatinga árvores e arbustos e suas utilidades* [*Caatinga trees and shrubs and their uses*] (1 ed.). São Paulo: D & Z.

Martinelli, G., & Moraes, M. A. (2013). *Livro vermelho da flora do Brasil.* Rio de Janeiro: Instituto de Pesquisas Jardim Botânico do Rio de Janeiro [*Red flora book of Brazil*].

Matos, M. de Q., & Felfili, J. M. (2010). Florística, fitossociologia e diversidade da vegetação arbórea nas matas de galeria do Parque Nacional de Sete Cidades [Floristic, phytosociology and diversity of tree vegetation in the gallery forests of the Sete Cidades National Park] (PNSC), Piauí, Brasil. *Acta bot. bras.*, 24, 483-496.

Mori, S. A., Silva, L., Lisboa, G. & Coradin, L. (1989). *Manual de manejo do herbário fanerogâmico.* Ilhéus: Centro de Pesquisas do Cacau [*Handbook of the phanerogamic herbarium.* Ilhéus: Cocoa Research Center].

Oliveira, L. D. S. D., Soares, S. M. N. A., Soares, F. D. A. R., & de Barros, R. F. M. (2007). Levantamento Florístico do Parque Ambiental Paquetá, Batalha, Piauí. *Revista Brasileira de Biociências* [Floristic of the Environmental Park Paquetá, Batalha, Piauí. *Brazilian Journal of Biosciences*], 5(S2), pg-372.

Oliveira, M. E. A., Sampaio, E. V. S. B., Castro, A. A. J. F., & Rodal, M. J. N. (1997). *Flora e fitossociologia de uma área de transição carrasco-caatinga de areia em Padre Marcos, Piauí* [*Flora and phytosociology of a Carrasco-Caatinga transition area of sand in Padre Marcos, Piauí.*]. Naturalia, São Paulo, 22, 131-150.

Ribeiro, J. F., & Walter, B. M. T. (1998). Fitofisionomias do bioma Cerrado [Phytophysiognomies of the Cerrado biome]. In: Sano, S. M. & Almeida, S. P. (Eds.). *Cerrado: ambiente e flora* (pp. 89-166). Planaltina: EMBRAPA.

Rizzini, C. T., & Mors, W. B. (1995). *Botânica econômica brasileira* [*Brazilian economic botany*] (2ed.). São Paulo: EPU-EDUSP.

Santos, F. S. F. (2009). *Composição florística e estrutural da vegetação de restinga doEstado do Piauí.* Tese de Doutorado, Programa de Pós-Graduação em Botânica, Universidade Federal Rural de Pernambuco, Recife, PE, Brasil [*Floristic and structural composition of the restinga vegetation of the State of Piauí.* PhD Thesis, Graduate Program in Botany, Federal Rural University of Pernambuco, Recife, PE, Brazil].

Silva, A. T., Muniz, C. F. S., Wanderley, M. G. L., Kirizawa, M., Sendulsky, T., Silva, T. S., Maluf, A. M., Silvestre, M. S. F., Chiea, Custódio, A. F., Mantovani, W., Jung, S. L., Barros, F. & Oliveira, L. C. A. (1989). Pteridófitas e fanerógamas. In: Fidalgo, O. & Bononi, V. L. R. *Técnicas de coleta, preservação e herborização de material botânico.* Série Documentos. São Paulo: Instituto de Botânica [*Techniques of collection, preservation and herborization of botanical material.* Series Documents. São Paulo: Institute of Botany].

Silva, G. A. R., Bastos, E. M., & Sobreira, J. A. R. (2014). Levantamento da flora apícola emduas áreas produtoras de mel no estado do Piauí. *Enciclopédia biosfera*, Centro Científico Conhecer-Goiânia [Survey of the bee flora in two areas producing honey in the state of Piauí. *Encyclopedia biosphere*, Centro Científico Conhecer-Goiânia], 10(18), 3305-3316.

Simpson, B. B., & Ogorzaly, M. C. (1995). *Economic botany: plants in our world.* (2ed.). NewYork: Ed. McGraw-Hill.

Sousa, F. D. C. D., Araújo, M. P., & Lemos, J. R. (2015). Ethnobotanical Study with Native Species in a Rural Village in Piauí State, Northeast Brazil. *Journal of Plant Sciences*, 3(2), 45-53.

Souza, M. L. D. E. R., Falkenberg, D. B., Amaral, L. G., Fronza, M., Araujo, A. C., & de Sá, M. R. (1991). Vegetação do Portal da Daniela, Florianópolis, SC, Brasil. I. Levantamento florístico e mapa fitogeográfico [Vegetation of the Daniela Portal, Florianópolis, SC, Brazil. I. Floristic survey and phytogeographic map]. *INSULA Revista de Botânica*, 21, 87-117.

Vaz, A. M. S. F., Lima, M. P. M., & Marquete, R. (1992). Técnicas e manejo de coleções botânicas. In: *Manual Técnico da Vegetação Brasileira* [Techniques and management of botanical collections. In: *Technical Manual of Brazilian Vegetation*] (pp. 55-75). [Manuais Técnicos emGeociências I]. Rio de Janeiro: IBGE.

Zickel, C. S., Vicente, A., Almeida Jr, E. B., Cantarelli, J. R. R., & Sacramento, A. C. (2004). Flora e vegetação das restingas no Nordeste Brasileiro [Flora and vegetation of restingas in the Brazilian Northeast]. In: Eskinazi, E. L., Neumann, S. L., Costa, M. F. *Oceanografia: um cenário tropical* [*Oceanography: a tropical setting*] (pp. 689-701). Recife: Bargaço.

BIOGRAPHICAL SKETCHES

Lucas Santos Araújo

Affiliation: Federal University of Piauí state, Brazil

Education: Undergraduate student in Biological Sciences

Research and Professional Experience: Floristics, Phytogeography and Ethnobotany.

Publications from the Last 3 Years:
http://lattes.cnpq.br/1156545441612504

Elizabete Cristina de Cerqueira

Affiliation: Federal University of Piauí state, Brazil

Education: Bachelor's Degree in Biological Sciences

Research and Professional Experience: Botanic and Education

Publications from the Last 3 Years:

Cerqueira, E. C.; Leite, A. V.; Carvalho, F V. M. (2016). O estágio como ferramenta motivadora para a carreira docente de licenciandos em ciências biológicas. *Anais do Congresso Nacional de Educação* [The internship as a motivating tool for the teaching career of graduates in biological sciences. *Proceedings of the National Congress of Education] -CONEDU*, Natal, RN, Brasil, 3.

Leite, A. V.; Cerqueira, E. C.; Sousa, F. C. C.; Nascimento, K. S.; Carvalho, F. V. M. (2016). As dificuldades encontradas pela professora de ciências no ensino de jovens e adultos (EJA) no sistema prisional de Parnaíba-PI. *Anais do Congresso Nacional de Educação* [The difficulties encountered by the science teacher in the teaching of youth and adults (EJA) in the prison system of Parnaíba-PI. *Proceedings of the National Congress of Education*].*-CONEDU*, Natal, RN, Brasil, 3.

Silva, G. O., Cerqueira, E. C., Vieira, I. R., Andrade, I. M. (2016). Organização e Diversidade vegetal no Herbário Delta do Parnaíba (HDELTA), Campus Ministro Reis Velloso, Parnaíba, Piauí. *Anais do Seminário de Extensão e Cultura da UFPI e II Fórum de Extensão e Cultura das IES do Piauí.* Tema: Curricularização da extensão e o protagonismo estudantil [Organization and Plant Diversity in the Herbarium Delta do Parnaíba (HDELTA), Campus Ministro Reis Velloso, Parnaíba, Piauí. *Proceedings of the Extension and Culture Seminar of the UFPI and II Forum of Extension and Culture of the IES of Piauí. Theme: Curriculum extension and student protagonism*], Teresina, PI, Brasil, 5.

Silva, M. S., Cerqueira, E. C., Mayo, S. J., Andrade, I. M. de (2017, maio). Estudo taxonômico das espécies de Cyperus l. (Cyperaceae) na costa piauiense. *Anais da 35ª Reunião Nordestina de Botânica: Ressurgir, Crescer e Evoluir.* Recife, PE, Brasil [Taxonomic study of the species of Cyperus l. (Cyperaceae) on the coast of Piauí. *Proceedings of the 35th Northeastern Meeting of Botany: Resurface, Grow and Evolve*. Recife, PE, Brazil], 35.

Cerqueira, E. C., Lemos, J. R. (2018). Levantamento florístico em trilhas naturais de um sítio com potencial turístico no norte do Piauí como subsídio à Educação ambiental e conservação da fitodiversidade. In: Lemos, J. R. (Org.). *Pesquisas Botânicas e Ecológicas no Piauí* [Floristic survey on natural trails of a site with tourist potential in the north of Piauí as a subsidy to Environmental education and conservation of phytodiversity. In: Lemos, J.R. (Org.). *Botany and Ecological Research in Piauí*]. (vol. 1, 1. ed. pp. 149-159). Curitiba-PR: CRV.

Jesus Rodrigues Lemos

Affiliation: Federal University of Piauí state, Brazil

Education: Doctorate in Biological Sciences – Botany

Research and Professional Experience: Floristics, Phytogeography and Ethnobotany.

Professional Appointments: Professor of Botany

Publications from the Last 3 Years:
http://lattes.cnpq.br/0603749727482775

INDEX

#

A

B

C

D

E

F

G

H

I

J

K

L

M

N

O

P